HERBERT E. ELLINGER

AUTOMOTIVE ENGINES
Theory and Servicing

Prentice-Hall, Inc., Englewood Cliffs, NJ 07632

Library of Congress Cataloging in Publication Data
ELLINGER, HERBERT E
   Automotive engines.
   Includes Index.
    1. Automobiles—Motors.   2. Automobiles—motors—Maintenance and repair.   I. Title.
TL210.E368   629.2'5   80-18154
ISBN 0-13-054999-1

Editorial Production/Supervision by Theodore Pastrick
Interior Design by Steven Bobker
Page Layout by Diane Koromhas
Manufacturing Buyer: Joyce Levatino

10 9 8 7 6 5 4

PRENTICE-HALL INTERNATIONAL, INC., *London*
PRENTICE-HALL OF AUSTRALIA PTY. LIMITED, *Sydney*
PRENTICE-HALL OF CANADA, LTD., *Toronto*
PRENTICE-HALL OF INDIA PRIVATE LIMITED, *New Delhi*
PRENTICE-HALL OF JAPAN, INC., *Tokyo*
PRENTICE-HALL OF SOUTHEAST ASIA PTE. LTD., *Singapore*
WHITEHALL BOOKS LIMITED, *Wellington, New Zealand*

CONTENTS

Contents   v

PREFACE

Automotive engines is always an interesting subject for those interested in automobiles and mechanics. Some are interested in how the engine is made and how it operates. Others are interested in how the engine is repaired and maintained. The 1974 edition of AUTOMOTIVE ENGINES was well accepted but it was weak in automotive service. In this book the theory has been updated to include the current production engines. The discussion of service of the automotive engine has been greatly expanded. In addition, the engine operating systems are covered. This book has been written to meet the needs of both those interested in theory and those interested in engine service.

The first section of the book describes the way the engine fits into the entire vehicle. It develops many of the common terms used throughout the book. The fourth chapter shows how the engine systems are related to the engine performance, economy, and emissions.

The second section discusses the engine parts in the same order in which an engine is disassembled and reassembled. With this manner of organization an engine can be disassembled and studied as the book is being studied. In each chapter the requirements of the parts in the engine are discussed first. This is followed by a discussion of how the parts are made to meet these requirements. The last chapter in this section is a brief discussion of engine assembly. It is assumed that the study engine will not be fitted with new parts so there is little discussion of operating clearances in this discussion of a study engine. Enough information is given so that the engine can be started and run for a short time without coolant. I have found that running a study engine is an excellent way to check the student's work on the study engine. It is also an excellent way to motivate the student to do good work.

The third section is written to be a guide for the repair of an engine that needs to be reconditioned. Like section two, this section follows the general order that would be used to repair a typical automotive engine. The approach used is to find out what is wrong with the engine, to determine why it got that way, and to decide what should be done to correct that condition. Stress is placed on the reason for the repair and the standards to which it should be reconditioned, regardless of the type of engine rebuilding equipment available for use. Illustrations show typical premature failure. These are accompanied by a discussion of what caused the condition and how it can be prevented in the reconditioned engine. Reconditioning procedures are discussed in detail. This requires several chapters. The final chapter of this section describes the procedures used to assemble an engine being used in a vehicle. It also includes typical engine repairs that do not require the engine to be completely disassembled.

The last section of this book deals with the engine operating systems that must be correctly installed after the engine has been reconditioned. These systems require service during the life of the engine. Preventative maintenance of these systems, usually called tune-up, is discussed. There is no discussion of system repair. Repair of these systems is beyond the scope of this book. Only enough information is given so that the reader will know how to keep the engine operating properly and to do preventative maintenance.

There has been an attempt to avoid discussion of specific engine makes and models. These keep changing from year to year, especially as the automobiles are downsized and the economy is increased. Obsolete features are purposely omitted unless they are needed to show the development of a current engine design feature. Experimental and limited production designs have also been avoided, except where high interest level dictates a need for a short discussion.

I wish to express my sincere thanks to the people who have provided material for use in this book. Many of my students have brought damaged engine parts that were photographed for illustrations in the book. Many graduates have sent up-to-date material to help me keep abreast of current engine technology. The unrestricted use of the automotive laboratories, automotive equipment, and training aids at Western Michigan University to take photographs is appreciated. Special recognition is given to Jack Trumbla, Automotive Instructor at the Michigan State Technical Institute and Rehabilitation Center for reading the rough draft of this book and for helping with some of the engine service photographs. Helen Horn has done an excellent job of typing the manuscript.

HERBERT E. ELLINGER

A NOTE TO THE STUDENT AND INSTRUCTOR ABOUT THE USE OF METRICS IN THIS BOOK

The metric system of measurement is starting to be used in domestic manufacturing. Cars imported into this country use metric standards. Manufacturers of domestic cars are switching over. Both customary and metric units of measurement are used throughout this book. The only exceptions are several complicated examples where the metric equivalents were omitted so that the examples would not be confusing.

Using metric units is no harder than using customary units. It is simply a matter of getting used to using them. When customary units used in this text were converted to metric units they were kept at the same level of precision. If the customary unit was "approximately 1 to 2 inches" the commonsense metric conversion would be "approximately 25 to 50 mm" or simply "approximately 1 to 2 inches (25-50 mm)." Of course, the exact conversion values are 25.4 mm to 50.8 mm but the customary unit used wasn't exact to begin with so measurements to tenths of millimeters are not necessary. Where high precision is called for conversion is made to high precision metric values.

It is important to keep in mind how big the basic length metric unit is; there are roughly 25 mm to 1 inch; thus, 0.100 inch = 2 ½ mm (2.5 mm) and 0.010 inch = ¼ mm (0.25 mm). If the specifications are given in hundreth of an inch, the specification of tenths of a milimeter is actually more exact.

Don't be alarmed if you try to convert a customary unit to a metric unit and find the value given in the text is somewhat different than your results. In the text the metric values were rounded to a commonsense value, whenever doing so would not affect the operation or description being discussed. If the results you get in making the conversion differ from that shown by more than 5 percent, recheck your calculations. Once you become accustomed to the metric system, you'll probably find it easier to use than customary units. The important thing is to practice using metrics.

ACKNOWLEDGEMENTS

A great number of individuals and organizations have cooperated in providing reference material and illustrations used in this text. The author wishes to express sincere thanks to the following organizations for their special contributions:

Auto Parts Distributors
Champion Spark Plug Company
Chrysler Corporation
Clayton Manufacturing Company
Dana Corporation
The Dow Chemical Company
Ford Motor Company
General Motors Corporation:
  AC Delco Division
  Buick Motor Division
  Cadillac Motor Car Division
  Central Foundry Division
  Chevrolet Motor Division
  Oldsmobile Division
Geo. Olcott Company
Greenlee Brothers and Company
Modine Manufacturing Company
Presolite Company
Sealed Power Corporation
Society of Automotive Engineers
Sunnen Products Company
TRW, Michigan Division

PART

I

ENGINE

INTRODUCTION

Before studying the details of engine design and repair, it is important to know the general working parts of the engine. It is also important to know what the engine is expected to do. This section of the book first discusses how the engine produces power and how the power is used in the automobile. It also includes a short review of alternative engines.

Engines are made in a number of different ways. For example, the engine may have the cam in the block or in the head. Differences in the design of the engine, such as the location of the cam, form engine classifications. The classifications of engine features used to describe engines are discussed in this section of the book.

Engines are further described by referring to the engine specifications. These are defined in this section of the book so that the terms will mean the same thing to all readers. Examples are given to show how the specifications can be used to compare the size, power, and efficiency of engines.

The modern automobile must meet local, state, and federal regulations. Automobiles have had emission controls since 1968. Fuel mileage of automobiles has been controlled since 1978. The engine is the major part that affects both emissions and fuel economy. In general, modifications made on engines to reduce emissions also reduce fuel economy. Engine systems have to be carefully designed and maintained to balance low emissions and good fuel economy. A discussion of engine emissions, their control, and their relationship to fuel economy is included in this section of the book.

1

CHAPTER

1

The Automobile Engine

Each summer, those interested in automobiles read newspapers and magazine articles describing new model automobiles, vans, and pickups. They visit their dealer showrooms when the vehicles are put on display. Here, they examine the styling features, the interior, and the power train. The engine is one of the most interesting parts of the power train.

The engine converts part of the fuel energy to useful power. This power is used to move the vehicle. There are two directions manufacturers can choose when designing an engine. They can design the engine to produce *maximum power* with no regard to the amount of fuel used. This is what is done to make racing engines. It is also what is done when building a hot rod engine. Other engines are designed to get *maximum fuel economy*. This is commonly called gas mileage. An engine that gets high gas mileage is often said to be efficient. Each engine designed for either maximum power or maximum economy must be individually hand-built using precision rebuilding procedures to fully meet the design requirements. An engine made in this way is called a *custom-built engine*.

The production engine is a compromise between the maximum power and the maximum economy that could be produced by the engine. This compromise is modified by government emission and fuel economy regulations. Within the production limits, some engines approach the maximum power limit. The best examples of this approach were the "muscle" cars of the late 1960s. During this period, fuel was cheap and emission control had just started. By the late 1970s, gasoline prices had more than doubled and fuel economy regulations were in force. Engines produced during this time were built for economy. Any individual production engine can be modified to improve either the power or the economy. Sometimes, both can be done at the same time. This is usually very expensive and causes the engine to produce higher emissions. These high emission levels are illegal in some states.

There are very few custom-built engines that do not start with standard parts from *mass-produced engines*. Mass-produced engines are the type found in typical automobiles. Standard parts are modified by the custom engine builder. One reason that mass-produced engine parts are used is that they have proven their strength and durability. Newly designed parts require a great deal of use to prove that they can do the job required of them. Another reason is that mass-produced parts are much less expensive than complete custom parts.

Each custom built engine is carefully and individually hand-built. This means that the labor cost is high. The mass-produced engine is built on complex machinery of the type shown in Figure 1-1. When the machinery can be kept in continuous operation, the cost of each individual part is very low when compared to a custom part. Mass-produced parts are nearly identical. Because of this, the parts can be interchanged between engines. This also helps to keep the cost low. Mass production gives the buyer the best compromise among maximum power, maximum economy, and low emissions for the dollar cost. Custom rebuilding produces small gains at a high cost.

1-1 Engine Power

Vehicle movement is caused by the power of the engine. Figure 1-2 shows the parts of an engine described in the following paragraph. In a gasoline-powered automotive engine, the correct amount of fuel is mixed with air by the carburetor or fuel-injection system to form the *intake charge*. This air/fuel charge is pushed by atmospheric pressure through the intake runner and an intake valve to fill the partial vacuum in the *combustion chamber*. The intake valve closes when the combustion chamber has taken as much charge as

Figure 1-1 A part of the machining line used to make mass-produced engine parts (Courtesy of Greenlee Brothers & Company).

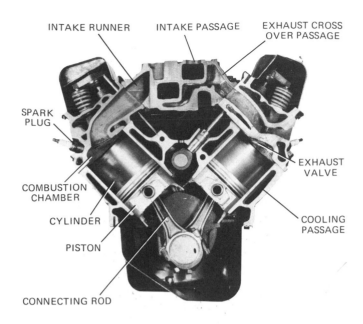

INTAKE RUNNER   INTAKE PASSAGE   EXHAUST CROSS OVER PASSAGE

SPARK PLUG

COMBUSTION CHAMBER

CYLINDER

PISTON

CONNECTING ROD

EXHAUST VALVE

COOLING PASSAGE

Figure 1-2 Names of engine parts.

form of heat. The heat increases the pressure of the gases within the combustion chamber. In a typical automotive engine, the high pressure caused by the heat forces a *piston* to move down in the *cylinder*. Piston movement and power are transferred to a rotating *crankshaft* through a *connecting* rod. When the piston approaches the bottom of its downward stroke, an exhaust valve opens, releasing the expanded combustion gases into an exhaust manifold and muffler system. The sequence of events then repeats itself.

Small utility engines usually have one cylinder. As more power is needed, more cylinders are used. Combustion alternates between cylinders to provide a smooth power output. Automobiles use reciprocating engines with pistons moving up and down in the cylinder. Inline engines have four, five, or six cylinders. V-type engines have two banks of three or four cylinders, each within the same *block* and they use the same crankshaft to form a V-6 or V-8 engine. These are illustrated in Figure 1-3.

The basic engine needs several systems. It requires a *fuel system* to transfer fuel from the tank to the carburetor or fuel injector. An *ignition system* is required to supply an electrical arc across the gap of the spark plug at the correct instant in each cylinder. A *cooling system* is needed to keep engine heat under control. Moving parts are kept

possible. The charge is compressed as the piston moves upward. The charge is ignited at the instant required to produce power. The charge will burn very rapidly. In an engine, the burning of the charge is called *combustion*. Combustion in the engine is so fast that it is often incorrectly called an explosion. To function properly, burning of the charge is controlled combustion. This combustion releases the fuel energy at a controlled rate in the

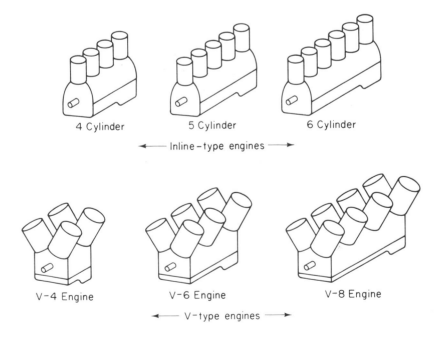

4 Cylinder   5 Cylinder   6 Cylinder

←— Inline-type engines —→

V-4 Engine   V-6 Engine   V-8 Engine

←— V-type engines —→

Figure 1-3 Automotive engine cylinder arrangements.

from touching each other by a cushion of oil supplied by the *lubricating system*. Failure of any of these engine systems will cause the engine to fail.

1-2 Where The Energy Goes

In a spark-ignited gasoline engine, only 25% of the energy in the fuel is changed to useful work as engine power at the crankshaft. In a diesel engine, this may be as high as 35%. The rest of the fuel energy is wasted as heat. About half of the wasted heat energy goes out of the engine with the exhaust gas. The other half leaves the engine through the cooling system. It may be observed that internal friction produces heat and most of this heat finds its way into the engine cooling system. In this way, the friction heat is removed from the engine.

Power Loss. Unfortunatley, all of the power produced at the crankshaft, called *gross horsepower,* is not usable to drive the vehicle. A number of power-absorbing accessories are mounted on the engine. These include a coolant pump, cooling fan, electrical charging system, fuel pump, air injection pump, air conditioner compressor, power steering, and air cleaner. Some power is also required to pull the intake charge into the combustion chamber and to push the exhaust out through a converter and muffler. When all of the power-absorbing accessories are being fully used, they will absorb about 25% of the power being produced by the crankshaft. The remaining 75% of the power at the crankshaft is usable power and it is called *net horsepower.*

The remaining useful power is further reduced as it goes through the drive line of the vehicle. The gears and bearings in the transmission and rear axle take their toll. A standard transmission uses about 2% of the net engine power, while a typical automatic transmission uses about 5%. Automatic transmissions with lock-up torque converters use less of the net engine power than the rest of the automatic transmissions.

The power needed to drive the vehicle is affected by the strength and direction of the wind and by the rolling resistance of the vehicle. Both of **these are affected by the vehicle speed, streamlining,** weight, type of tires used, road condition, road grade, humidity, and temperature.

Available Power. The total power actually available to drive the vehicle may be as little as 50% of the power delivered to the crankshaft. Reducing this loss helps both maximum power and fuel economy. There are a number of things being done to modern engines to reduce these losses. First, the engine is being modified and finely adjusted by using electronic components to precisely control fuel and ignition. Power-absorbing accessory and drive losses are being minimized. **Minimized means they are being reduced as much as possible.** Typical examples include cycling the air conditioner compressor, electric cooling fans, reduced weight, lock-up torque converters, overdrive transmissions, streamlining, and radial tires.

1-3 Drive Lines

The first automobiles were built using buggy and bicycle parts. Most of these used front wheel steering, as buggies did. The rear axle was stationary, so it was not difficult to connect the power drive to the rear wheels. The first automobiles had the engines under the driver's seat. This placed the engine close to the drive wheels so that a simple drive line could connect between them. A chain drive was generally used. As engines became more powerful, cooling the underseat engine became a problem. It was logical to move the engine forward so it could be cooled by the air hitting the front of the automobile. At the same time, the drive line was improved by using shafts and gears rather than chains.

Rear-Wheel Drive. The front-engine, rear-wheel drive became the standard drive line arrangement, as shown in Figure 1-4. A *clutch* or *torque converter* is connected to the engine crankshaft. Its purpose is to provide a means to effectively disconnect the engine from the drive line. This allows the engine to idle while the vehicle is stopped. It is engaged to drive the vehicle.

A *transmission* is located directly behind the clutch or torque converter. It is needed to provide gear reduction, which will produce high torque at the drive wheels to start the automobile moving and to drive it up steep grades. The transmission also provides a reverse gear for backing the vehicle. Gear range selection is either manual or automatic.

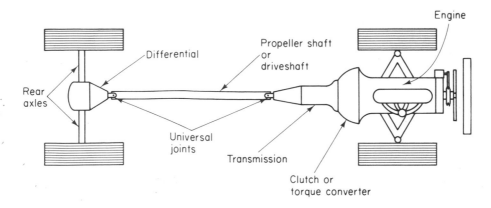

Figure 1-4 Typical front engine-rear wheel drive line.

In front-engine, rear-wheel-drive vehicles, the transmission is located under the front floor pan. A *propeller shaft* or *drive shaft* is used to carry engine power to the *rear axle*. The propeller shaft has *universal joints* on each end to provide flexibility as the suspension springs compress and expand.

Ring and pinion gears, along with a differential on the *rear axle,* splits the incoming power to each drive wheel. This also allows the drive wheels to turn at different speeds as they go over bumps and around corners.

Front-Wheel Drive. Vehicles with front-wheel drives, illustrated in Figure 1-5, usually bolt the transmission case to the differential case or they are in the same case. They do not have a propeller shaft. Short axle shafts, called *half shafts,* between the differential and the drive wheels have universal joints on each end. Front-wheel-drive drive lines are complicated because they must steer as well as drive. Two engine mounting arrangements are used on front-wheel-drive vehicles. Some have the engine mounted with the crankshaft and cylinders from back to front (longitudinal). Sometimes this is called north and south. These require the use of a ring and pinion gear set, similar to a rear axle, to split the power going to the drive wheels. A more popular arrangement is to mount the engine with the crankshaft and cylinders from side to side between the front wheels (transverse). Sometimes this is called east and west. Simple spur gears or a silent drive chain can be used in the drive line of front-wheel-drive automobiles when the engine is transverse-mounted. With transverse engine mounting, the vehicle can be designed with the maximum interior space for the vehicle size. It is, therefore, most commonly used for small vehicles.

1-4 Driveability

The operator expects the engine to start immediately when cranked and to run smoothly during idle, acceleration, and cruising. It should operate this way when the engine is either cold or hot. This type of operation is called driveability. It is observed any time the engine is operated. Therefore, driveability is more important to the operator than is

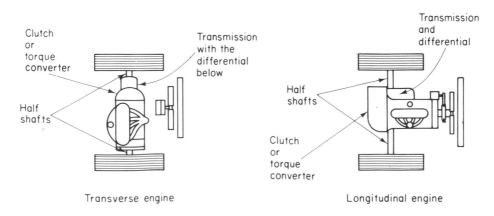

Figure 1-5 Two types of front engine-front wheel drive lines.

maximum power or maximum economy. The only way an engine can give good driveability while producing low emissions and good fuel economy is to have all the parts in good working order and adjusted correctly. When an engine does not give good driveability, the operator usually has it tuned up, even if it only requires a carburetor adjustment. This wastes time and the operator's money. In place of a tune-up, the problem should be diagnosed and the faulty component repaired, adjusted, or replaced. Then the engine should be rechecked to make sure the repair actually corrected the problem.

Improper engine operation may have an effect on other vehicle units. For example, an engine may have a faulty idle adjustment that causes a transmission shift problem. The automatic transmission shift points are controlled by movements of the throttle linkage or by manifold vacuum. If the engine does not operate correctly, the transmission may not shift correctly. Both affect driveability.

To run, the engine needs the correct air/fuel mixture charge, combustion chambers that seal properly for good compression, and good ignition at the correct instant. If one is *missing*, the engine will not start or run. If one is *faulty*, the engine will not run properly. The air and fuel are mixed in the correct ratio using a carburetor, fuel injector, or fuel metering system. Compression results from the mechanical condition of the engine. In most cases, at least partial disassembly is required to correct a compression problem. Basic ignition is timed in relation to the crankshaft angle. This is based on the piston position. Ignition timing is advanced from the basic timing using either mechanical or electronic advance mechanisms. If these do not operate properly, they are replaced. The spark plug must be in good condition to ignite the charge in the combustion chamber.

1-5 Engine Cycles

Engine cycles are identified by the number of piston strokes required to complete the cycle. **A piston stroke is a one-way piston movement between the top and bottom of the cylinder. A cycle is a complete series of events that continually repeat.** Most automobile engines use a *four-stroke cycle* (Figure 1-6).

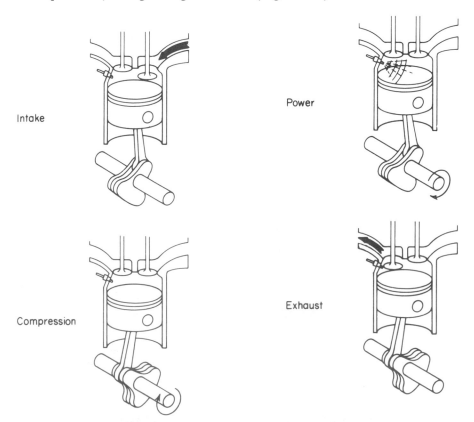

Intake

Power

Compression

Exhaust

Figure 1-6 Typical four-stroke cycle of a spark-ignited gasoline engine.

Four-Stroke Cycle. The four-stroke cycle starts with the piston at the top of the stroke. An intake valve opens as the piston moves down on the first, or *intake stroke*. The open intake valve allows the intake charge to enter the cylinder. The intake valve closes after the piston moves past the bottom of the stroke. The piston then moves up on the second stroke, called the *compression stroke*, to squeeze the intake charge into the small space of the combustion chamber. Near the top of the compression stroke, the spark plug ignites the charge and fuel burns. The heat released from the burning fuel raises the charge pressure. This pressure pushes the piston down on the third, or *power stroke*. The exhaust valve opens near the bottom of the stroke to release the spent exhaust gases. The piston moves up on the fourth stroke, called the *exhaust stroke*, to push all of the exhaust gas from the engine. This completes the 720° four-stroke cycle. The piston is then in a position to start the next four-stroke cycle with another intake stroke. The four-stroke cycle is repeated *every other* crankshaft revolution.

Two-Stroke Cycle. Some small engines use a *two-stroke cycle*. This cycle starts with the piston at top center on the power stroke. As the piston nears the bottom of the power stroke, the exhaust valve or port opens to release the spent gases. **A port is an opening in the side of the cylinder.** The intake valve or port opens very shortly after the exhaust opens and an intake charge is forced into the cylinder under pressure. This aids in pushing the exhaust gases from the cylinder. Both intake and exhaust valves or ports close as the piston starts up on the compression stroke. The two-stroke cycle engine has a power stroke *each* crankshaft revolution.

1-6 Alternative Engines

In 1961, California laws required all automobiles to have positive crankcase ventilation systems. In 1966, hydrocarbon and carbon monoxide exhaust emissions were first controlled in California. Controls of these emissions in the exhaust were established for the rest of the states in 1968, following California's lead. Nitrogen oxides were first controlled in California on 1972 automobiles. Nitrogen oxides were controlled by federal regula-

tions starting with the 1973 automobiles. Allowable emissions have been reduced every year or two since they were first controlled. To meet these tightening emission regulations, the automotive industry did two things. First, they modified the existing engine to meet the new regulations. In general, these modifications reduced both maximum power and fuel economy. The second thing the automotive industry did was to examine the potential of alternative engine types. Alternative engines continue to be under study, but none is able to solve the emission and economy problems at a competitive cost.

Fuel Economy. In 1973, the crude-oil-producing nations restricted the fuel supply and readjusted the price of crude oil. This caused a sudden shortage of gasoline and the price nearly doubled within a year. The shortage led the Congress, in 1975, to set fuel economy standards that would reduce fuel consumption 40% from the 1974 average. Regulations called for 18 mpg corporate average fuel economy (CAFE) in 1978. In 1979, the average was 19 mpg, and 20 mpg in 1980. Legislation at the time this was written raises the corporate average fuel economy to 27.5 mpg by 1985.

With manufacturing technology for mass-produced automobiles, the automobile manufacturers have been able to meet both the emission and economy regulations by producing engines that operate as economically as possible and by reducing power losses in the drive line. The most common way this is done is to use electronic engine controls, reduce vehicle weight so that a smaller engine can be used, using low-friction lubricants, installing radial tires, and streamlining the vehicle. Small engines working in their mid-power range, are more economical than large engines working at the low end of their power range. A small engine has little reserve power for acceleration into traffic or for passing. A *turbocharger* added to some small engines with good fuel economy will provide power for acceleration. At the same time, the small turbocharged engine will have good fuel economy at part throttle operation.

Diesel. The diesel engine has been commonly used in heavy vehicles and on stationary machinery. The diesel engine has high thermal efficiency, so it provides good fuel economy. The diesel engine produces low hydrocarbon and carbon monoxide exhaust emissions. These reasons make it a good alternative to the reciprocating gasoline engine for use in automobiles. Mechanically, the two engines are very much alike. The diesel engine is somewhat heavier and it is more expensive. Their greatest difference is in the fuel and ignition systems. Two factors limit the application of the diesel engine in passenger cars. First is the high price, and second is the fact that it is difficult for the diesel engine to meet the very low nitrogen oxide emission standards in the original 1981 regulations. This may change with new technology or by modifying the standards that applied to diesel engines.

Rotating Combustion Chamber. A second successful alternative engine is the rotating combustion chamber or Wankel engine. It has some advantages over a piston engine. The rotating combustion chamber engine runs very smoothly and it produces high power for its size and weight. It runs quietly because it does not use a noisy valve train. It operates on low-octane gasoline as a result of the large cooling surface around the combustion chamber. This surface causes the engine to produce low nitrogen oxide emissions. This same cooling surface causes the engine to have poor fuel economy and to produce large amounts of hydrocarbon emissions. In 1972, it looked as if the rotating combustion chamber engine was going to be one of the major alternative engines for automobiles. However, it could not compete with the piston gasoline engine on economy, durability, and exhaust emission control. There may still be a future for this engine in small vehicles when the engine technology improves. The fact that it is small and light for the amount of power it produces makes it attractive for use in small automobiles. Using this small lightweight engine, the vehicle size and the weight could also be reduced. This means that a smaller engine in a small vehicle may allow the same space for passengers and luggage, while it still produces the same performance as a larger engine in a larger vehicle.

The basic rotating combustion chamber engine consists of a triangular-shaped rotor turning in a housing. The housing is a geometric figure called a two-lobed epitrochoid. A seal on each corner, or apex, of the rotor is in constant contact with the housing, so the rotor must turn with an eccentric motion. This means that the center of the rotor moves around the center of the engine. The eccentric motion can be seen in Figure 1-7. This eccentric movement makes expanding and contracting volumes between the flat portions of the rotor and the housing. As the volume expands, an air/fuel intake charge is drawn in through an intake port. In Figure 1-7 the port is shown in the housing. When the volume reaches its largest size, the port is closed as the apex seal moves past it. Continued rotor rotation reduces the volume, compressing the charge. Spark plugs ignite the charge. The high-pressure gases developed during combustion will force the volume to expand. This pulse of power provides the engine rotating force called *torque*. When the volume again reaches its largest size, one of the apex seals moves past an exhaust port, allowing the spent high-pressure gases to escape from the engine. Continued rotation reduces the size of combustion chamber volume to force the remaining exhaust gases from the engine. This completes a cycle similar to the four-stroke cycle of the reciprocating engine. Continued rotation of the rotor starts the next cycle with the next intake charge.

While the one chamber is going through its cycle, the other two chambers formed between the rotor and housing also go through similar cycles. This produces three power pulses each time the rotor makes one revolution.

Power produced by the rotor forces an eccentric shaft to turn. The action is similar to the connecting rod and crankshaft. The eccentric shaft turns three revolutions for each revolution of the rotor. This places the eccentric in the correct position to be pushed or rotated by each power pulse. An internal gear within the rotor meshes with an external tooth gear on one of the side housings. The purpose of the gear is to keep the rotor correctly indexed to the eccentric and housing. The gears do not carry any of the torque load.

Intake and exhaust ports are located in the rotor housing on some engines and in the side housings on others. A depression in the rotor flats forms the combustion chamber. Because the combustion chamber is relatively long, some engines use two spark plugs to ignite the charge for rapid,

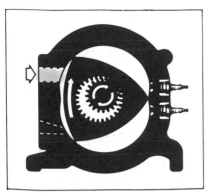

1. Chamber at minimum space volume. Intake cycle begins.

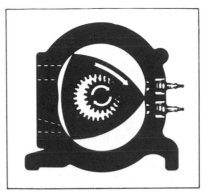

5. Start of compression cycle. Next chamber begins intake cycle).

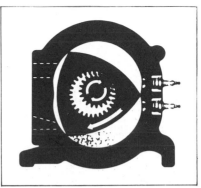

9. Power cycle continues. (Third chamber begins intake cycle).

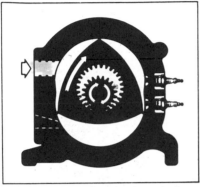

2. Rotor revolves, fuel/air mixture is drawn through carburetor (arrow).

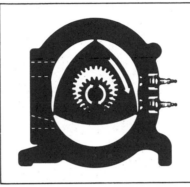

6. Fuel/air mixture is nearly compressed.

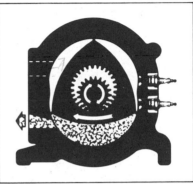

10. Expanded gases have reached maximum volume and begin exhaust.

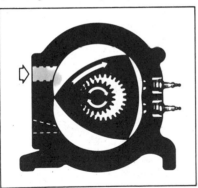

3. Chamber nearly filled with fuel/air mixture.

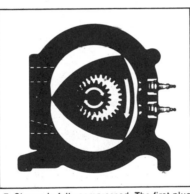

7. Charge is fully compressed. The first plug fires. Slightly later, second plug fires.

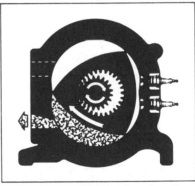

11. Burned gases discharge through port.

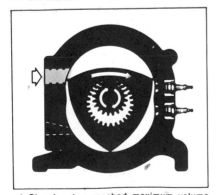

4. Chamber has reached maximum volume. Carburetor intake is almost closed and compression starts.

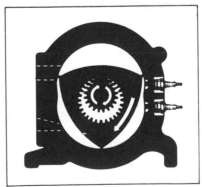

8. Gas expansion and power cycle.

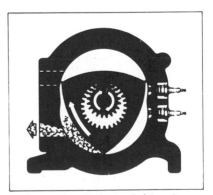

12. Exhaust cycle nears end. All cycles are then repeated on this rotor face.

Figure 1-7 Rotating combustion chamber engine cycle (Courtesy of Champion Spark Plug Company).

complete combustion. Two complete ignition systems are required when two spark plugs are used.

The rotor operates in a hot environment, so auxiliary cooling is needed. Two methods are in common use. One is to cool the rotor by running large quantities of oil through the inside of the rotor. The other is to bring the inlet charge through the rotor on its way to the combustion chamber; this will effectively air cool the rotor.

The rotor seals are a critical part of the rotating combustion chamber engine. Apex rotor seals slide on the inner housing surface to separate the combustion chambers from each other. Side seals on the rotor slide on the side housings to seal the combustion chamber from the lubricating system, bearings, and indexing gears. The most critical sealing spot in the engine is the corner seal between the apex seal and the side seal. All of the seals must fit exactly to operate correctly.

The rotating combustion chamber engine has accessory systems like those of reciprocating engines. The engine is supplied with a typical gear-type oil pump, a centrifugal coolant pump and cooling fan, carburetor, starter, electrical charging system, ignition system, and emission control devices.

Turbine. A gas turbine engine is another alternative engine. By the mid-1950s, each of the big three automobile manufacturers had gas turbine engines running in experimental automobiles. Ford and General Motors concentrated on development of turbine engines for trucks, while Chrysler concentrated on automobile applications. By 1963, Chrysler had turbine engines in a number of specially built automobiles that were loaned to the general public for evaluation. At the end of this test period, the Chysler turbine project was put back on an experimental basis.

High cost and poor fuel economy were the major reasons that the turbine engine was not put into production. A great deal of research effort has been put into the development of ceramic materials for the high-temperature sections of the engine. Parts from ceramic materials should be less costly to produce than the same parts made from special high-temperature metals. Ceramics are expected to allow the use of higher temperatures in the hot section of the gas turbine engine. This will result in both better fuel economy and greater power for the same-size engine. A schematic drawing of a typical automotive gas turbine is shown in Figure 1-8.

The power-producing member of the gas tur-

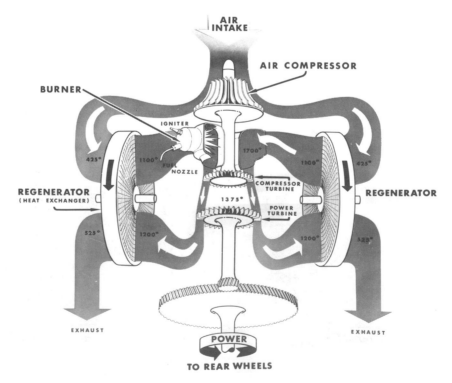

Figure 1-8 Schematic view of an automobile turbine engine (Courtesy of Chrysler Corporation).

bine has one rotating shaft with an air *compressor* on one end and a *compressor turbine* on the other. In operation, the shaft spins, turning the compressor. The compressor pulls outside air into the engine and forces it into a *burner*. Here, fuel is mixed with the air and the mixture is ignited. After initial ignition, the mixture burns with a continuous flame as long as fuel and air are supplied with a combustible air/fuel ratio. The heat developed in the burner gives increased energy to the gases. As the gases leave the burner, they are ducted to the compressor turbine. The compressor turbine, its shaft, and the air compressor are spun by these high-energy gases. A great deal of energy remains in the exhaust gases after they spin the compressor–turbine assembly. If this were a jet engine, these remaining gases would be ducted to the atmosphere to provide thrust for the aircraft. In a gas turbine, these high-energy gases are directed to a second turbine called the *power turbine*. It is geared to a transmission and drive line to power the vehicle.

The gas turbine just described is not very efficient because there is still a good deal of usable heat energy remaining in the exhaust gases. Modern experimental automotive-type turbines direct the exhaust gas through a rotating mesh *regenerator*. The latest regenerators are made from ceramic materials. The exhaust coming from the power turbine heats a rotating regenerator. The heated portion of the regenerator rotates into the air that is flowing from the air compressor to the burner. This flowing air removes heat energy from the regenerator and adds it to the compressed air. Using the exhaust heat reduces the amount of fuel required to develop the necessary engine operating temperature. Regeneration, then, increases the thermal efficiency of the engine.

The gas turbine fuel system is much more complicated than the fuel system of the piston or rotating combustion chamber engine. Fuel flow rates must be adjusted for changes in engine temperature and atmospheric pressure. A governor is required to control the maximum compressor turbine speed. A lubrication system, with an oil pump, supply tank, and cooler, forces the oil through the engine to lubricate and cool the engine bearings. The engine has accessory drives to operate the charging system, power steering pump, vacuum pump, and so on. An ignition system

operated by an electronic start controller is used during the automatic start cycle. In an automobile, large air filters and silencers are needed to handle the large volume of air required to operate the engine. Even though the actual compressor and gasifier turbine is about the size of a large starter motor, the entire engine with its accessories and air ducting fills an automobile engine compartment.

Exhaust emissions from turbine engines are very similar to those produced by diesel engines. Both engine types use excess air. This keeps hydrocarbons and carbon monoxide emissions low. Both engine types operate at high temperatures. This causes nitrogen oxides to be high. As with diesel engines, either the turbine technology will have to change or the emission standards will have to be modified before turbine engines can be used in passenger vehicles.

The big push for alternative engines occurred when tough emission standards were set. When it became evident that energy conservation was just as important as clean air, most of the research funds for alternative engines were restricted. The reason for this was that the possible alternative engines either used too much fuel and/or were too complex.

Electric. Electric motor drives produce no vehicle emissions. However, most of the electricity is produced by power plants burning fossil fuel, such as coal and oil. This moves the production of emissions from the vehicle to the electric power plant. It takes more energy at the power plant to produce a vehicle mile than would be required if the fuel were burned within the vehicle engine. One of the big drawbacks for the expanded use of electric vehicles is the battery. Present battery technology can store only a relatively small amount of energy. These batteries charge slowly and release their energy slowly. They are also very heavy. The battery weight must be carried by the vehicle at all times and this consumes energy. Electric vehicles are commonly used in lift trucks, golf carts, and so on. A breakthrough in battery technology will have to occur before the electric automobile will come into common use.

Hybrid. Some engineering is being done on hybrid vehicles. These are essentially electric vehicles with smaller batteries and a small onboard piston engine operating a large generator. The engine is sized so that it operates at its most efficient speed while driving the generator. The onboard piston engine will operate at its most efficient speed near peak torque at all times. In this way the engine speed is not affected by the vehicle speed. Using this system the hybrid vehicle is not entirely dependent on the batteries. The hybrid power gives it a much greater driving range than electric vehicles. Still the hybrid vehicle has to carry batteries, as well as the engine–generator system.

Steam. Automobiles have been operated with steam. It was used on some of the earliest automobiles in the late 1800s and early 1900s. Steam engines were dropped in favor of the gasoline engine. In the 1970s, modern steam engines were again used in some experimental vehicles. These engines require a great deal of plumbing and they must carry a large quantity of water. Steam engines have a slow cold-start. The operator must wait for heat to build up steam pressure before the vehicle will move. Steam engines do produce low emissions, but their thermal efficiency is low. This means their fuel mileage is very poor. Steam engine vehicles are heavy, so they require more energy to move than a lightweight vehicle.

Stirling. A Stirling engine has the highest potential thermal efficiency. Because it has external combus-

tion, it has very low emissions. It can burn any kind of fuel. The Stirling engine has a complicated operating mechanism and it must have very close operating clearances. It requires special high-temperature materials on the hot sections of the engine. The Stirling engine is still an advanced engineering concept.

Other Types. Other alternative automotive engine concepts that have had some laboratory testing include fuel cell engines, free piston engines, ammonia engines, hydrogen engines, and thermonuclear power. These have not been developed to the point that they could be considered as replacements for the piston gasoline engine, however.

1-7 Summary

The automobile engine has been used for approximately 100 years. During this time, many new designs have been developed. Still, the standard gasoline automotive engine is the dominant type. It has been refined until it can be manufactured at low cost, is very reliable, and can be easily repaired.

In spite of the development, a great deal of the power in the gasoline cannot be used to drive the vehicle. Heat energy is lost through the exhaust and cooling systems. It is also used up as it drives the mechanical systems in the vehicle. The mechanical systems have friction, and this is lost as wasted heat energy.

REVIEW QUESTIONS

1. What two directions can the manufacturer choose when the engine is being designed? [INTRODUCTION]
2. How do custom-built engines differ from mass-produced engines? [INTRODUCTION]
3. How does combustion produce power to turn the crankshaft? [1-1]
4. Name four engine systems. [1-1]
5. Where does the heat of combustion go out of the engine? [1-2]
6. Where does the part of the gross horsepower that is not included in the net horsepower go? [1-2]
7. Where is power lost between the net engine horsepower and the available power? [1-2]
8. What is the purpose of the clutch or torque converter? [1-3]

9. Name the major parts of the drive line on a vehicle with a front engine and rear drive wheels. [1-3] *Torque Converter, Trans - Drive Shaft U joints differential*

10. How does the drive line of a front engine with front drive wheels differ from a drive line used with a front engine and rear drive wheels? [1-3]

11. Why are front wheel drive lines more complicated than rear wheel drive lines? [1-3]

12. How do longitudinal and transverse engine mountings differ when used with front wheel drives? [1-3] *Do not have a propeller shaft*

13. What is driveability? [1-4] *Operator expects when to start - ride the end run smooth*

14. What is needed for an engine to run? [1-4]

15. Name the strokes of a four-stroke cycle. [1-5] *Intake - compression - power - exhaust*

16. What direction does the piston move on each stroke? [1-5] *up + down*

17. How many revolutions does the crankshaft turn to complete the four-stroke cycle? [1-5] *two*

18. How do the strokes of a two-stroke cycle engine differ from those of a four-stroke cycle engine? [1-5] *has a power stroke each crank shaft revolution*

19. What two things were done by the automotive industry to meet the emission regulations? [1-6] *Electronic eng control - reduced weight - radial tire - better oil*

20. List six ways manufacturers changed the automobile to improve fuel economy. [1-6] *use smaller engine + streamlining vehicle*

21. What is the advantage of adding a turbocharger to a small engine that has good fuel economy? [1-6] *will provide good acceleration*

22. What makes a diesel engine a good alternative engine for use in an automobile? [1-6] *it provide good gas economy*

23. What are the advantages of the rotating combustion chamber engine? What are the disadvantages of this type of engine? [1-6] *it use a lot of gas hi hydro carbon emission*

24. Describe the combustion cycle of a rotating combustion chamber engine. [1-6] *for each revolution there is three power strokes*

25. What are the advantages and disadvantages of turbine engines for use in automobiles? [1-6] *hi cost and poor fuel economy*

26. How does the gas turbine regenerator increase the thermal efficiency of the engine? [1-6] *regeneration or recirculated hot energy efficiency*

27. What type of exhaust emissions are a problem for a turbine engine? [1-6] *it will not meet emission control tests*

28. Why is the battery one of the drawbacks for the expanded use of electric vehicles? [1-6] *there are heavy - battery charge slowly - radiates slowly*

29. What is a hybrid vehicle? [1-6] *electric car with a smaller battery and*

30. What are the advantages and disadvantages of steam engines? [1-6] *a small piston operation & a large generator hybrid power gives much greater range than electric cars.*

CHAPTER
2

Automotive Engine Classification

Energy is required to produce power. Natural energy, such as water power, is converted to useful rotating mechanical power through mechanical devices such as water wheels and turbines. Chemical energy in fuel is converted to heat by burning the fuel at a controlled rate. This process is called *combustion*. When engine combustion occurs in a space separated from the power-producing chamber, the engine is said to be an *external combustion engine*. If engine combustion occurs within the power chamber, the engine is called an *internal combustion engine*. Engines used in automobiles are internal combustion heat engines. They convert chemical energy of the gasoline into heat within a power chamber that is called a *combustion chamber*. Heat energy released in the combustion chamber raises the temperature of the combustion gases within the chamber. The increase in gas temperature causes the pressure of the gases to increase. The pressure developed within the combustion chamber is applied to the head of a piston or to a turbine wheel to produce a usable *mechanical force*. This force is converted into useful *mechanical power*.

The internal combustion engine was used in some of the first self-propelled vehicles. Its development and use since that time has proven it to be a most reliable and economical engine. Recently, there has been a great deal of concern over the damage being done to our environment by the emissions (exhaust gases) of automotive engines. A great amount of engineering time and money has been

Rectolinear Direction one Direction

spent to develop an external combustion engine with a very low level of harmful vehicle emissions. There are a large number of complicated problems using external combustion engines that will have to be solved. They include problems with the heat-transfer medium, warm-up time, fluid sealing, boiler, condenser, transfer pumps, engine weight, and efficiency.

Most of the modern automotive engines are gasoline-powered reciprocating piston engines. Some diesel reciprocating piston engines are gradually being put in automobiles to improve the corporate average fuel economy (CAFE) required by law. The basic internal parts of the gasoline and diesel engine are similar. The parts have the same function and names in both engine types. Automobile engine parts are built as light as possible but strong enough to provide good durability. Heavy-duty gasoline engine parts, such as those used in vans and trucks, are made heavier than automotive engine parts. Diesel engine parts are made still heavier. The following discussion will primarily cover automotive engines. Reference to heavy-duty, diesel, and performance engine parts will be made occasionally for clarification.

2-1 Basic Engine Construction

Figure 2-1 shows a cutaway section of a basic automotive engine. A *piston* that moves up and down, or reciprocates, in a *cylinder,* can be seen in this illustration. The piston is attached to a *crankshaft* with a *connecting rod.* This arrangement allows the piston to reciprocate (move up and down) in the cylinder as the crankshaft rotates. The combustion pressure developed in the *combustion chamber* at the correct time will push the piston downward to rotate the crankshaft.

2-2 Engine Design Features

Internal combustion engines are described by referring to a number of their different design features. These can be broken down into the following, readily recognized features.

Operating Cycles. Four-stroke cycles and two-stroke cycles, as previously described, are the most common engine operating cycles. A third cycle, the turbine cycle, is beginning to appear in some

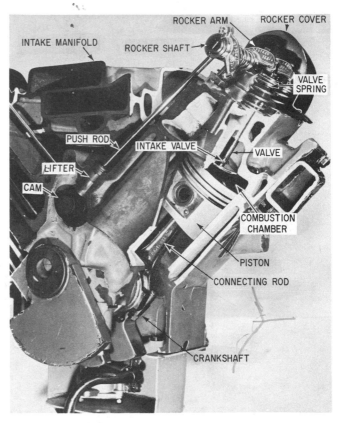

Figure 2-1 Cutaway view of a typical V-type automotive engine.

ground vehicles. This cycle is continuous, with a constant quantity of compressed air and fuel being supplied to the combustion chamber. A constant flow of high-pressure gas from the combustion chamber is directed through a turbine to produce a usable rotating force.

Ignition Types. Engines are also classified according to their type of ignition. Most automobiles use gasoline mixed with correct proportions of air, as an intake charge mixture. The charge is ignited with a spark plug at the correct instant in the cycle. It is, therefore, called a *spark-ignited,* or *SI engine.* Heavy-duty vehicles may use less expensive diesel fuel. In these engines, the air is compressed more than in spark-ignited engines. The compressed air therefore becomes quite hot. Near the end of the compression stroke, diesel fuel is injected under high pressure into this hot compressed air. The high temperature air ignites the fuel, releasing heat, which further increases the combustion chamber pressure. An engine with this type of ignition is called a *compression ignition, CI engine,* or *diesel.*

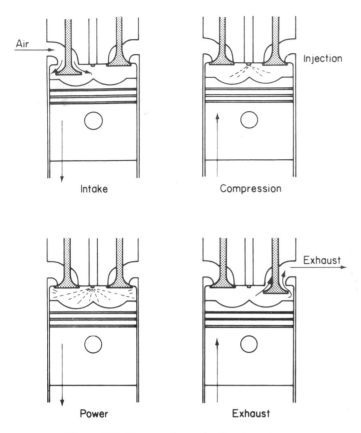

Figure 2-2 Four-stroke cycle diesel engine.

The four-stroke diesel cycle is shown in Figure 2-2 and the two-stroke diesel cycle is shown in Figure 2-3.

Cooling Methods. Most automobile engines use a liquid coolant, usually water plus an antifreeze, to maintain the engine at a constant operating temperature. This is done by transferring heat through the cylinder wall and head metal to the coolant. The coolant passage around the cylinder

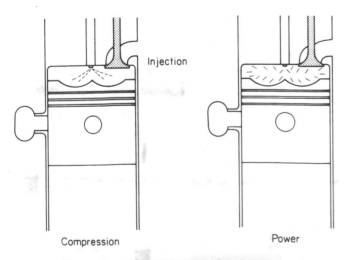

Figure 2-3 Two-stroke cycle diesel engine.

and head is called the water jacket. The coolant flows from the water jacket to a radiator. The radiator removes the heat by running the coolant through thin-walled tubes that are exposed to a flow of atmospheric air. This system is called a *liquid cooling system*. The external part of a typical cooling system is shown in Figure 2-4.

Some automobile engines maintain a constant operating temperature by transferring heat from the metal around the combustion chamber and cylinder directly to the air. This is done by directing air across fins that surround the combustion chamber and cylinder. Cooling the engine by this method is called *air cooling*. One example of air

Figure 2-4 Liquid cooling system.

cooling is shown in Figure 2-5. Air-cooled automobile engines are only found on some small imported vehicles.

Engine Configuration. The power that an engine produces is in direct proportion to the *mass* or weight of air that it uses. Engine designers can change this mass of air by (1) altering the piston diameter and stroke, (2) by changing the number of

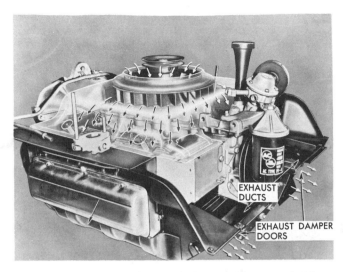

Figure 2-5 Air cooling system (Courtesy of Chevrolet Motor Division, General Motors Corporation).

cycles per minute, (3) by improving the air flow design, or (4) by increasing the number of cylinders used. Materials, manufacturing processes, and performance requirements are used to design the operating characteristics of the engine for each application requirement.

The modern automobile uses four, five, six, or eight cylinders. Generally speaking, as more power is required from the engine, more cylinders are used. More cylinders also provide smoother engine operation. Some modern V-8 engines have less displacement than older inline six-cylinder engines. In the evolution of the internal combustion spark-ignited engine, the four- and six-cylinder engines have used an inline cylinder arrangement; that is, the cylinder center lines are parallel, one next to the other. They are usually located within a single casting, called a block. A typical inline engine is shown in Figure 2-6.

Eight-cylinder engines are usually made in a form that might be considered two four-cylinder inline blocks set at a 90° angle to each other, using a common crankshaft. The two blocks are in one casting in this engine, forming a V-type engine as shown in Figure 2-7. New six-cylinder engines are also being designed as V-type engines. When compared to an inline engine, the V-type engine has a number of advantages. First, it has a compact overall size. This allows it to be used in smaller and lighter vehicles. Second, it is a rigid, compact design. This makes it much stronger for the material used and it makes less mechanical noise.

Figure 2-6 Inline engine.

Figure 2-7 V-type engine.

For the same displacement, the V-type engine can be built lighter while still producing the same power. The V-6 design is more readily adaptable to small, front-wheel-drive vehicles than either the V-8 or the inline six. It operates more smoothly than a four-cylinder engine of equal power.

Several rear engine drive automobiles use a four- or six-cylinder engine with half of the cylinders at 180° to the other half. The cylinder centerlines are usually horizontal and they are, therefore, called *horizontally-opposed engines*. A horizontally-opposed air-cooled engine is shown in Figure 2-5.

Notable examples of still different combinations in production are V-4, V-12, V-16, and inline five and eight. These are usually limited-production engines and are built to serve special operational requirments.

Valve Arrangement. Engines may be classified according to the location and type of the valve system employed (Figure 2-8). In one type, the valves are placed in the block right next to the cylinder. This allows the inlet and outlet passages to be short. The head is a simple top for the cylinder. It includes a passage for coolant, an opening for a spark plug, and a space that forms the top of the combustion chamber. With both valves located on one side of the cylinder, a cross-sectional view has an L-shape. This type of valve arrangement is, therefore, called an *L-head* or flat-head engine.

A modification of this type of engine has one valve on each side of the cylinder. It is called a *T-head* engine because of its appearance in a cross-sectional view.

Most current automotive engines have both valves in the cylinder head. This reduces the cost of the engine block and it allows better engine breathing with large inlet and exhaust passages. These passages are called *ports*. Both ports may be on one side of the head or one on each side. The head for the valve-in-head engine is a large complex casting. It has openings for the valve ports, coolant, valve operating mechanism, and lubricant. The cost of this complex type of cylinder head is offset by the reduced cost of the simple block design. Engines having the valves in the head are called an *overhead-valve* engine or an *I-head* engine.

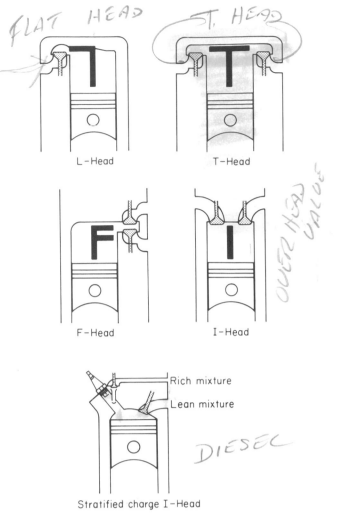

Figure 2-8 Valve arrangements.

An engine combining the features of both the L-head and the I-head engines has been produced. One valve is in the head and the other valve is in the block. This engine type is called an *F-head*. It has many of the advantages of both L-head and I-head engines. It also has most of their disadvantages as well. Very few engines have been built with an F-head.

A modification of the I-head engine includes a third small valve. It is located with the spark plug in a *precombustion chamber*. This chamber is connected by a small passage to the combustion chamber. On the intake stroke, a fuel rich intake charge is drawn into the combustion chamber through the small valve. At the same time, a lean intake charge mixture is taken in through the normal intake valve. On the compression stroke, some of the lean mixture is pushed into the precombustion chamber. Ignition occurs easily in the fuel rich charge in the precombustion chamber. The hot burning gases rush from the precombustion chamber into the lean charge in the main combustion chamber, igniting it. In this way, a very lean charge can be burned in the engine to minimize unwanted exhaust emissions. This method of combustion has two different combustion charge mixtures: rich and lean.

One modification of this engine brings fresh air in through a small third valve in the engine. This valve is operated by the intake valve mechanism. At low engine speeds, a high-velocity air stream enters through the small valve. This stream of air cleans the spark plug. It also thoroughly mixes the intake charge for efficient combustion. At high engine speeds, the normal flow of the intake charge through the standard intake valve is high. This normal flow causes the charge to thoroughly mix. At the same time, the air flow through the small valve is low at high engine speeds.

Combustion Systems. A great deal of engineering effort has gone into the study of engine combustion. As a result of these studies, refinements have been made to improve engine efficiency while maintaining low emission levels. The standard automotive engine thoroughly mixes the entire air and fuel charge before burning. The thoroughly mixed charge is called a *homogenous* mixture. The total charge mixture must be in the combustible air/fuel mixture range. The charge is ignited directly by the spark plug. Other types of engines ignite a rich, homogenous charge mixture in the precombustion chamber attached to the main combustion chamber. The burning fuel rushes out of the precombustion chamber to ignite a lean charge in the main combustion chamber.

When the charge is not thoroughly mixed before combustion, the charge is said to be *stratified,* or in layers. Part of the charge is too rich to burn and part is too lean to be ignited by a spark. The total charge may have the correct mixture or it may be lean. The spark plug is located in the part of the charge where the mixture has a combustable air/fuel ratio. After ignition, burning spreads to the rich and lean parts of the charge. A diesel engine is one type of the stratified charge engines. Spark-ignited engines can also have stratified charges. Some of these were described in the preceding section.

Camshaft Location. In engines that have valves in the block, the valves are operated by a camshaft located directly below the valves. An L-head engine uses this arrangement. In a valve-in-head engine, the camshaft is usually located in the block. An alternative location in some engines places the camshaft above the valves on the head. This is called an *overhead-cam* engine.

When the camshaft is located in the block, the overhead valves are driven through a lifter, pushrod, and rocker arm assembly. This type of engine will be called a *pushrod engine*. When the camshaft is located on the head, the valves are opened by some type of cam follower. Typical camshaft arrangements are shown in Figure 2-9.

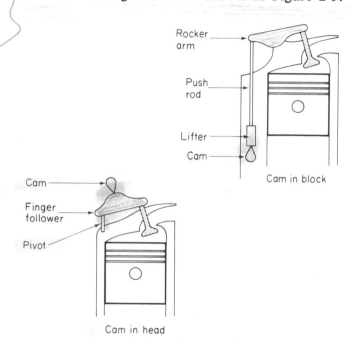

Figure 2-9 Camshaft locations.

2-3 Summary

A number of engine design characteristics are used to describe and identify automotive engines. The most common are the four-stroke cycle, spark ignition, liquid-cooled, V or inline cylinders, and overhead valves with either pushrods and rocker arms or overhead-cam mechanisms. New design characteristics are continually being developed. These include the stratified charge, CVCC, and MCA combustion chambers; counter-rotating balance shafts; revised fuel metering and fuel injection; electronic engine controls; and so on. Many of these features will be discussed in the chapters that follow.

REVIEW QUESTIONS

IN LINE V 4, V-6, V8, X, Y, DIESEL

1. List seven readily recognized engine design features. [2-2]
2. Name two types or ways used for each of the design features listed in question 2. [2-2]
3. What are the advantages of a V-type engine over an inline-type engine? [2-2] *IT WEIGHT LESS, IT IS STRONGE - IT IS SMALLER - IT MAKE LESS NOISE*
4. How does a precombustion chamber engine differ from other types? [2-2]
5. How does a stratified charge differ from a homogenous charge? [2-2]

STRATIFIED charge HAS LAYER - PART LEAN - PART RICH.

HOmogenous = IS EQUaly DISTRIBUTION.

TAKES A THIRD VALVE WITH RICH MIXTURE IT MINIMIZE EXHAUST EMISSION

CHAPTER
3

Automotive Engine Specifications

When comparing engines or discussing performance trends and engine modifications, it is important to have a clear understanding of terms and calculations. Calculations involve the use of mathematics. If terms and calaculations are not thoroughly understood, any discussion of engine performance and modification will lead to confusion.

A number of physical terms are used when talking about engines. These will be reviewed to help clarify the meaning of calculations that are used to describe engines and the way they operate. Many terms that describe engines are carelessly used because they are not thoroughly understood.

A numerical value for each physical term is used in the calculation of engine specifications. Many of these specifications are used to describe and compare engines. Therefore, it is important to review them.

Until the last few years, all engineering measurements in the United States have been in *customary units,* such as the inch, foot, and pound. Most of the rest of the world has been using some form of the metric system of measurement. Measurements throughout the world are being standardized under an international system called the *SI metric* system. Metric terms are based on units of 10. For instance, the basic length measurement is the meter. Long distances are measured in kilometers (1000 meters). Small distances are measured in millimeters (one thousandth of meter). **To minimize the chance of confusion, the** following examples are given primarily in customary units. Where applicable, metric examples are also shown.

The prefixes used for metric values are as follows:

Multiplication Factor	Prefix	Symbol
$1000 = 10^3$	kilo	k
$100 = 10^2$	hecto	h
$10 = 10$	deka	da
$0.1 = 10^{-1}$	deci	d
$0.01 = 10^{-2}$	centi	c
$0.001 = 10^{-3}$	milli	m
$0.000\ 001 = 10^{-6}$	micro	μ

3-1 Work, Torque, and Power

The power produced by an engine is usually considered to be the most important operating characteristic. The characteristic is the special property being discussed. It is interesting to note that power is not a basic measurement. **Power is the amount of work done in a specified time.** For example, it takes more power to accelerate a vehicle from 0 to 60 mph in 6 seconds than to do it in 10 seconds. Time in seconds is fully understood. Work may not be understood.

Work is the result of a force acting through a distance. In the customary units, force is measured in pounds and distance in feet. In the metric system, force is measured in kilograms and distance in meters.

It requires work to lift an engine out of a vehicle. In the following example, assume that an engine weighs 440 lb (200 kg). It has to be lifted 4.92 ft (1.5 m) to raise it high enough to clear the chassis. The work required to lift the engine would be **2165 ft/lb (300 kg•m).**

Force	×	distance	=	work (3-1)
engine weight	×	lifted	=	work done
400 lb	×	4.92 ft	=	2165 ft/lb
200 kg	×	1.5 m	=	300 kg•m

The engine crankshaft does not lift weight to do work, but it rotates. The twisting effort of the shaft is called torque. **Torque is a rotating force being produced at a distance.** The numerical values of torque are similar to those for work—ft/lb and kg•m. Technically, torque should be expressed, in the customary units, as pound feet, to distinguish it from work. However, it is commonly called *foot pounds*. In the metric system, torque is measured in newton meters. Torque is somewhat different from work in the metric system. Torque in the metric system is specified in newton meters (N•m). Kilogram meters are often used for torque in the metric system (kg•m = 9.8 N•m).

The following conversion factors are useful when working torque problems:

MULTIPLY:	BY:	TO GET:
ft lb	1.356	N•m
N•m	0.737	ft lb
ft lb	0.138	kg•m
kg•m	7.246	ft lb
N•m	0.102	kg•m
kg•m	9.807	N•m

These same torque values are used when tightening nuts and cap screws during engine assembly. A tool called a *torque wrench* is used to properly tighten fasteners at the correct torque or tightness.

It would take the *same* amount of work to gradually lift the engine from the chassis in 1 hour or to lift it in 1 minute. It takes 60 times more power to lift the engine in 1 minute than it does to do it in 1 hour. This example shows that power is the amount of work done in a given period of time.

The power an engine produces is called horsepower. *Horsepower* is the power measurement used in the customary unit system. The term *watt* is used for power measurement in the metric system. One horsepower is the power required to move 500 ft lb/min in 1 second (33,000 ft lb/min). One watt is the power to move 1 newton meter per second. This value is too small to be useful in measuring engine power, so the term *kilowatt* is used (kW = watts × 1000). To relate the value of the power units, 1 horsepower equals 0.746 kilowatt. One horsepower is, therefore, just over three-fourths of a kilowatt. To say it another way, 1 kilowatt equals 1.34 hp.

The actual horsepower produced by the crankshaft of an engine is measured with a *dynamometer* connected to the engine crankshaft. This is illustrated in Figure 3-1. A dynamometer is a piece of engineering equipment that places a load on the engine. This causes the engine to work. The load

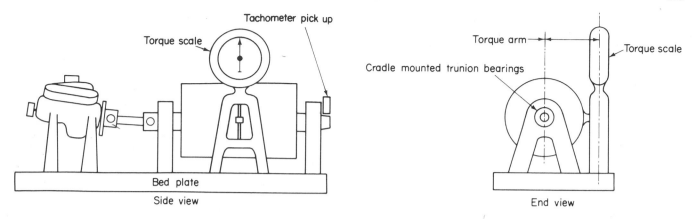

Figure 3-1. Line drawing of an engine dynamometer.

may be mechanical, electrical, or hydraulic, depending on the design of the dynamometer. The load holds the engine speed so it is called a brake. The horsepower measured on a dynamometer is called brake horsepower. The dynamometer has some means to control the amount of load on the engine. It also has a means to measure the amount of twisting force the engine crankshaft places against the load. The engine speed will slow as the load is increased. It will speed up as the load is decreased.

The load on the engine is a *force*. It is measured in pounds, kilograms, or newtons. An arm is attached to the housing of the dynamometer. A measuring device, called a torque scale, is placed on the end of the arm. This can be seen in Figure 3-1.

Here it is necessary to go back to some basic fundamentals of geometry. In Figure 3-2, the length of the diameter of a circle is placed on the surface (circumference) of the circle as many times as possible. This could be repeated three times on the circle with a small arc remaining. This small remaining arc is 0.1416 of the diameter. The distance around the surface of the circle (circumference) is equal to 3.1416 diameter lengths. The Greek lower-case letter pi, π, is used to express this relationship (π = 3.1416).

Each turn of the crankshaft equals 3.1416 times the diameter. The scale arm length is the radius. The diameter used in this discussion is twice the length of the scale arm. In one revolution, the *work* done equals the scale reading (force) times the distance moved (2 π times the arm length). Power can be determined by knowing how quickly the work is being done. On engines, the number of revolutions the crankshaft makes in 1 minute is

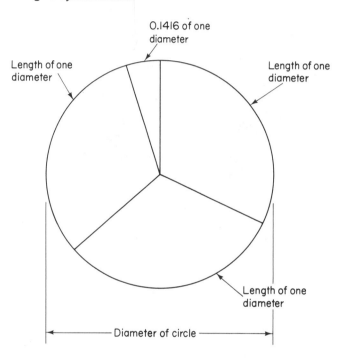

Figure 3-2. The meaning of pi (3.1416).

used for the time base. One horsepower is required to do 33,000 ft/lb of work in 1 minute.

The equation used to determine engine horsepower is based on the preceding facts. The force being applied is the engine torque (scale reading times the arm length as shown in Figure 3-3). The distance traveled in one revolution of the crankshaft is π times twice the radius (arm length). The distance traveled in 1 minute equals $2\pi R$ times the engine rpm. Multiplying these values gives the work done in 1 minute. This numerical answer must be divided by the work equivalent that 1 horsepower can do in 1 minute (1 hp = 33,000 ft lb/min).

The equation for determining engine brake

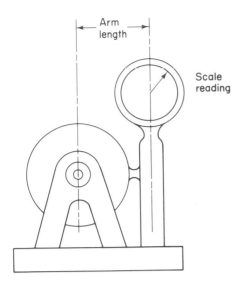

Figure 3-3. Engine dynamometer readings.

horsepower using values measured with a dynamometer is expressed as follows:

$$bhp = \frac{2\pi R \text{ ft} \times S \text{ lb} \times rpm}{33,000 \text{ ft lb/min}} \qquad (3\text{-}2)$$

where

$$R = \text{arm length (radius)}$$
$$S = \text{scale reading}$$
$$R \times S = \text{engine torque } (T)$$

This equation can be simplified as follows:

$$bhp = \frac{2\pi}{33,000 \text{ ft lb/min}} \times T \text{ ft lb} \times rpm \qquad (3\text{-}3)$$

$$= \frac{1}{5252} \times T \times rpm \qquad (3\text{-}4)$$

The form of the equation that is usually used for brake horsepower is

$$bhp = \frac{\text{torque} \times rpm}{5252} \qquad (3\text{-}5)$$

On most dynamometer systems, the scale arm length is made so that the effective length of the arm is equal to 5.252 ft (63.025 in.). This produces a simple-to-use dynamometer constant of 1000. The simplified equation to use with this type of dynamometer system becomes

$$bhp = \frac{\text{scale reading} \times rpm}{1000} \qquad (3\text{-}6)$$

3-2 Displacement and Compression Ratio

Assuming that all other factors are equal, the power an engine produces is directly proportional to the mass of the air used to burn the fuel. Mass is specified in pounds or kilograms, so it is usually equated to weight. Weight, however, is only the effect of gravity on an object. This can best be understood when one sees television pictures of an astronaut moving a large object in the zero gravity of space. The object has all of the same material it had when it was on earth. Its mass remains the same, but its weight keeps becoming less as it is moved away from the earth in the spacecraft. In our study of automotive engines, the mass of air will be considered the same as the weight of air.

A number of factors control the amount or mass of air that can enter an engine. One of the most easily understood is engine speed. With a wide-open throttle, the amount of air entering an engine each minute keeps increasing as the engine speed increases. At some point the engine reaches the speed at which it produces maximum power when no more air can enter the engine. Any increase in speed would cause a loss in the power produced. This can be seen at the right edge of Figure 3-8 (See page 29). A turbocharger can force more air into an engine to increase the power produced. The amount of air entering the gasoline engine is restricted by a throttle valve to control engine power. Other restricting factors are designed into the engine. These are beyond the control of the operator. They include engine displacement, passage size of the intake and exhaust system, and the opening and closing of the valves.

Engine size is described as displacement. **Displacement is the cubic inch or cubic centimeter volume displaced or swept by all of the pistons in one revolution of the crankshaft.** Engine displacement is not directly related to the number of cylinders or to the cylinder arrangement. Figure 3-4 illustrates the displacement of one cylinder.

The displacement of any engine can be determined easily. The cross-sectional area of the cylinder is multiplied by the stroke and the number of cylinders. Because cylinders are round, the cross-sectional area of the cylinder can be calculated using the familiar equation for the area of a circle, πr^2, where r is the radius of the circle. The cylinder bore (diameter) is given, so it is common to replace r with

26

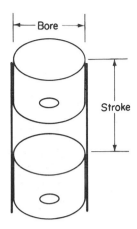

Figure 3-4. Dimensions used to determine the displacement of one cylinder.

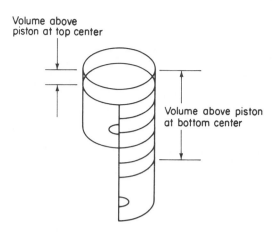

Figure 3-5. Dimensions used to determine the compression ratio of one cylinder.

the bore divided by 2. The equation becomes

$$\text{Area} = \pi \left(\frac{\text{bore}}{2}\right)^2 \text{ or } \frac{\pi}{4} \text{ bore}^2 \qquad (3\text{-}7)$$

The equation can be simplified to the form

$$\text{Area} = \frac{3.1416}{4} \text{ bore}^2 \text{ or } 0.785 \text{ bore}^2 \qquad (3\text{-}8)$$

The equation for engine displacement becomes

$$\text{Displacement} = 0.785 \times \text{bore}^2 \times \text{stroke} \times \text{number of cylinders} \qquad (3\text{-}9)$$

Example: (customary units): What is the displacement of an eight-cylinder engine that has a 3.5-in. bore and a 3.75-in. stroke?

$$\text{Displacement} = 0.785 \times 12.25 \text{ in.}^2 \times 3.75 \text{ in.} \times 8 \text{ cylinders}$$
$$= 288.5 \text{ in.}^3$$

Example: (metric units): What is the displacement of a four-cylinder engine that has an 8-cm bore and a 10-cm stroke?

$$\text{Displacement} = 0.785 \times 64 \text{ cm}^2 \times 10 \text{ cm} \times 4 \text{ cylinders}$$
$$= 2009.6 \text{ cm}^3$$

Compression ratio is often confused with displacement because each is related to piston position and to cylinder volume. In most engines, when the piston is at the top of its stroke, the top of the

piston is even or level with the top of the block. The combustion chamber volume, above the piston when it is at the top of the stroke, is the cavity in the cylinder head. This is modified slightly by the shape of the head of the piston. The combustion chamber volume, *added to* the displacement of one cylinder, will give the volume above the piston when it is at the bottom of the stroke. **The compression ratio is the ratio of the volume in the cylinder above the piston when the piston is at the bottom of the stroke to the volume in the cylinder above the piston when the piston is at the top of the stroke.** Compression ratio of one cylinder is illustrated in Figure 3-5. The equation for compression ratio is

$$\text{Compression ratio} = \frac{\text{volume in cylinder (piston at bottom center)}}{\text{volume in cylinder (piston at top center)}} \qquad (3\text{-}10)$$

where the volume in the cylinder (piston at bottom center) equals the cylinder displacement plus the combustion chamber volume.

Example: (customary units): What is the compression ratio of an engine with a 50.3-in.³ displacement in one cylinder having a combustion chamber cavity of 6.7 in.³?

$$\text{Compression ratio} = \frac{(50.3 + 6.7) \text{ in.}^3}{6.7 \text{ in.}^3}$$
$$= \frac{57.0}{6.7} = 8.5$$

27

Example: (metric units): What is the compression ratio of an engine with a 322.5-cm³ displacement in one cylinder having a combustion chamber cavity of 43 cm³?

$$\text{Compression ratio} = \frac{(322.5 + 43)\ cm^3}{43\ cm^3}$$

$$= \frac{365.5}{43} = 8.5$$

An increase in compression ratio increases both compression pressure and temperature of the compressed intake charge. When the spark ignites the highly compressed charge, a higher combustion pressure is produced. Higher compression ratios gives the engine more turning effort or torque. However, they are limited by abnormal combustion (knock or detonation). Fuels with high-octane ratings are used to give knock-free operation. High-octane fuels do not decompose and cause knock as rapidly at high combustion chamber temperatures and pressures as low octane fuels.

3-3 Types of Horsepower

The combustion pressure on the piston develops the force that gives torque to the crankshaft. On a given engine, the only way that torque can be changed is to change the combustion pressure. This can be done by changes in the amount of charge taken into the engine, changes in the ignition timing, changes in the coolant temperature, and changes in the air/fuel mixture ratio of the intake charge.

The actual power produced by an engine is called brake horsepower. It is measured with a dynamometer, as discussed in Section 3-1. Theoretically, more power is produced in the cylinder than reaches the crankshaft. **The theoretical power produced by an engine is called indicated horsepower.** The basis for this horsepower is calculated from the engine size, operating speed, and the pressure developed in the cylinder. The combustion pressure in the cylinder is shown on an oscilloscope. It uses a cylinder pressure indicator attached to the spark plug (Figure 3-6). The term "indicated horsepower" comes from this cylinder pressure indicator. The

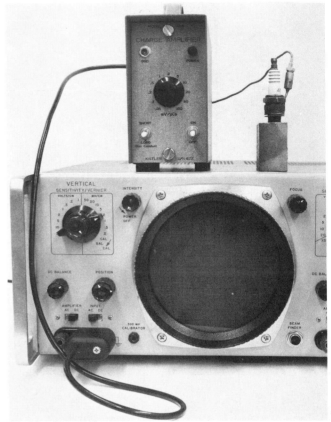

Figure 3-6. Equipment used to measure cylinder pressure as it changes during the compression and power strokes.

indicated pressure continually changes throughout the cycle. First, the average or mean pressure is determined from the indicator. The mean pressure is a constant pressure that would give the same torque as the changing pressure in the combustion chamber. This constant pressure is called the *mean effective pressure* (mep). Figure 3-7 shows the indicated mean effective pressure on a graph of the typical pressures in the combustion chamber during a four-stroke cycle.

Engine torque results from the mep acting on the top of the piston (area of the cylinder cross section) during the entire stroke. The pressure of combustion forces the piston down on the power stroke in each cylinder. Engine power results from the total number of power strokes per minute. In the automotive engine, there is one power stroke in each cylinder for each *two* crankshaft revolutions. The total power developed in 1 minute must be divided by 33,000 ft lb/min to give indicated horsepower (ihp). Indicated horsepower is easily

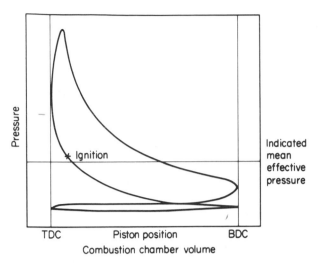

Figure 3-7. A pressure-volume indicator graph showing the line of equal pressure called indicated mean effective pressure (IMEP).

calculated using the following equation:

$$\text{Indicated horsepower} = \frac{\text{PLANK}}{33,000} \quad (3\text{-}11)$$

where P = average or mean indicated pressure (mep)
$\quad L$ = length of stroke, in feet
$\quad A$ = area of cylinder cross section
$\quad N$ = number of power strokes per minute
$\quad K$ = number of cylinders

Example: (customary units): If an eight-cylinder engine with a 3.5-in. bore and a 3.75-in. stroke produces its maximum horsepower at 4400 rpm, what is its indicated horsepower when the average indicated pressure is 131 lb/in.²?

$$\text{Indicated horsepower} = \frac{\begin{array}{c}131 \text{ lb/in.}^2 \times 0.3125 \text{ ft} \times \\ 9.62 \text{ in.}^2 \times 2200 \times 8\end{array}}{33,000 \text{ ft lb/hp}}$$
$$= 210 \text{ ihp}$$

The metric SI unit for power is the watt. Horsepower multiplied by 746 equals power in watts. In the example above, the engine would produce 156,660 W (210 × 746) or 156.66 kW.

The brake horsepower (bhp) of this same engine would be quite a bit less than the indicated horsepower. Assume that this engine produced 221

lb-ft of torque on a dynamometer when it was run at 4400 rpm. Substitute these values in the brake horsepower equation (3-5):

$$\begin{aligned}\text{Brake horsepower} &= \frac{\text{torque} \times \text{rpm}}{5252} \\ &= \frac{221 \text{ lb ft} \times 4400 \text{ rpm}}{5252} \\ &= 185 \text{ bhp}\end{aligned}$$

A typical torque and horsepower curve is shown in Figure 3-8.

There is an apparent loss of horsepower between the indicated horsepower developed in the cylinder and the horsepower delivered by the crankshaft. This loss is defined as friction horsepower. **Friction horsepower (fhp) is indicated horsepower less brake horsepower.**

$$\text{Friction horsepower} = \text{ihp} - \text{bhp} \quad (3\text{-}12)$$

Using the same engine conditions as in the preceding examples, the friction horsepower of this engine will be

$$\text{Friction horsepower} = 210 \text{ ihp} - 185 \text{ bhp} = 25 \text{ fhp}$$

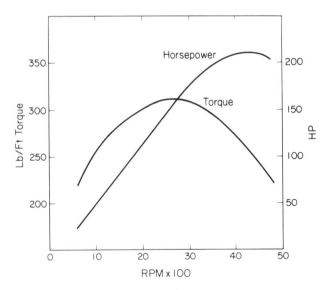

Figure 3-8. The typical shape of torque and horsepower curves of a spark ignited engine operating at full throttle.

3-4 Engine Efficiency

Efficiency is the output divided by the input. Both output and input must be in the same terms. It is commonly said that one engine is more efficient than another engine. Any expression of engine efficiency must be related to some common input and output values.

Horsepower values may be used to compare engines. Brake horsepower divided by indicated horsepower gives the *mechanical efficiency* of the engine. It is expressed as

$$\text{Mechanical efficiency} = \frac{\text{bhp (output)}}{\text{ihp (input)}} \quad (3\text{-}13)$$

Example: What is the mechanical efficiency of an engine that produces 210 indicated horsepower and 185 brake horsepower?

$$\text{Mechanical efficiency} = \frac{185 \text{ hp}}{210 \text{ hp}} = 88\%$$

Volumetric efficiency relates the actual air used by an engine compared to the maximum possible air that could be used at that speed.

$$\text{Volumetric efficiency} = \frac{\text{actual air used (output)}}{\substack{\text{maximum possible air} \\ \text{used (input)}}} \quad (3\text{-}14)$$

Example: What is the volumetric efficiency of an engine that will displace 367 ft³/min when it is using 330 ft³/min?

$$\text{Volumetric efficiency} = \frac{330 \text{ ft}^3/\text{min}}{367 \text{ ft}^3/\text{min}} = 90\%$$

Any change in throttle position, engine speed, or engine load will change the volumetric efficiency of an engine. At slow speeds and full throttle, there is enough time to fill the combustion chamber with air at atmospheric pressure. As engine speed in- creases, there is less time for the air to move through the intake valve so the volumetric efficiency is less. Volumetric efficiency also drops as the throttle is closed to reduce the air flow into the engine. The speed and torque that an engine pro- duces are controlled by changing the engine's volumetric efficiency with the throttle.

A third type of engine efficiency is *thermal effi- ciency*. This relates the maximum available heat energy in the fuel to the heat equivalent of the brake horsepower that the engine produces. One horsepower is equivalent to 42.4 Btu/min in customary units. Gasoline has approximately 110,000 Btu/gal. Using these figures, the thermal efficiency of an engine can be calculated by the for- mula

$$\text{Thermal efficiency} = \frac{\text{bhp} \times 42.4 \text{ Btu/min}}{\substack{110,000 \text{ Btu/gal} \times \text{gal} \\ \text{used/min}}} \quad (3\text{-}15)$$

Example: What is the thermal efficiency of an engine developing 15 hp at 50 mph using 2 gal of gasoline per hour (25 mpg or 1/30 gal/min?)

$$\text{Thermal efficiency} = \frac{15 \text{ hp} \times 42.4 \text{ Btu/min}}{110,000 \text{ Btu/gal} \times \frac{1}{30} \text{ gal/min}}$$
$$= 17.3\%$$

Maximum gasoline engine thermal efficiency is approximately 25%. The rest of the heat energy is used to overcome friction or it is expelled with the exhaust or through the cooling system. This was discussed in Section 1-2.

3-5 Summary

A number of terms are used to compare engines. The most common are torque, power, efficiency, displacement, and compression ratio. There are several types of power and efficiency. These terms must not be confused for a clear discussion of engines and their characteristics.

REVIEW QUESTIONS

1. How do work and power differ? [3-1]
2. How does torque differ from work? [3-1]
3. What is the purpose of a dynamometer? [3-1]

4. How does the mass of the air used by an engine affect the power produced by the engine? [3-2]
5. What limits the mass of air an engine can use? [3-2]
6. Calculate the displacement of a 4-cylinder engine that has a 4 in. bore and a 3 in. stroke. [3-2]
7. Calculate the compression ratio of the engine in question 6 when the combustion chamber volume is 5 in.^{3} [3-2]
8. On a given engine, what is done to change the crankshaft torque? [3-3]
9. How does brake horsepower differ from indicated horsepower? [3-3]
10. How does mechanical efficiency differ from friction horsepower? [3-4]
11. What causes changes in volumetric efficiency? [3-4]
12. How is volumetric efficiency used to change the engine speed and torque? [3-4]
13. What common terms does the average automobile operator use to describe the thermal efficiency of the engine? [3-4]

CHAPTER
4

Engine Emissions
and Economy

The automotive electrical system provides the electrical power for cranking the engine and for ignition. An operating engine produces the mechanical power necessary to operate the charging system and to move the vehicle.

To start an engine, an air/fuel intake charge is drawn into the engine as the starter motor cranks. This charge is ignited at the correct instant by the ignition system and the engine begins to run. The engine runs by converting part of the energy of the fuel into useful work.

All of the energy used to operate an engine comes from the fuel. In a spark-ignited engine, the fuel is usually gasoline. In some cases, a small amount of alcohol is added to the gasoline. Gasoline and alcohol are almost entirely composed of relatively volatile hydrocarbon molecules. **Volatile fuels evaporate easily and quickly.** Molecules that make up the fuel have many different physical and chemical properties. Motor gasoline is blended from selected parts of the refined petroleum to meet the operating conditions found in spark-ignited reciprocating engines. Alcohol is often added to gasoline to reduce the amount of petroleum required to operate engines.

Many hydrocarbons in gasoline start to combine with oxygen at temperatures below 600°F (320°C). This temperature is encountered in the combustion chamber on the compression stroke before ignition takes place. After ignition the products of combustion are mostly gases and a large quantity of heat energy. It is this heat energy that increases the pressure of the gases in the combustion chamber.

4-1 Normal Combustion

Liquid gasoline must be changed to a vapor to burn in an engine. This happens when the gasoline evaporates. In engines using a carburetor, or throttle body injection, vaporization of the gasoline is done in one-third of a second at idle speeds. It only takes one-thirtieth of a second at normal operating speeds. In fuel-injected engines, fuel vaporization must occur much faster. Vaporization occurs rapidly because liquid gasoline is broken into a sudsy foam that will rapidly mix with the intake air. For ideal combustion, the molecules of fuel must combine and mix with the correct number of molecules of oxygen in the air. **The chemically correct mixture is said to be stoichiometric and it has an equivalence ratio of 1.**

Air Density. At sea level, the air is heavy and dense. A relatively small quantity of sea level air supplies enough oxygen for a given amount of gasoline. The air becomes lighter and less dense at high altitudes and at high atmospheric temperatures. For a given volume the high altitude air contains less oxygen molecules than sea level air. Using the same volume the light air will cause the intake charge to become richer than if it were sea level air. The effect of air density becomes critical on some emission-controlled engines. Automobiles used high up in the mountains have engines that require leaner carburetor settings than those used at sea level. Automobiles are frequently operated in both high mountains and at sea level. Because of this some carburetors are made with altitude compensation devices. They automatically prevent overly rich mixtures when operated at high elevations.

The combustion process takes place after the intake valve closes and before the exhaust valve opens. The intake charge is trapped in the combustion chamber. Here the molecules of oxygen in the air come into contact with the hydrocarbon molecules of the gasoline. Because they are together they burn rapidly.

Combustion Products. When a gallon of gasoline is completely burned, it produces nearly a gallon of water. In addition, it produces sulfur dioxide in an amount that depends on the sulfur content in the gasoline being used. At normal operating temperatures, the water produced during combustion is in a vapor form. It leaves the cylinder as part of the exhaust gas. Condensed water vapor is visible coming from the tail pipe when the engine is first started in cold weather. Condensed moisture with sulfur dioxide makes the water acidic and this is corrosive. When the engine is cold, much of the moisture is condensed inside the engine, especially during low-temperature operating conditions such as suburban driving. The combination of corrosion and wear under these low-temperature operating conditions is the major reason for excessive wear of the top ring area of the cylinder wall.

Combustion. A spark plug ignites the charge in the combustion chamber at the correct instant near the end of the compression stroke. The spark must have enough energy to start the charge burning. From this point, a flame front moves smoothly across the combustion chamber during normal combustion. Normal combustion is illustrated in Figure 4-1.

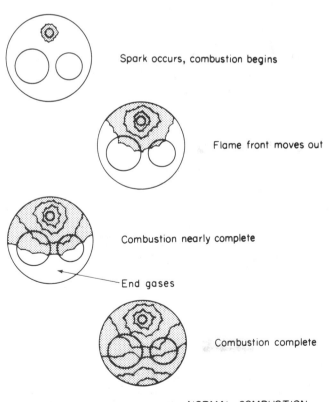

Spark occurs, combustion begins

Flame front moves out

Combustion nearly complete

End gases

Combustion complete

NORMAL COMBUSTION

Figure 4-1. Flame front movement during normal combustion.

Actual combustion is much more complex than it appears from this simplified description. During combustion the gases go through many steps or phases. For this discussion, the combustion is divided into two steps, *preflame reactions* and *combustion*.

A simple example is helpful in understanding *preflame reactions*. If one were to light a piece of paper with a match, the paper would first turn brown. This is caused by preflame chemical reactions. After turning brown, the paper would ignite, producing a flame. The flame is combustion. The charge in the combustion chamber reacts in a similar way. As the gases are compressed during the compression stroke, the temperature rises. Preflame chemical reactions take place in the high-temperature compressed charge. This changes the character of the charge. Preflame reactions prepare the charge for burning.

After ignition takes place, the flame front moves out in a spherical fashion modified by combustion chamber turbulence. The heat energy released behind the flame front increases the temperature and pressure in the combustion chamber. This higher pressure and temperature increase the rate of preflame reactions in a portion of the charge ahead of the flame front, called the end gases. Preflame reactions become more rapid at higher engine compression ratios. When preflame reactions increase too rapidly, they cause abnormal combustion.

4-2 Abnormal Combustion

Abnormal combustion is divided into two main types—knock and surface ignition. Each of these result in loss of power and in high combustion chamber temperature. Continued operation under either type of abnormal combustion will result in physical damage to the engine.

Knock. Engine knock is the result of rapid preflame reactions within the end gases. A pressure–crank angle trace of knocking combustion is shown in Figure 4-2. The reactions become so rapid that ignition occurs without a spark or flame in the end gases ahead of the flame front. This results in very rapid rates of combustion within the end gases that produce high-frequency pressure waves. These waves hit against the combustion chamber walls and cause

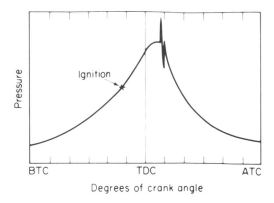

Figure 4-2. Pressure-crank, angle trace showing a high frequency pressure wave shortly after top center (ATC) during detonation.

a vibration noise that is called knock, detonation, or ping. This type of abnormal combustion is illustrated in Figure 4-3.

Reducing Knock. The tendency for an engine to knock with a given fuel can be reduced by any method that will lower the combustion *pressure*

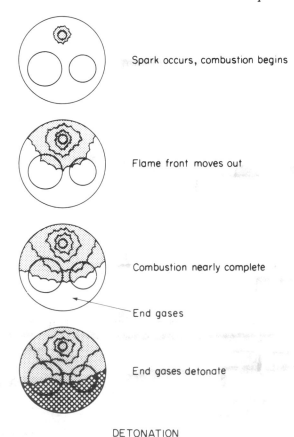

Spark occurs, combustion begins

Flame front moves out

Combustion nearly complete

End gases

End gases detonate

DETONATION

Figure 4-3. Flame front movement and end gas reaction during detonation.

and/or *temperature*. It can also be reduced by any method that will reduce the *time* the end gases are in the high pressures and temperatures. In addition, a change to a gasoline grade that will not cause rapid preflame reactions will reduce engine knock. **Octane rating is an indication of the antiknock properties of gasoline.** A gasoline that has high antiknock characteristics has a high-octane rating. The compression ratio of the engine has a major effect on compression pressure. Turbocharging also increases the pressure of the compressed charge. As the compression pressure is increased, the engine power and economy will also increase. This occurs because the increase in compression pressure results in higher combustion pressures. High combustion pressures, accompanied by higher combustion temperatures, cause a greater tendency to knock. Fuels with high-antiknock properties (high octane gasoline) are used in high-compression-ratio engines to allow the engine to run knock-free while developing increased power. Lower compression ratios are used in emission-controlled engines using catalytic converters so they will be able to run knock-free on low-octane no-lead gasoline.

Combustion chamber design also affects engine knock. Combustion chambers with a squash or quench area tend to have low knocking tendencies. This occurs because the end gases are thin and close to a cool metal surface in the quench area. Cooling the end gases causes a slowing of preflame reactions in them. This reduces the tendency of the engine to knock. Quenching of end gases is the main reason that a rotating combustion chamber engine will run knock-free on low-octane gasolines. Combustion chamber turbulence is also useful in reducing the tendency to knock by mixing cool and hot gases within the combustion chamber. This prevents a concentration of static hot end gases in which rapid preflame reactions can take place. Turbulence caused by the quench area is shown in Figure 4-4.

Smooth, efficient engine combustion is controlled by careful design of the combustion chamber quench areas. Unfortunately, these quenching areas cool the combustion gases so much that the flame goes out. This leaves some unburned hydrocarbons to be expelled with the exhaust gases. Combustion chamber quenching is the major source of hydrocarbon exhaust emissions when the

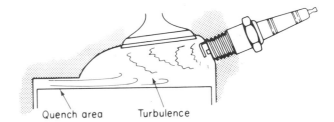

Figure 4-4. Gas turbulance in the combustion chamber at the end of the compression stroke.

carbon monoxide is low. These hydrocarbons are combined with oxygen in the catalytic converter to remove them from the exhaust gas.

Surface Ignition. Surface ignition is a broad term that indicates combustion has started at a source of ignition that is not the electrical arc at the spark plug. One type of surface ignition, called preignition, is shown in Figure 4-5. The result of surface ignition is to complete the combustion process sooner than normal. This causes the maximum combustion pressure to occur at the wrong time in the engine cycle. A pressure–crank angle trace of surface ignition

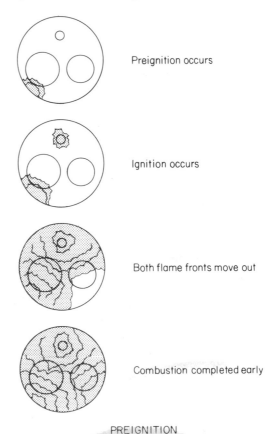

Preignition occurs

Ignition occurs

Both flame fronts move out

Combustion completed early

PREIGNITION

Figure 4-5. Flame front movement during preignition.

is shown in Figure 4-6. Surface ignition causes the engine to develop less power, less economy, and more exhaust emissions.

The source of secondary ignition is a hot spot. A spark plug electrode, a protruding gasket, a sharp valve edge, and so on, can become a hot-spot ignition source. These items can become red hot during engine operation. At this temperature they have enough heat energy to become a second source of ignition. These ignition sources seldom occur in modern engine designs as long as the engines are operated properly and given the proper maintenance.

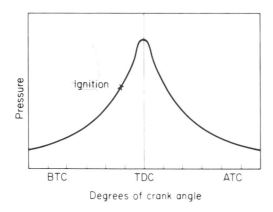

Figure 4-6. The shape of a typical pressure-crank angle trace during preignition.

Another secondary ignition hot spot is that of combustion chamber deposits. These deposits result from the type of fuel and oil used in the engine. They also result from the type of engine operation. Fuel and lubricant suppliers have been doing a great deal of research to make products that minimize *deposit ignition*. A deposit ignition source may be a hot loose carbon deposit flake that ignites only one charge. It is then exhausted from the engine with the spent exhaust gases. This is called a *wild ping*. Sometimes, the flake will remain attached to the combustion chamber wall. When this happens, it will ignite a number of charges until the deposit is burned up or the engine operating conditions are changed.

Names are given to many specific abnormal combustion conditions that are caused by surface ignition. If surface ignition occurs before the spark plug fires, it is called *preignition*. It may be heard or it may be silent. It may be a wild ping or it may be a continuous *runaway surface ignition*. If sur-

face ignition occurs after the ignition switch is turned off, it is called *run-on* or *dieseling*.

4-3 Vehicle Emissions

In a vehicle with no emission controls, crankcase vapors account for 25% of the total vehicle emission. The exhaust emits 60% of the emissions. The remaining emissions evaporate from fuel tank and carburetor vents.

Vehicle emissions come from the engine crankcase, gasoline tank, carburetor, and exhaust. The crankcase vapors from the engines have been completely controlled since 1968. Evaporative vapors from the gasoline tank and carburetor have been controlled under almost every condition in automobiles built since 1971. Vapors from the fuel tank and carburetor are vented to an activated carbon charcoal canister, where they are trapped and stored when the engine is not running. In order to run, the engine will use large amounts of air and will get rid of a large quantity of exhaust. Federal standards require that the exhaust gases be modified so that they will contain only a very small amount of harmful exhaust products. This will minimize their pollution of the atmosphere.

Exhaust Emissions. Harmful exhaust emissions consist of carbon monoxide (CO), hydrocarbons (HC), and oxides of nitrogen (NO_x). Control devices on modern automotive engines are designed to keep the exhaust emissions at a minimum. Figure 4-7 shows a typical graph of the exhaust emissions from an engine without emission controls.

The carbon monoxide level is lowered by operating the engine with a lean air/fuel mixture. The lean mixture will ignite easier when it is heated. This reduces the tendency to misfire with lean air/fuel mixtures. Most of the carbon monoxide forms during idling, acceleration, and deceleration. Rich mixtures are used under these operating conditions to compensate for poor fuel distribution within the manifold. Rich mixtures also form when a cold engine is being choked.

Most of the unburned hydrocarbons are produced by the surface quenching of combustion caused by the cooling effect of the metal surface of the combustion chamber. This is impossible to avoid. The quenching effect is reduced by designing

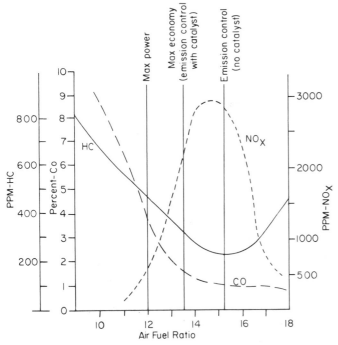

Figure 4-7. The shape of typical HC, CO, and NOx exhaust emission curves from an uncontrolled engine.

volume. It naturally produces large amounts of unburned hydrocarbons.

The last controlled emission produced in the combustion chamber is oxides of nitrogen (NO_x). NO_x forms when air, which consists of approximately 80% nitrogen and 20% oxygen, chemically reacts to combine these elements in the high-temperature combustion. A graph of typical combustion temperature is shown in Figure 4-8. Engine thermal efficiency and power will increase the production of oxides of nitrogen. The higher the peak cycle temperature, the greater the production of NO_x. The production of NO_x also increases as the mixture is leaned until an excessively lean mixture occurs. This, of course, causes misfiring, which results in an increase in unburned HC. This can be seen on the right side of the graph in Figure 4-7. Because the rotary combustion chamber engine operates at low thermal efficiencies, it produces small amounts of NO_x. Diesels and turbines operate at high temperatures with large quantities of air, so they produce large amounts of NO_x.

4-4 Exhaust Emission Control

Exhaust emission control is accomplished by (1) carefully controlling the air/fuel mixture being sent

the engine to have a low combustion chamber surface-to-volume ratio. To do this, the engine cylinder bore is small and the stroke is long. The rotary combustion chamber engine has a very high combustion chamber surface area compared to its

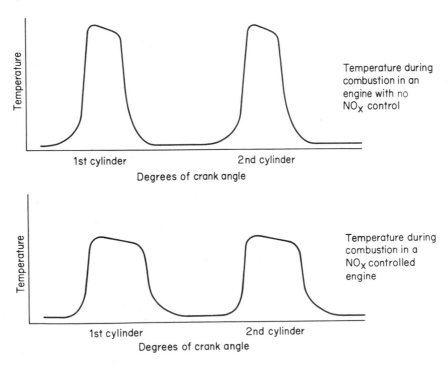

Figure 4-8. A graph of typical temperatures within the combustion chamber as the engine runs.

to the combustion chamber, (2) by controlling the combustion process, and (3) by eliminating any harmful emission products still remaining in the exhaust gases.

Unburned hydrocarbons and carbon monoxide emissions are most critical at idle, during acceleration, and during deceleration. At these operating conditions, the engine will run with a rich mixture. Oxides of nitrogen are produced at cruising speeds when the engine runs with high thermal efficiencies, accompanied by high peak combustion temperaatures.

Control of the Charge. Exhaust emission control starts with the carburetor. Emission control carburetors are very carefully calibrated and adjusted. They operate on as lean a mixture as possible while still providing each cylinder with a combustible mixture. With mixtures on the lean side of their best combustible range, the engine may have rough idling and may surge during cruising operation. Surge is an increase and decrease in speed with no change in the accelerator pedal position. Careful carburetor adjustment to specifications will minimize this undesirable engine operation.

When the engine is cold, it must be partially choked to operate. Choking gives the engine a rich air/fuel mixture that produces a large amount of hydrocarbons and carbon monoxide. Emission control engines have their chokes designed to open as rapidly as possible and still maintain vehicle driveability. Driveability is starting, idling and running that satisfies the operator. This is done by heating the thermostatic choke spring rapidly or by using a more sensitive thermostatic choke spring. Some engine models use an electric heater to open the choke more rapidly than is possible using engine heat alone. Production of hydrocarbon and carbon monoxide exhaust emission products will reduce as the choke opens.

Control by the Combustion Chamber. Many factors affect combustion once the intake charge gets into the combustion chamber. The combustion chamber shape greatly affects the amount of unburned hydrocarbons remaining in the exhaust. The part of the charge that is close to the combustion chamber surface is kept so cool that it does not burn. These unburned gases form much of the un-

burned hydrocarbons in the exhaust. The areas within the combustion chamber that quench the charge have been kept small in emission-controlled engines. Figure 4-9 shows quench and low-quench combustion chambers.

The combustion chamber temperature must be high to help burn lean mixtures. Increasing the operating temperatures of the cooling system thermostat helps to provide the high combustion temperatures. Temperatures are also kept high by restricting distributor vacuum advance that normally occurs on engines that have no emission controls. High engine temperatures cause more rapid hydrocarbon oxidation. Oxidation occurs during both preflame reactions and combustion. This ox-

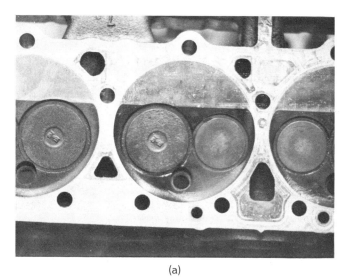

(a)

(b)

Figure 4-9. Typical combustion chambers. (a) High quench wedge shape and (b) modified low quench shape.

idation lowers the amount of hydrocarbons remaining in the exhaust. While idling in traffic, coolant flow through the engine is low because the coolant pump is turned slowly. The fan is also driven slowly. Operating the engine like this will often lead to excessive engine temperatures. The distributor vacuum advance system is provided with a temperature-operated bypass valve to prevent excessive engine temperatures. At an engine coolant temperature of approximately 220 °F (104 °C), the bypass opens to apply manifold vacuum to the distributor vacuum advance unit. This causes the engine to run more efficiently, and this, in turn, increases the engine speed. The resulting higher engine speed increases the cooling pump and fan speed to move the coolant faster and to pass more air through the radiator, which results in a lower engine operating temperature. A section view of a typical temperature-operated bypass valve is shown in Figure 4-10.

Engine emission control systems using the features just described to clean up the exhaust are given a number of different names by the automobile manufacturers. They all have essentially the same operating units and go by names such as IMCO (Improved Combustion), CAP (Clean Air Package), CAS (Clean Air System), CCS (Controlled Combustion System), and Engine Modification.

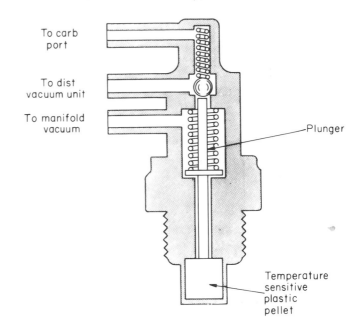

Figure 4-10. Section view of a typical temperature-operated bypass valve showing the internal parts.

Control by Ignition Timing. Basic ignition timing is retarded from the best timing on emission-controlled engines that do not use catalytic converters. This makes it necessary to have the throttle open further to provide the same engine idle speed, as shown in Figure 4-11. The higher idle speed allows the engine to be run with a leaner mixture. Lean mixtures, together with the higher temperatures that result from retarded timing, help to

IDENTICAL ENGINES IDLING AT 700 RPM

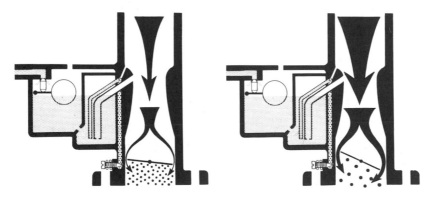

INITIAL TIMING 8° B.T.D.C.
Relatively little air flow past carburetor throttle plate which is in nearly closed position.

INITIAL TIMING 4° B.T.D.C.
Larger throttle plate opening is needed to allow more air into the fuel to achieve 700 R.P.M.

Figure 4-11. Emission controlled engine operates with the throttle opened more than an engine that has no emission controls (Courtesy of AC Division, General Motors Corporation).

reduce hydrocarbon and carbon monoxide emissions in the exhaust. The amount of these undesirable exhaust emissions produced by an engine is the result of carburetion and the ignition timing effect on combustion. To operate properly, basic engine timing must be set to specifications.

Standard ignition systems have two advance systems: mechanical centrifugal and vacuum. The mechanical centrifugal advance is designed to provide the required advance for full-throttle *maximum power* engine operation. Vacuum advance provides the added advance needed at part throttle to give the engine *maximum economy*. At full throttle, there is no vacuum on the distributor port of the carburetor. This can be seen in Figure 4-12.

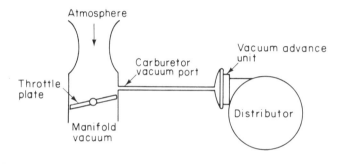

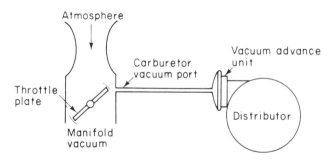

Figure 4-12. Location of the distributor vacuum port in the carburetor.

Therefore, the only ignition advance being used at full throttle is the mechanical centrifugal advance. The vacuum advance unit has no effect on the engine's full throttle power. The vacuum advance control is modified on emission-controlled engines so that during part throttle operation the engine will produce low exhaust emissions.

Distributor vacuum advance on most emission-controlled engines is connected to the distributor vacuum port in the carburetor through a temperature-operated bypass valve that may also be called a thermostatic vacuum switch (TVS), ported vacuum switch (PVS), or distributor vacuum control valve. It provides the distributor with a modified distributor vacuum signal from the carburetor port. The connecting hose may also include an advance delay valve that slows the reaction as ported vacuum increases. Figure 4-13 shows a typical engine installation that also includes a transmission controlled spark (TCS) solenoid valve.

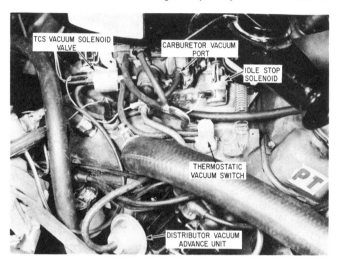

Figure 4-13. Emission controls on a typical V-type engine.

Control by Valve Timing. Valve timing also has a great effect on combustion. Before emission-controlled engines, the valves were timed to give the engine efficient operation at its design speed. Low-speed engines had a short valve opening duration and very little valve overlap. *Overlap* occurs when both valves are partly open between the exhaust and intake strokes. High-speed engines had a long valve-opening duration and a large valve overlap. These are illustrated in Figure 4-14. Most emission-controlled engines have a short valve-opening duration with a large valve overlap. The relatively short valve opening is normal for the low-speed automobile engine. The large valve overlap allows some of the exhaust gas to return to the engine with the fresh intake charge. The exhaust gas takes the place of some of the intake charge to minimize the amount of charge used. This lowers the peak combustion charge temperature. Lowering the peak combustion temperature results in less NO_x. Unfortunately, long valve overlap causes rough engine idle, poor low-speed operation, and poor fuel economy.

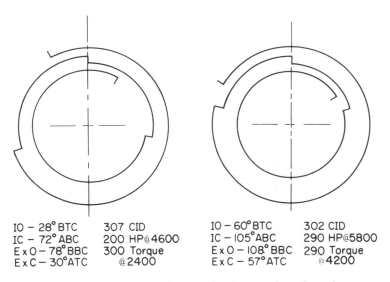

IO – 28° BTC	307 CID	IO – 60° BTC	302 CID
IC – 72° ABC	200 HP@4600	IC – 105° ABC	290 HP@5800
ExO – 78° BBC	300 Torque	ExO – 108° BBC	290 Torque
ExC – 30° ATC	@2400	ExC – 57° ATC	@4200

Figure 4-14. Valve timing spiral showing the degree when the valves open and close on two engines of the same make having similar displacements but entirely different performance characteristics.

Control by Exhaust Gas Recirculation. A second, more positive means of diluting the intake charge with exhaust gas is with the addition of an exhaust gas recirculation (EGR) system. This system cools and recirculates a portion of the exhaust gas back through the engine intake manifold. The exhaust gas takes up space in the charge and it will not burn. This leaves less space for the fresh intake charge to get into the combustion chamber. A smaller amount of combustible charge results in a lower peak combustion temperature and less NO_x production. This is shown as a graph in Figure 4-15. On

V-type engines, the intake manifold exhaust crossover is an easy place to interconnect the intake and exhaust passages. An example of these passages can be seen in Figure 4-16. Inline engines often use external lines (Figure 4-17). EGR systems are equipped with a recirculating valve which limits the amount of exhaust gas recirculation. A typical EGR valve is shown in Figure 4-18. It will only recirculate exhaust gas when the engine operating conditions promote the formation of NO_x. This will occur under light acceleration and cruising speeds when the air/fuel mixture is lean. The exhaust gas recir-

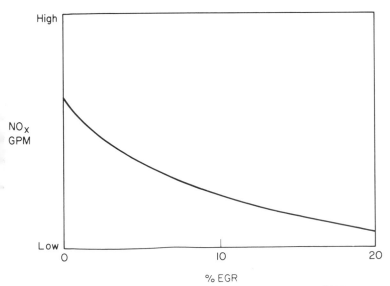

Figure 4-15. A graph showing the reduction of NOx as the amount of EGR is increased.

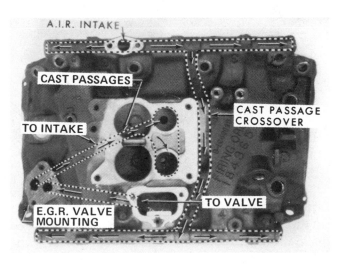

Figure 4-16. Typical internal intake manifold passages used for EGR and air injection (Courtesy of Buick Motor Division, General Motors Corporation).

Figure 4-17. Typical external line connecting the exhaust manifold to the EGR valve (not visable) on the intake manifold. The air injection system check valve can be seen.

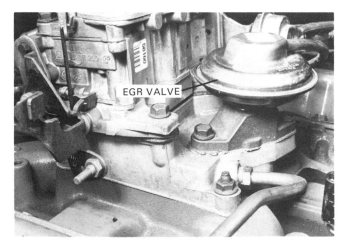

Figure 4-18. A typical EGR valve located on an adapter between the carburetor and the intake manifold.

culating valve is closed at idle and at low speeds. It gradually opens as the throttle is opened and the velocity of the charge increases in the manifold. It closes again at full throttle, so it does not affect full power operation.

The EGR system may also be equipped with a

valve that blocks EGR operation at low engine temperatures and at low vehicle speeds. This is accomplished by a temperature-sensing unit in the engine cooling system and a speed-sensing unit in the speedometer cable. Some EGR valves have a back-pressure transducer that senses pressure in the exhaust manifold to change the amount of opening.

The first models of the exhaust gas recirculation systems reduce fuel economy and increased the formation of HC. The economy can be improved with EGR systems if a very high ignition advance is used. This, in turn, increases the amount of HC produced. The high HC can be cleaned up after it leaves the engine.

Control of the Exhaust Gas. Much of the HC remaining in the exhaust can be cleaned up. The first method used was to pump fresh air into the exhaust manifold. The fresh air supplied additional oxygen that helped burn the HC and CO remaining in the exhaust gases. This method of cleaning the exhaust does not affect engine efficiency. It uses only a small amount of power, which is required to operate the air injection pump. A schematic drawing of an air injection system is shown in Figure 4-19.

Starting with the 1975 model automobiles, the majority of domestic automobiles have been

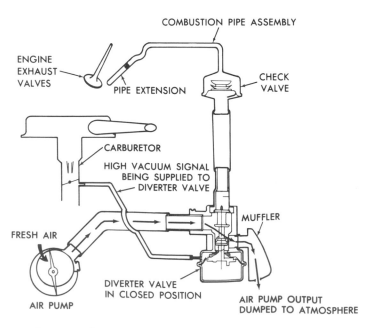

Figure 4-19. A schematic drawing of an air injection system (Courtesy of the AC Division, General Motors Corporation).

equipped with *catalytic converters.* A catalytic converter is made of a bed of catalyst pellets or of a honeycomb grid of catalyst. These are shown in Figure 4-20. It is installed in the exhaust system. As the exhaust gas passes through the catalyst bed or grid, most of the remaining HC and CO exhaust contaminants are oxidized. The action of the catalyst causes this to occur at a much lower temperature than in the combustion chamber. An air injection system is sometimes used to provide additional air for more oxygen. The life of the catalyst will be shortened by contamination, especially with lead contamination. Vehicles using a catalytic converter *must* be run on lead-free gasoline for the maximum life of the catalyst.

4-5 Fuel Economy Improvement

The power the engine produces is proportional to the mass of air going into the engine when the air/fuel mixture is correct and ignition occurs at the proper instant. When the throttle of a spark-ignited engine is closed only a small mass of air can get into the engine. This allows the engine to produce only enough power to idle. As the throttle is opened, more air mass can get into the engine to produce more power. This causes the engine to speed up or do more work as it moves the vehicle. The engine produces maximum power when the throttle is wide-open at maximum rpm. Design changes can be made to modify an engine to allow more air mass to go into the engine. The engine dis-

placement can be increased by increasing the cylinder bore and/or the piston stroke. The air mass can be increased by changing the valve timing to provide better engine breathing at high engine speeds. This, of course, gives a valve timing that produces poor engine breathing at part throttle operation. A turbocharger can be installed to increase the mass of the air going into the engine, and this will cause the engine to produce more power. The maximum power delivered to the engine crankshaft is also modified by the engine temperature, lubricant viscosity, and internal engine friction.

Power and Economy. Any specific engine is capable of producing either maximum power or maximum economy. It cannot produce both at the same time. The fundamental trade-off here is between acceleration and fuel economy, regardless of engine size or gear ratio. Maximum power requires a rich air/fuel mixture so that all of the oxygen molecules in the air have a good chance of combining with fuel molecules. This makes the best use of the *air* that is in the combustion chamber. It does, however, leave unburned hydrocarbons and carbon monoxide. An engine designed for economy, on the other hand, uses a lean air/fuel mixture so that all of the fuel molecules have a good chance of combining with oxygen molecules in the air. This makes the best use of the *fuel* to produce useful power. If the mixture in some of the cylinders becomes too lean to burn, misfiring will occur. Misfiring leaves

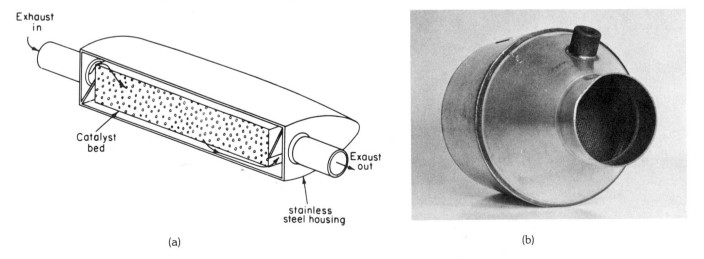

(a) (b)

Figure 4-20. Catalytic converter types. (a) Pellet-type and (b) honeycomb grid type.

unburned hydrocarbons that are expelled from the cylinders with exhaust gases. This will increase the exhaust HC.

Engine Mapping. For any operating conditions, the engine will operate most economically with a specific air/fuel mixture and a specific timing. Any change from these values, either greater or less, will decrease the fuel economy. In engineering terms, fuel economy is specified as thermal efficiency and it is described as *brake specific fuel consumption* (BSFC). This is a fundamental measurement of engine efficiency. **BSFC is determined by dividing the pounds of fuel consumed per hour by the brake horsepower produced** (lb fuel/bhp hour). Manu-

facturers use a process called *mapping* to determine an engine's requirements.

For mapping, the engine is operated on a dynamometer while only one variable is changed at a time and the data are plotted on a graph that becomes a map. The engine variables include the air/fuel ratio, engine torque, EGR, ignition timing, and rpm. An engine map would look similar to Figure 4-21. The engineer is able to precisely design the engine operating systems from the engine maps.

Computer Controls. Mechanical and vacuum controls are too slow to respond to the precise air/fuel ratio, EGR, and timing requirements of the engine as the speed and load change. Therefore, modern

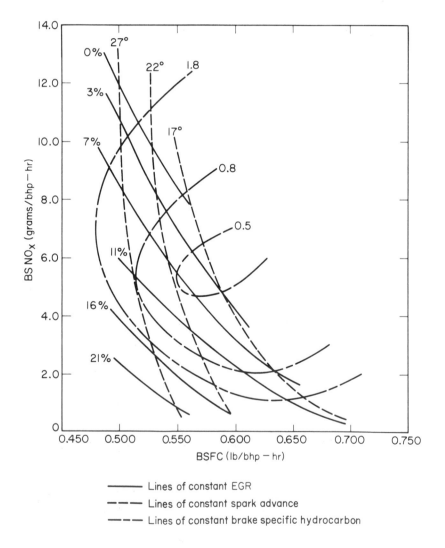

Figure 4-21. A typical engine map of the effects of brake specific NOx, brake specific HC, percent of EGR, and ignition spark advance on brake specific fuel consumption.

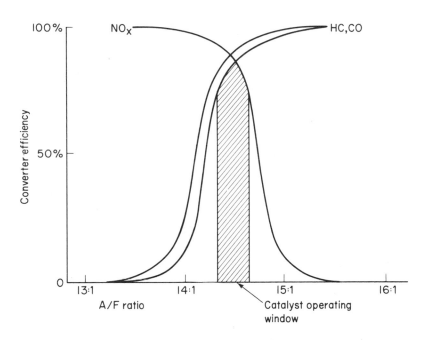

Figure 4-22. Three way catalytic converter efficiency in relation to the air/fuel ratio of the intake charge.

on-board computer systems have been developed to take over control of these engine operations. Computer controls were first applied to ignition advance by Chrysler in 1976 as lean-burn controls in some model automobiles. In 1977, Oldsmobile first used the MISAR ignition computer on the Toronado. Ford first adapted computer timing ignition controls and also included and EGR control on the Versailles in 1978. These computer controls have been updated to control EGR, early fuel evaporation, and carburetor air/fuel ratios. In some systems, they even retard timing if the engine begins to detonate and knock.

As reductions in NO_x emissions have been mandated by law, it was necessary to use a new type of catalytic converter to chemically *reduce* or separate NO_x to nitrogen and oxygen. Three-way converters have been developed using rhodium for reducing the NO_x remaining in the exhaust. Platinum with palladium are used in the converter for *oxidizing* the remaining HC and CO. These three-way converters are most efficient when the air/fuel mixture is *stoichiometric*. This chemically correct stoichiometric air/fuel ratio is also said to have the *equivalence ratio* of 1. Because the three-way converter will only be efficient very near the stoichiometric air/fuel ratio, a feedback system has been made that will continually adjust the fuel metering system during engine operation. It determines the correct air/fuel ratio by measuring the

amount of oxygen remaining in the exhaust. If too much oxygen remains, the air/fuel ratio will be enrichened. If too little oxygen remains, the air/fuel ratio will be leaned. The slight change in the air/fuel ratio is called a catalyst operating *window,* as shown in Figure 4-22. Fortunately, this is exactly the air/fuel ratio needed for best fuel economy. Computer-controlled fuel metering systems have been developed to maintain air/fuel ratios close to the ideal.

The computer-controlled fuel metering system uses an exhaust gas oxygen (EGO) sensor made of zirconium dioxide with platinum electrodes. A simplified sketch of the sensor is shown in Figure 4-23. The EGO sensor is placed in the exhaust

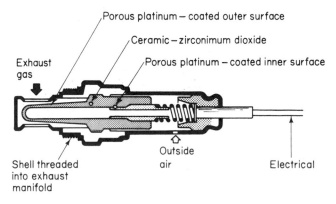

Figure 4-23. Simplified section of a zirconium dioxide exhaust gas oxygen sensor used with a three way converter and feedback metering system.

manifold ahead of the three-way converter to measure the oxygen remaining in the exhaust. This sensor changes its electrical conductivity at the stoichiometric air/fuel ratio. The electrical signal from the EGO sensor is fed into the engine control computer. If the air/fuel ratio is incorrect, the computer adjusts a control on the carburetor or injection system to correct the air/fuel ratio. This is why this system is called a feedback system. It continually and rapidly tunes the engine while the engine is running. A typical feedback loop is illustrated in Figure 4-24. Changes in the exhaust oxygen are sensed by the EGO and the signal is again fed back to the fuel metering system to close the loop.

4-6 Summary

In general, engine modifications that are required to reduce harmful exhaust emissions also reduce fuel economy. By careful design of the engine and its systems, the reduction in fuel economy is kept at a minimum. To do this, the air/fuel ratio is precisely metered. Engine timing is advanced when the EGR system is in operation and retarded when the catalytic converter is not warmed up.

Fuel economy improvements have been most dramatic when the manufacturers reduced the vehicle weight and the drive line friction. Aerodynamic body design has also reduced the power required to operate the vehicle.

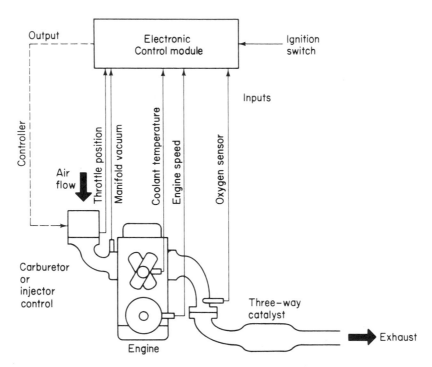

Figure 4-24. A block diagram of a feedback loop controlling the fuel metering to hold the air/fuel ratio within the catalyst operating window.

REVIEW QUESTIONS

1. Why does operating a vehicle at high altitude affect the intake charge mixture? [4-1]
2. What causes most of the wear in the top ring area of the cylinder? [4-1]
3. How does knock differ from preignition? [4-2]
4. Name four things that can be done to reduce knock. [4-2]
5. When is it necessary to use high octane gasoline in an engine? [4-2]
6. What causes surface ignition? [4-2]
7. What can be done to minimize surface ignition? [4-2]

8. Where do vehicle emissions come from? [4-3]
9. How is the carbon monoxide level lowered? [4-3]
10. What causes most of the unburned hydrocarbons in the exhaust? [4-3]
11. What causes oxides of nitrogen (NO_x) to form in the combustion chamber? [4-3]
12. What exhaust emissions are a problem in diesel engines? [4-3]
13. Name three ways in which exhaust emissions are controlled?   [4-4]
14. Where does exhaust emission control start? [4-4]
15. Why is the choke opened rapidly? [4-4]
16. In what way do quench and low-quench combustion chambers differ? [4-4]
17. How does the temperature-operated bypass help to prevent excessive engine temperatures? [4-4]
18. How does ignition timing help reduce hydrocarbons and carbon monoxide emissions? [4-4]
19. Under what operating conditions does the mechanical centrifugal advance control the ignition timing? [4-4]
20. Under what operating conditions does the vacuum advance control the timing? [4-4]
21. What is the advantage of large valve overlap? [4-4]
22. Under what engine operating conditions do oxides of nitrogen form? [4-4]
23. How can EGR improve fuel economy? [4-4]
24. How does a catalytic converter reduce exhaust emissions? [4-4]
25. Why is it necessary to use lead-free gasoline in engines with catalytic converters? [4-4]
26. What engine design changes can be made to allow the engine to use a larger mass of air? [4-5]
27. How does the air/fuel ratio differ for power than for economy? [4-5]
28. Why have computer controls been adapted to engines? [4-5]
29. Why is it important to have the intake charge mixture at stoichiometric? [4-5]
30. What is a feedback fuel metering system? [4-5]

PART

II

ENGINE

CONSTRUCTION

This part of the book is a study of the design details of reciprocating engines. The discussion of each engine part starts with the purpose and function of the part. This is followed by a description of how the parts differ among the most common engines.

Overview

The approach used in this part is to discuss the operation of an entire system. A discussion of the individual parts follows the normal order of engine disassembly. This makes this section of the book very usable as a guide for beginning students as they work on a study engine.

Chapter 5 reviews the basic operating systems of the engine. First, the purpose of the system is discussed. This is followed by a brief description of how each system functions. This chapter can be used as the engine accessories are being removed from the engine.

Chapters 6 through 9 discuss the engine parts in the same order the engine is being disassembled. It starts with the manifolds, then the head and valves. The lower engine discussion starts with the piston and rod assemblies. This is followed by a discussion of the shafts and bearings. While the engine is disassembled, the seals, block, internal lubricating system, and cooling system are studied.

A study of an engine is not complete until it is properly reassembled. Chapter 12 describes the engine assembly details.

Many schools find that students are much more careful working on an engine when they know it is going to be run after assembly. Section 12-7 covers the information necessary to give the engine a brief run to make sure that it is assembled and timed correctly. It is assumed that the engine will *not* be fitted with new parts, so detailed clearances are not specified. These are discussed in detail in Chapter 18.

Shop Safety

As you go into the shop to work on engines, you should be reminded of safety in the shop. It is up to each person in the shop to make sure the shop is safe at all times.

Personal Dress. Safety in the shop starts with each individual. Loose-fitting clothing can get caught in moving equipment. This is also true of loose, long hair. Work trousers and either short sleeve or buttoned long-sleeve shirts should be worn. Long sleeves will protect the arms from becoming burned around hot engine parts. Loose, long neckties should never be worn in the auto shop. It is advisable to remove rings and wristwatches. This is done so they do not catch in parts and injure your hands.

Canvas and rubber shoes should never be worn in the auto shop. They provide no foot protection if something drops on them. Shoes should be heavy leather. Safety shoes provide the best protection.

Safety glasses should *always* be worn in the shop. It is easy to understand that they should be worn while using a grinder or wire wheel. It is also easy to see why they should be worn to protect the eyes from splash and chips as parts are being cleaned. It is not as easy to understand that it is important to wear safety glasses when inspecting parts. It may be that a fellow student is working on a part that could fly into your eyes. An excellent example of this is a **valve retainer or split lock flying through the air** when a valve spring compressor slips as the spring is compressed.

Some shops require hard hats where there is some chance of head injury. In these shops, hard hats are used when working under vehicles and under hoods. In engine work, they would be used while removing and installing the engine. Hard hats should be used when handling material overhead. This is often done in storerooms.

Shop Cleanliness. A safe shop is a clean shop. The technician can stumble over parts that are piled carelessly in the shop. Piled parts can become damaged and even lost. Parts should always be stacked and stored properly in neat order.

The floors must be kept free from fluids. You can slip and fall on spilled fluids, especially oil. This is one of the major causes of serious injury in the shop. Any spilled fluid should be cleaned up before going on with more work. If the fluid is not cleaned up, it will be tracked throughout the shop. This will make the cleanup take much more time and it will use more cleaning materials.

Air hoses and electric extension cords should be kept rolled up when they are not in use. Fellow students could stumble over them.

The service shops should be well ventilated. Ventilation will remove fumes that endanger your health. It will also remove fumes that could become a hazard, such as gasoline vapors. Exhaust gases are removed with a shop exhaust system. If an engine is to be run more than 30 seconds, it should be connected to the shop exhaust system. In 30 seconds, a vehicle can be moved or an engine started to make a quick check. Room ventilation should remove these and other shop fumes.

Fire is always a hazard in the auto shop. Gasoline vapors are very flammable. They ignite very quickly. Oils, greases, plastics, and rubber products used in the shop will also burn when they get hot enough. This is also true of the automobile interior and electric wire insulation. The best fire protection is to eliminate conditions that could cause a fire. **Fire occurs only when air and fuel are mixed in the correct air/fuel ratio.** A part of this mixture has to be heated to the ignition temperature. Removing vapor fumes and eliminating sparks from the auto shop do a good job in minimizing the fire hazard. This includes care to prevent gasoline spilling. If gasoline is spilled, it should be cleaned up immediately. The cleaning material should be removed to a safe place. Sometimes, a welding torch is used in automobile service. All flammable materials should be removed from the area where the flame is to be used. When this is not possible, the flammable parts should be covered with soaking-wet shop towels while another person holds a fire extinguisher, ready for immediate use.

If a fire occurs, immediately take steps to put it

out with appropriate fire extinguishers. A small fire can be easily extinguished. If it is allowed to become a large fire, it will be difficult to put out. Fires in the carburetor intake can easily be put out by cranking the engine. This will draw the flame into the engine. These flames can also be snuffed out by closing the choke plate. Fire on open oil or fuel will not cause an explosion. An explosion will only occur from an enclosed combustible mixture. Never have an open flame around an empty gasoline tank or around the top of a battery. Both contain an explosive gas mixture.

Tools and Equipment. Most accidents that occur while using tools are the result of improper or careless use of the tools. Accidents involving tools usually result from the tool slipping, causing damage to the part or injury to the technician. It is important, therefore, to use good-quality tools, maintained in proper condition, and they must be used correctly.

Inexpensive tools that are readily available at shopping centers are seldom satisfactory for use in automotive service. A technician who is using tools to earn a living should invest in first-line high-quality tools. These tools are designed for easy use, are strong, and fit the fasteners properly. Quality tools are guaranteed. They will be replaced at no charge if they break while being used correctly.

Fender covers and seat covers should be used to protect the vehicle from scratches and soil that could occur when working with tools and dirty vehicle parts. It is much easier to cover the vehicle than it is to clean it after it has become dirty or to repair scratches.

Engine handling is best done using a sling designed for hoisting engines. They can be handled with a securely fastened chain. Rope should *never* be used to hoist an engine. It can break and the engine will fall. The engine should never be allowed to hang on a hydraulic hoist. If the hydraulic fluid develops a leak, the engine will fall.

Always keep the protective shields in place when using a bench grinder or wire brush. Wear eye protection and keep a firm grip on the part. If the piece is small, it can be firmly clamped in a locking pliers. This will hold it while grinding or wire brushing. Never grind on the side of a grinding wheel. This can produce unnatural wheel loading that may cause the grinding wheel to fly apart.

Some engine parts are quite heavy. If a part is too heavy to handle easily, ask one of your fellow students to help you handle it. Straining to move a part can cause back injury. If the part is dropped, it will likely be damaged and it could even damage you if it lands on your foot or rolls into your leg. Parts that are too heavy to lift should be handled with a hoist.

An air gun is a common automotive shop tool. The air gun is not a toy. Dirt blown by high air pressure can penetrate the skin. Shop air guns should operate on low air pressure. Be sure to direct the air so that it does not blow on anyone. Also, direct the air so it does not raise dust in the shop. The dust could settle on partly assembled engine parts to cause early wear of the reconditioned engine.

Specific safety precautions are discussed throughout the text. They are described where they apply to serve as a reminder while the service procedures are discussed.

Care of Parts. Safety involves the care of the parts as well as protection against personal injury. Disassembled engine parts should be placed in neat order. In this way, the operating location of each part can be identified. This is necessary to locate the cause of premature part failure. It also prevents handling damage that would have to be corrected before the engine is assembled. The repair of parts damaged while being handled will add cost to the engine repair.

Some engine parts have sharp edges. These should be handled with care to prevent cuts.

What to Do in Emergencies. Report *all* accidents to your instructor. The instructor will follow prescribed school procedures for the particular emergency.

Damage to the parts should be reported so that the replacement parts can be ordered or the damage corrected. It would be unusual for a semester to be completed without someone damaging a part by dropping it.

Personal injury should also be reported. Small cuts or bruises can often be cared for with first-aid materials kept in the auto shop. More serious injury will require professional medical help. If all personal injury accidents are reported, the instructor will decide how it should be cared for.

CHAPTER
5

Engine Operating Systems

It is important to have a general understanding of the engine operating systems. This is necessary to properly understand the purpose and function of the internal engine parts. Some of the components of the operating systems attach directly to the engine. The interaction between the operating systems and the engine can be studied as the engine is being worked on in the shop.

Two external systems are necessary to allow the engine to run. These are the *fuel metering system* and the *ignition system*. An internal *lubrication system* is required to supply oil which will prevent metal-to-metal contact between moving parts, to minimize friction, and to provide durability. The *cooling system* removes about half of the rejected heat to keep the engine from seizing. Seizing would keep the engine from rotating. A *cranking system* is required to rotate the crankshaft to get the engine started. A *battery* is needed to store energy so that it is available for the starter and ignition. A *charging system* is required to provide electricity for the electrical running load and to recharge the battery after the engine starts. The electrical *running load* consists of lights, instruments, windshield wiper, heater and air conditioner motors, radio, and so on, that can be operated continuously as the vehicle is driven. Intermittent loads, such as power door locks, power windows, power seats, turn signals, and so on, may require more current than that supplied by the charging system. The intermittent power is taken from the battery. It is replaced by the charging system after the intermittent load is discontinued.

A number of emission control components are installed on the engine. Many of these have interconnecting vacuum hoses and electrical wires. It is most important that they be connected properly to operate correctly. Care in disassembly will aid in correct reassembly. Safety items placed in the engine compartment include power steering and power brake units. The power steering pump is mounted on the engine. It is driven by a belt from the front of the crankshaft. The power brake unit is operated either by vacuum from the engine intake manifold or from hydraulic pressure provided by the power steering pump.

Comfort and convenience systems have components under the hood. The largest is the air conditioning pump attached to the engine. It has hoses, tubing, and wires that connect to body-mounted parts of the system. When a speed control is used, it connects to the throttle control. Some of the components of the speed control system are mounted to the body and some to the engine. They also require attaching hoses and wires.

There are only three basic things that are necessary for the engine to run as it is cranked for starting. First, the *air/fuel intake charge* that is inducted into the engine must be a combustible mixture. Second, the engine must be mechanically sound so that the intake charge will be compressed. Third, the ignition system must flash an *arc across the spark plug* at the correct instant in the cycle. The arc must have enough energy to start the compressed charge to burn. The pressure that builds up in the combustion gases will force the piston down. The up-and-down motion of the piston is converted to rotary motion through a connecting rod and crankshaft. All other parts within the engine support these basic functions.

5-1 Fuel Metering

Most automobile engines use a carburetor for metering the fuel. A fuel pump transfers fuel from the gasoline tank through a filter to the carburetor. The carburetor has a linkage connecting from the accelerator pedal which the driver uses to control engine power. A number of vacuum hoses are connected to the carburetor. Some carburetors have electrical solenoids attached to the linkages and operating vacuum valves. An installed carburetor with these attachments is shown in Figure 5-1.

Float System. Gasoline enters the carburetor *float bowl* through an inlet valve. When the gasoline reaches the required level in the float bowl, a float

Figure 5-1. Typical carburetor mounted on a V-type engine.

system closes the *needle valve* to stop the flow of fuel. The float system allows gasoline to flow into the bowl at the same rate that it is used so that the fuel level remains nearly constant.

Air flow through the carburetor is controlled by a *throttle plate* located at the outlet side of the carburetor passage. As the engine runs, the intake strokes of the cylinder forms a partial vacuum in the manifold below the throttle plate. When the throttle plate is closed for engine idling speeds, the intake strokes pull a high vacuum in the manifold. As the throttle is slightly opened for driving speeds, the amount or mass of air flowing through the carburetor into the manifold will increase. This decreases the manifold vacuum. At full throttle (wide open), the air flow through the carburetor is so rapid that there is almost no vacuum remaining in the manifold. *Manifold vacuum is high when the throttle is closed and low when the throttle is open.* This manifold vacuum is used for a number of operating units. For example, heater and air conditioning doors as well as power brakes.

The systems within the carburetor mix the correct amount of fuel with the air flowing through the carburetor to produce an intake charge that has the proper air/fuel mixture. A narrow section in the air passage, called a *venturi,* reduces the pressure of the air as it passes through the narrowest part. The faster the air flows the less pressure is in the venturi. The low pressure produced at this point is called venturi vacuum. The main fuel discharge nozzle is located here. A section view line drawing of a simple carburetor is shown in Figure 5-2.

Main System. Fuel leaves the float bowl through a *main metering jet* and enters the *main well.* Air from a *main air bleed* also enters the main well to break up the solid stream of fuel. This mixture of air and fuel flows through a passage to the main discharge nozzle where it enters and mixes with the large air flow at the venturi. This forms the intake charge. The air/fuel mixture of the intake charge is controlled by the size of the main metering jet and the size of the air bleed. It is enriched with a larger main jet to let more fuel into the main well or by restricting the air bleed. A main system of a simple carburetor is shown in Figure 5-3.

Idle System. The main fuel system of the carburetor is used during steady highway speeds. A separate system, called the *idle and transfer system,* is required for low-speed operation. Fuel for the idle system is taken from the main well in most carburetors. After fuel enters this system, it is taken up above the fuel level through the idle tube where it meets an *idle air bleed.* The air and fuel flows through a fixed idle restriction within the passage, then down the idle passage. It passes openings in the passage above the throttle plate, called *transfer ports.* These allow more air to bleed into the fuel in

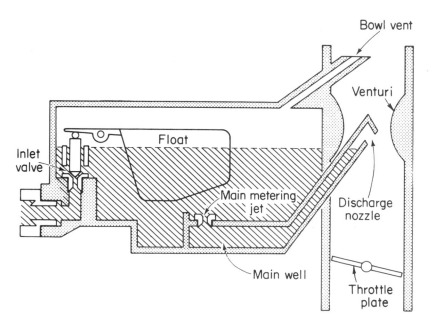

Figure 5-2. Section view line drawing of a simple carburetor.

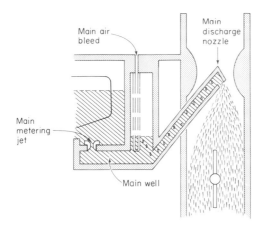

Figure 5-3. A simple carburetor main system.

the idle system for better air/fuel mixing. Finally, the mixture passes an idle mixture screw to be discharged at the *idle port* below the throttle. Here it mixes with the air for idling. This system is shown in Figure 5-4.

When the throttle is gradually opened past the transfer port, the port is exposed to manifold vacuum. This causes fuel to be delivered through the transfer port as well as through the idle port. Both ports supply enough fuel to the increased air flow to maintain a combustible air/fuel mixture in the intake charge. As the throttle is opened still further, the main fuel system begins to deliver fuel. The combination of these systems delivers the correct air/fuel mixture until the throttle is nearly wide open.

Power System. A power system comes into operation as the throttle approaches wide open. Either or both manifold vacuum and throttle position are used to open an auxiliary carburetor system, called a *power system.* This system supplies extra fuel that is needed for full power. A measured quantity of extra fuel is allowed to flow from the float bowl to the main well to be delivered to the air flow through the main discharge nozzle. One type of power system is illustrated in Figure 5-5.

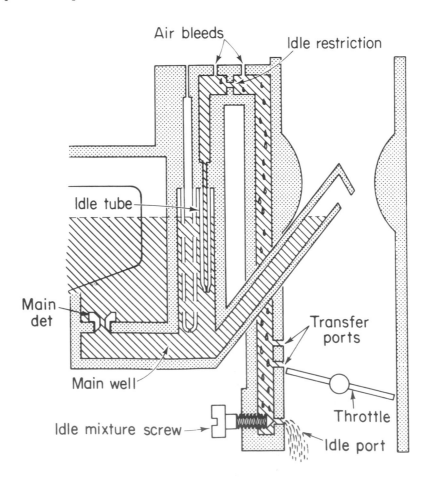

Figure 5-4. A typical carburetor idle system.

Acceleration System. At the moment of rapid throttle opening, the air flowing through the carburetor will start to move more quickly than the fuel. This increased air flow would cause the air/fuel mixture of the intake charge to lean at a time when a rich mixture is needed. An *acceleration pump system* is provided to correct this problem. The acceleration pump pulls fuel from the float bowl through an *inlet check valve*. It is pushed out of the pump through an *outlet check valve* to a special *discharge nozzle* located on the air cleaner side of the venturi. A spring system in the operating mechanism provides fuel delivery for a few seconds' duration. The fuel flow will last until fuel begins to flow in the main and power systems. No fuel will flow through the acceleration system at any other time. A typical acceleration system is shown in Figure 5-6.

Choke System. Carburetors have a *choke* to aid in starting a cold engine. The choke is a plate in the *air horn* on the air cleaner side of the venturi. When the engine is cold, the choke plate closes to limit the incoming air. This lets manifold vacuum pull fuel from the main discharge nozzle as well as the idle and transfer ports while cranking. Fuel delivered from all three of these systems provides a very rich mixture that is needed to start the engine. Immediately after a start the choke is opened slightly. As the engine becomes warm, the choke mechanism

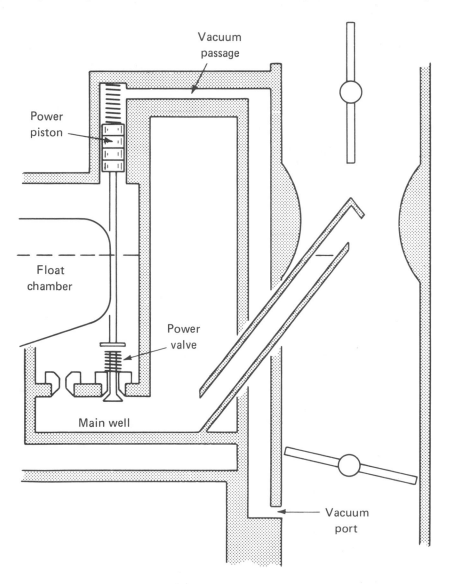

Figure 5-5. One type of power system.

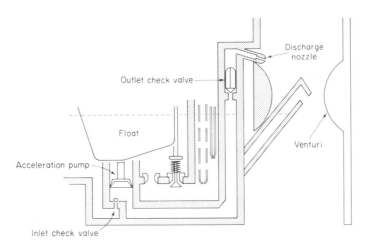

Figure 5-6. Parts of an acceleration system.

gradually opens the choke. The choke is fully open when the engine is at its normal operating temperature. The parts of a typical choke system can be seen in Figure 5-7.

Secondary System. The carburetor just described applies to the *primary barrel* of a single barrel (one bore and one throttle plate) carburetor. A carburetor with two primary barrels (two bores with a throttle plate in each) duplicates all of the parts of the single-barrel carburetor, except that the carburetor uses only one float bowl, choke, accelerating system, and in some cases, one power system.

The greatest amount of air flow is needed for high-speed engine operation. In addition to the two primary barrels, large-capacity carburetors have two *secondary barrels* to allow the engine to get enough air. Some small engines use a carburetor with one primary barrel and one secondary barrel for this same purpose. The secondary system may use the same float bowl as the primary system, or it may have a separate float bowl. Secondary systems have a main fuel system designed to operate at the air/fuel ratio needed for full power. Some secondary systems also contain an idle and transfer system for smooth engine operation as the secondary throttle plates begin to open. Secondary systems do not use a choke, acceleration system, or a power system. Figure 5-8 shows the manifold side of a four-barrel carburetor with the primary throttles nearly open as the secondary throttles begin to open.

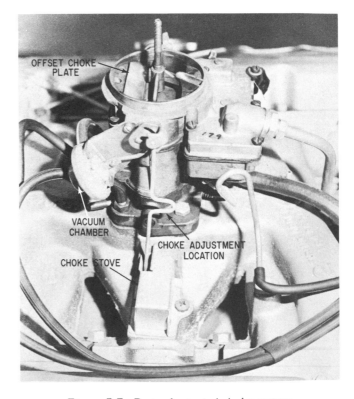

Figure 5-7. Parts of a typical choke system.

Auxiliary Controls. Carburetors are provided with a number of additional devices to improve driveability, economy, and low emission levels. Among these devices are the fast idle cam, unloader, throttle return check, secondary lock out, idle stop solenoid, and vacuum kick. In addition, the carburetors include vacuum ports for the air silencer, heat damper, EGR valve, carbon canister, and the ignition vacuum advance.

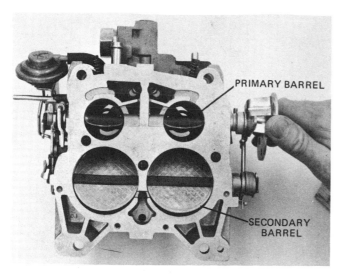

Figure 5-8. Four barrel carburetor with the large secondary barrels just beginning to open as the primary barrels are nearly open.

During the study of the ignition sytem, it is important to understand how the distributor vacuum advance port is connected into the carburetor. The port uses the carburetor body, but it does not interconnect into any of the carburetor operating sytems. This port can be seen in Figure 5-9. There is a direct passage from the external port nipple to an opening in the carburetor bore. The opening is located just above the high edge of the throttle plate when it is in the closed position. In this location, it senses atmospheric pressure when the throttle is closed. As the throttle opens slightly, the edge of the throttle plate moves across the opening, exposing it to manifold vacuum. The vacuum on this opening is

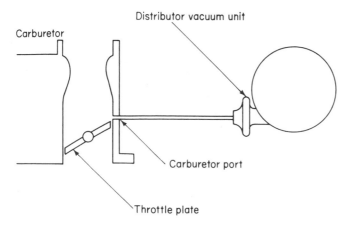

Figure 5-9. Carburetor port above the high side of the throttle plate for the control of the distributor advance.

connected to the distributor vacuum unit. When the throttle plate becomes nearly wide open, the manifold vacuum in the carburetor bore is reduced to nearly atmospheric pressure. At wide-open throttle, there is not enough vacuum at the distributor vacuum port to advance the timing. Vacuum from this port is called *ported vacuum*. To summarize, there is no effective vacuum on the port when the throttle is fully closed against the idle stop or when it is fully open. At part throttle, ported vacuum is high enough to advance the ignition.

5-2 Ignition System

An air/fuel intake charge, compression, and ignition are required to make an engine start and run. Of the three, ignition is most critical. The correct amount of fuel mixed with the air is required for proper engine operation. The engine will start, however, even with no carburetor, if a small amount of gasoline is poured into the intake manifold. High compression is required for maximum power and economy, but the engine will start and run with low compression. In all cases, the ignition system must be able to produce a strong spark at the correct instant to ignite the intake charge after it has been compressed in the cylinder.

Ignition Requirements. Ignition requirements change as engine operating conditions change. Higher compression pressures require a higher ignition voltage. **Voltage is a name given to the amount of electrical pressure.** Cold combustion chambers require a higher ignition voltage while starting than the voltage for a warmed-up chamber. Both rich and lean intake charge air/fuel mixtures require higher ignition voltage than the voltage required for chemically correct mixtures. The ignition timing must advance as engine speed increases to provide proper combustion that leads to engine efficiency. Part throttle operation requires more ignition advance than either full throttle operation or idle. Exhaust emission is affected by ignition timing. Some emission controls change ignition timing to minimize polluting exhaust emissions. These changes usually reduce the thermal efficiency of the engine and reduce the gas mileage.

The ignition system provides a voltage that is high enough to form an arc between the spark plug electrodes. It must do this at the correct instant in

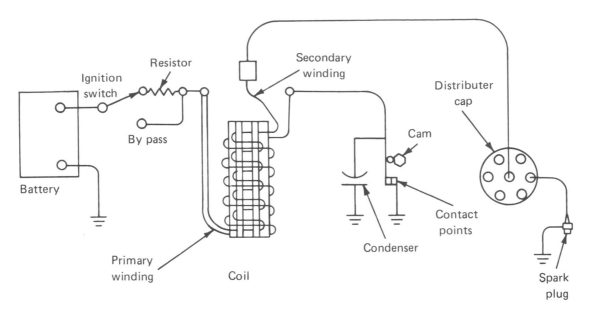

Figure 5-10. Schematic drawing of a typical breaker point ignition system.

the engine cycle. This is called *ignition timing*. The correct ignition timing allows the engine to produce the required power, economy, and emission level demanded of modern automobiles.

The ignition system consists of two parts: a low-voltage primary circuit and a high-voltage secondary circuit. The primary circuit of a standard ignition system includes the battery, ignition switch, ballast resistor, cam-operated breaker points, condenser, and the heavy primary windings in the coil with their connecting wiring. The secondary circuit includes the large number of fine coil secondary windings, the distributor rotor, distributor cap, ignition cables, and spark plugs. Figure 5-10 is a schematic diagram of a typical breaker-point ignition system.

Ignition Operation. When the ignition switch and *breaker points* are closed in a typical ignition system, the primary circuit is completed, through ground. This allows electrical current to flow in the primary circuit. This current flow builds a *magnetic field* in the coil (Figure 5-11). As the engine rotates, it turns the breaker-point *cam* within the distributor housing. The cam pushes against the breaker-point *rubbing block,* forcing the points apart. Breaker-point separation interrupts and stops the primary current flow. When the current flow stops, the magnetic field in the coil collapses through the secondary windings (Figure 5-12). The *condenser,*

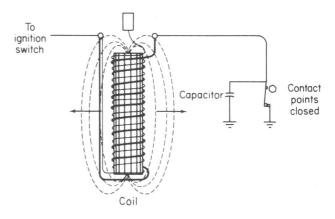

Figure 5-11. Magnetic field build up in the coil while the contact points are closed.

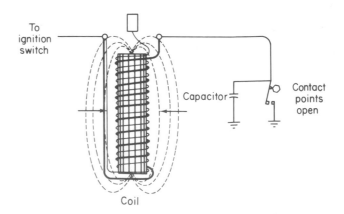

Figure 5-12. Magnetic field collapse at the instant the contact points open.

sometimes called a capacitor, within the distributor minimizes contact point arcing as it helps control the rapid collapse of the magnetic field. Field collapse induces a momentary high-voltage surge in the coil secondary windings. At this instant, the *rotor tip* is lined up with the proper distributor *cap electrode*. This can be seen in Figure 5-13. The high

voltage is impressed through the *secondary cables* to the spark plug in the cylinder to be fired. This high voltage causes an electrical arc to form across the spark plug electrode gap. The arc ignites the compressed intake charge in the combustion chamber at the correct instant. The resulting combustion increases pressure above the piston to push it down in the cylinder.

Solid-State Ignition. Starting in 1971, Chrysler made solid-state ignition systems standard on some automobile models. By 1975, all domestic automobiles were being equipped with solid-state ignition. These ignition systems do not require a regular tune-up. Replacing spark plugs is the only periodic service required on the ignition system.

In solid-state ignition systems, a *pulse inductor* is located in the distributor in place of breaker points. A signal produced by the pulse inductor is sent to an *electronic control unit* (ECU). The ECU momentarily stops the primary current flowing through the coil, just as breaker points do. From this point on, the ignition system produces a high secondary voltage just as in a breaker-point system. Figure 5-14 is a schematic drawing of a solid-state ignition system.

Computer-Controlled Ignition. The next development in ignition systems was the *computer-controlled ignition advance*. The computer takes the place of the mechanical and vacuum advance mechanisms of the ignition systems just described. It provides more exact ignition timing for low emissions and good fuel economy. The basic wiring of a

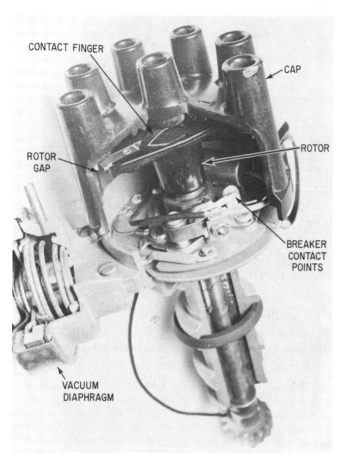

Figure 5-13. A cut away distributor showing the rotor tip aligned with a cap electrode at the rotor gap.

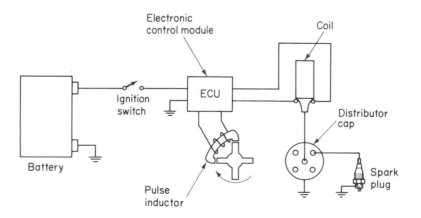

Figure 5-14. A schematic drawing of a solid state ignition system.

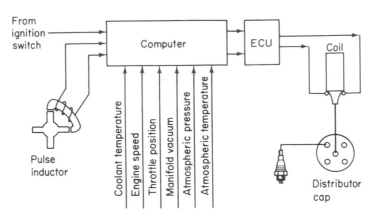

Figure 5-15. A schematic diagram of the basic wiring of a computer controlled ignition advance.

computer-controlled ignition advance is shown in Figure 5-15.

The computer-controlled ignition advance uses a solid-state pulse inductor in the distributor. The pulse indicator signal at full advance is sent to the computer. At the same time, the computer also receives signals from engine coolant temperature, engine speed, throttle position, manifold vacuum, atmospheric pressure, and so on. The computer compares these signals to the program built into the computer. The computer then delays the signal the correct amount, then it triggers an ECU, which, in turn, triggers the coil primary. A distributor rotor and cap sends the coil secondary discharge through secondary cables to the spark plugs.

5-3 Electrical System

The automotive engine electrical system provides electrical energy to supply the ignition system as well as all of the other electrical loads of the vehicle. The starting and charging systems are the primary electrical systems on the engine. The battery is connected electrically between these systems. A typical engine electrical system is illustrated in Figure 5-16.

The Battery. The lead–acid type of battery is the primary source of electricity for starting modern engines (Figure 5-17). It also serves as a reserve source of electricity for the electrical running load of the vehicle. The battery size depends on its use.

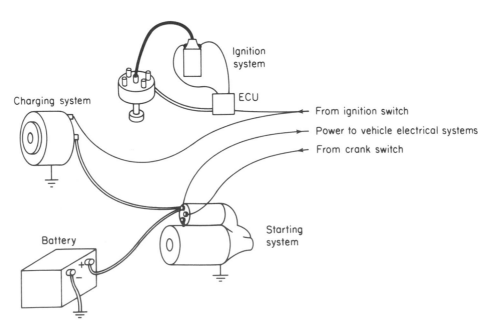

Figure 5-16. The battery connected into the engine electrical systems.

Figure 5-17. A typical side terminal lead acid automobile battery.

Vehicles with large engines require a greater cranking power and so they use a large battery. Large batteries are also used in vehicles with a large number of electrically operated accessories. Small batteries are found in vehicles with small engines and light electrical loads.

A properly maintained lead–acid battery of the type used in automobiles will give from 3 to 4 years of trouble-free service. Proper battery maintenance involves keeping the case and terminals clean, charged, full of water, and well supported in the carrier. When a battery fails to start the engine, the automotive service technician must be able to check the battery and the rest of the electrical system. This is necessary to determine the cause of failure in order to properly correct the problem. It could be that the battery has failed or it could be that other electrical system parts have failed.

The Cranking Motor Circuit. Half of the main cranking motor electrical circuit consists of a heavy cable connecting the positive battery post to the starter terminal through a heavy-duty switch, relay, or solenoid. In the other half of the circuit, the negative battery post is connected to the engine block through a battery ground cable. This circuit carries high current while the engine is being cranked, so it must have low resistance.

The rest of the vehicle electrical system is con-

nected to the battery-starter insulated circuit at a junction between the battery and starter relay or solenoid. This junction might be the battery cable clamp, the battery junction of the solenoid, or a special junction block. The ignition switch gets its electricity from this junction. When the ignition switch is turned to the start position, it energizes the relay or solenoid to operate the starter and crank the engine. This part of the electrical system is called the starter switch circuit.

While cranking, the starter is mechanically connected to the engine ring gear by the starter drive mechanism. The drive mechanism engages with a stationary ring gear while cranking. It is also designed to protect the starter motor from overspeed when the engine starts. Figure 5-18 shows a typical starter mounting location on an engine.

The Charging System. The *alternator,* or *diode-rectified generator,* is the source of electrical energy used to operate all of the electrical devices on the automobile while the engine is running. In addition, the alternator has extra capacity to recharge the battery. The charging circuit consists of wiring (called the insulated circuit) that interconnects the alternator, the regulator, the battery, and the vehicle electrical system.

The main portion of the insulated circuit consists of a wire between the alternator BAT terminal and a *junction* where the alternator can feed both the vehicle electrical system and the battery. A second wire between this junction and the battery completes the insulated portion of the charging circuit.

Figure 5-18. A typical starter mounted on a V-type engine.

The battery uses this second wire to feed the vehicle electrical system when the alternator does not supply adequate electrical power by itself. Vehicle manufacturers use a number of different junction points. The junction may be the BAT terminal of the starter solenoid, a junction block on the radiator support or inner fender pan, or the BAT terminal of the horn relay. The grounded portion of the charging circuit is the metal-to-metal contact between the alternator case and the engine. The engine is connected to the negative battery post through the battery ground cable. This completes the main charging circuit. The ground cable is the same cable that completes the cranking motor circuit.

The *voltage regulator* is an important part of the charging system. It is connected in series with the alternator rotor field winding. The regulator controls electrical current going through the field. This controls the effective strength of the magnetic field. In this way, it controls the alternator output voltage. When mechanical voltage regulators are used, one end of the regulator-field circuit is connected to the insulated circuit through a switch (this can be called a B circuit). The other end of the field is connected to ground. Solid-state regulators may be either internal integrated regulators or external regulators. They are located electrically between the rotor field and ground (this can be called an A circuit). Figure 5-19 shows the typical mounting of an internally regulated alternator on an engine.

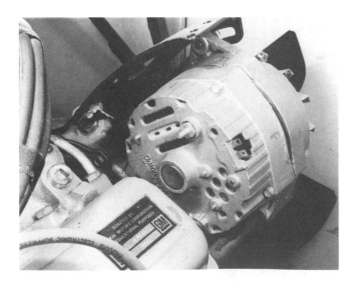

Figure 5-19. A typical alternator having an internal regulator mounted on a V-type engine.

A switch is needed in the regulator-field circuit to open the circuit when the engine is not running and to close the circuit when the engine operates. The *ignition switch* serves this function. Some vehicles use a heavy-duty ignition switch to directly feed electrical power into the regulator-field circuit. Other systems use a standard-duty ignition switch to signal a *field relay* located in the same case with the voltage regulator. The relay connects the regulator-field circuit to the charging circuit. This causes the charge indicator lamp to go out when the alternator is charging.

The charging circuit includes an indicator, either an ammeter or and indicator lamp, to show battery charge or discharge. The ammeter, when used, is connected in the charging circuit between the junction and the battery. It will show charge when the alternator is producing enough output to charge the battery. It shows discharge when the alternator cannot supply enough electrical current by itself for the electrical demand. The current shown as discharge is coming from the battery to help supply the electrical demand.

5-4 Lubrication System

Lubricating oil is often called the life blood of an engine. It circulates through passages in the engine that carry it to all of the rubbing surfaces. The main job of the oil is to form and maintain a film between these surfaces. The film is thick enough to keep the surfaces from touching. This action *minimizes friction* and wear within the engine. The lubricant has useful secondary functions. Cool lubricant picks up heat from the hot engine parts. It carries the heat to the oil pan. The oil is cooled as air moves past the outside surface of the pan. This helps to *cool* the engine. The oil flow also carries wear particles from the rubbing surfaces to the pan, where they will be drained from the engine during an oil change. This helps to keep the engine clean. Oil between the engine parts *cushions* the parts from shock as the combustion charge forces the piston down.

The lubricant used for motor oil must have properties that meet the engine requirements. The most important property of the motor oil is its thickness at normal engine operating temperatures. The oil thickness is called *viscosity*. If the oil is too thin, it will rapidly leak from the bearing clearances, allowing the parts to contact. This will result in scoring of the engine parts. **Scoring is heavy scratches on the rubbing surfaces.** When the oil is too thick, the oil will drag between the rubbing surfaces. This requires excessive engine power. The oil drag characteristic is most noticeable when one compares the cranking speed of a cold engine to a warm engine.

The motor oil producers put additives into the oil. These help the oil to clean the engine, minimize scuffing, reduce rusting, resist oxidation, and maintain correct viscosity characteristics as the oil temperature changes. The oil should be replaced when it gets dirty or when its properties no longer protect the engine.

Figure 5-20 is a phantom view of a typical lubrication system. The oil rests in the bottom of the oil pan when the engine is not running. When the engine is started, the *oil pump* pulls the oil through a pickup screen and pushes it through the *oil filter* into the drilled main engine oil passage that is called a *main oil gallery*. A pressure regulator or *relief valve* built into the oil pump limits the maximum pressure in the main gallery. Drilled oil passages connect between the main oil gallery and the *engine bearings*. In some engines the *lifter bores* intersect the main oil gallery to supply the hydraulic lifters with pressurized oil. After the oil has lubricated the parts, it runs back to the oil pan, completing the lubrication cycle. Either cast openings or drilled holes lead from the rocker assemblies through the head and block. The oil can readily flow back from the overhead valve mechanism to the oil pan through these holes.

5-5 Cooling System

The primary function of the automotive engine cooling system is to maintain the normal operating temperature of the block and head. Coolant flow is held at a minimum during warm-up until the normal engine temperature is reached. The coolant flow is then increased, as required, to maintain the normal engine operating temperature. If the engine operating temperature is too low, scuffing and wear rates will increase within the engine. **Scuffing is the transfer of metal from one surface to the surface it is rubbing on.** If temperature is too high, hard deposits will form that can cause part sticking and passage clogging. Operating the engine at normal temperatures will minimize these problems. It will help increase the service life of the engine.

Most automobiles use a liquid cooling system. The cooling liquid enters the cooling pump, also called a *water pump*. The *fan* is attached to the pump, so both the fan and pump are driven by the *fan belt*. The pump delivers coolant to the *coolant passages* in the engine block, where it flows around the cylinders. Coolant flows from the block to the head. Here, the coolant is collected at a common point where the *thermostat* is located. A small amount of coolant is bypassed around the thermostat and returned to the engine during warm-up to maintain equal temperatures all through the engine. When coolant becomes warm, the thermostat opens, allowing coolant to flow to one of the *radiator* tanks. Coolant flows through small tubes in the radiator. These tubes are soldered to metal fins forming the radiator *core*. Heat transfers from the coolant to the metal fins. Air picks up the heat from the radiator core. The coolant is a transfer medium that carries heat from the combustion chamber to the air and out of the vehicle. Cooled liquid coolant returns from the other radiator tank to the water pump to be forced back through the cooling system circuit. This completes the cooling cycle. Figure 5-21 shows a phantom view of the internal cooling system.

To function properly, the coolant must remain as a liquid. If the coolant boils, the resulting water

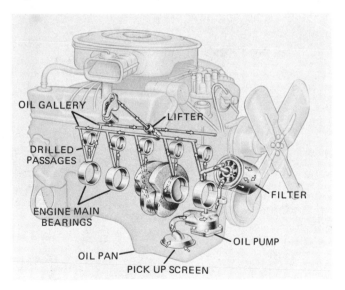

Figure 5-20. A phantom view of a typical lubrication system of a V-type engine (Courtesy of Ford Motor Company).

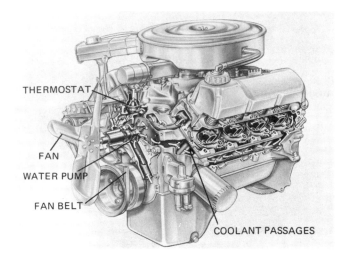

THERMOSTAT

FAN

WATER PUMP

FAN BELT

COOLANT PASSAGES

Figure 5-21. A phantom view of a typical internal cooling system of a V-type engine (Courtesy of Ford Motor Company).

vapor will not pick up the required heat. If the coolant freezes, it will not flow through the system, so it cannot pick up heat. Glycol–water coolant mixtures are used to lower the freezing temperature and increase the boiling temperature to give the coolant a large usable temperature range.

REVIEW QUESTIONS

1. What are the three basics needed for an engine to run? [INTRODUCTION]
2. What is the purpose of the carburetor float system? [5-1]
3. What happens to venturi vacuum and manifold vacuum at idle? [5-1]
4. What happens to venturi vacuum and manifold vacuum at full throttle? [5-1]
5. How is the air/fuel mixture of the intake charge controlled? [5-1]
6. Describe the flow of fuel in the idle system. [5-1]
7. How does the operation of the transfer port differ when the engine is idling than when it is running faster than idle? [5-1]
8. What does the power system do and when does it do it? [5-1]
9. Why is it necessary to have an acceleration pump? [5-1]
10. Why is it necessary to have a choke to start a cold engine? [5-1]
11. When does the carburetor secondary system operate? [5-1]
12. What happens to the coil primary current that produces a high secondary voltage in the coil? [5-2]
13. How does the operation of a solid-state ignition system differ from a breaker-point ignition system? How are they the same? [5-2]
14. How does the computer-controlled ignition advance differ from the solid-state ignition? [5-2]
15. How is a battery properly maintained? [5-3]
16. Describe the electrical connections in the starter circuit. [5-3]
17. When does a voltage regulator actually regulate the voltage? [5-3]
18. In what circuit does the voltage regulator do the regulating? [5-3]
19. When is a field relay used in a regulator-field circuit? [5-3]
20. What is the purpose of lubrication in an engine? [5-4]
21. Why is the viscosity of the oil important? [5-4]
22. When should the oil be replaced? [5-4]
23. What is the primary function of the cooling system? [5-5]
24. What is the purpose of the thermostat? [5-5]
25. Why must the coolant remain liquid? [5-5]

66

6

Intake and Exhaust Manifolds

Automotive engines are required to run smoothly. Smooth operation can only occur when each combustion chamber produces the same pressure as every other chamber in the engine. To do this, each cylinder must receive a charge exactly like the charge going into the other cylinders in quality and quantity. The charges must have the same physical properties and the same air/fuel mixture.

The air coming into an engine will flow through the carburetor or injector throttle body. The carburetor or injector is the device that provides the charge quality. It does this by mixing fuel into the incoming air in the correct proportions. The intake manifold directs an equal quantity of the mixed intake charge to each intake valve. Each intake valve is timed the same as the others to allow an equal quantity of the intake charge to enter each combustion chamber. If individual port injectors are used, they have injector nozzles that inject the fuel into the air just before it enters the intake valve. Throttle body injection injects the fuel just above the throttle plate. An ignition distributor must be timed to send a spark across the spark plug gap when each piston has compressed the charge the same amount. When all of these requirements are met, the pressure in the combustion chambers during the power stroke will be equal.

It is unfortunate that an engine has not, as yet, been designed to meet these ideal requirements under all operating conditions. For minimum manufacturing cost, tolerances must be quite large. Because of this, the valve and ignition timing between cylinders is not exactly the same. The manifold passages have different sizes, angles, temperatures, and flow rates. All of these tend to make the quality and quantity of the intake charge somewhat different between cylinders. This is especially true at low engine speeds. Intake manifolds that are carefully designed will supply a more uniform charge to the cylinders. This helps to make the engine run smoothly, but these manifolds are costly to manufacture.

Students generally find that they can understand automotive components better if they work on them right after they have finished reading about them. This chapter and those immediately following have been written with this in mind. An engine can be disassembled as the chapters are studied.

After the engine accessories have been taken off the engine, the manifolds can be removed. It is necessary to drain the coolant before removing any intake manifold that has a coolant passage. The manifolds can be lifted from the engine after all of the assembly bolts and hold-down clamps are removed.

6-1 Intake Manifold Criteria

The carburetor delivers finely divided droplets of liquid fuel into the incoming air in a combustible air/fuel ratio (Figure 6-1). These particles start to evaporate as soon as they leave the carburetor. With a carburetor engine operating at its best volumetric efficiency, about 60% of the fuel will evaporate by the time the intake charge reaches the combustion chamber. This means that there will be some liquid

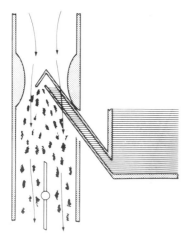

Figure 6-1. Finely divided droplets of a liquid fuel mixed into the incoming air by the carburetor.

droplets in the charge as it flows through the manifold. The droplets stay in the charge as long as the charge flows at high velocities. At maximum horsepower, these velocities may reach 300 ft/s. Separation of the droplets from the charge as it passes through manifold occurs when the velocity drops below 50 ft/s. Intake charge velocities at idle speeds are often below this value. When separation occurs, extra fuel must be supplied to the charge in order to have a combustible mixture reach the combustion chamber at low engine speeds.

Manifold sizes are a compromise. They must have a cross section large enough to allow enough charge flow for maximum power. The cross section would look like a pipe cut in two. The cross section must be small enough so that the flow velocities of the charge will be fast enough to keep the fuel droplets in suspension. This is required so that equal mixture reaches each cylinder. Manifold cross sectional size is one of the reasons that engines designed especially for racing will not run at low engine speeds. Racing manifolds must be large enough to reach maximum horsepower. This size, however, allows the charge to move slowly so the fuel will separate from the charge at low engine speeds. Fuel separation leads to poor accelerator response. Passenger car engines are primarily designed for economy at light-load, part-throttle operation. Their manifolds, therefore, have a much smaller cross-sectional area than racing engines. This small size will help keep high flow velocities of the charge throughout the normal operating speed range of the engine.

It should be noted that fuel separation problems do not exist on engines that have fuel-injected into the intake near the inlet port. Engines with port injection can operate satisfactorily at low speeds, even with large intake manifold cross sections.

In a four-stroke cycle, the inlet stroke is approximately one-fourth of the cycle. On a single-cylinder, four-cycle engine, the carburetor would be working one-fourth of the time on the intake stroke. Four cylinders can, therefore, be attached to the same-size carburetor when cylinders are timed so that each cylinder takes a different quarter of the 720° four-stroke cycle. Using this technique, one carburetor will satisfy the requirements of four cylinders just as well as it will satisfy one cylinder. This principle is also used in modern V-8 automotive engines. In these engines, the intake manifold is divided into two sections or branches. Each branch supplies four cylinders. When a two-barrel carburetor is used on a V-type engine, one barrel supplies each branch of the manifold. The cylinder cycles are timed so that only one cylinder draws an intake charge from the carburetor barrel at a time.

Six-cylinder inline engines usually have a single-barrel carburetor to supply all the cylinders. In these engines, the single carburetor barrel will be supplying some intake charge to two cylinders at the same time. Therefore, a carburetor used on six-cylinder inline engines requires a larger barrel. Some six-cylinder inline engines use a two-barrel carburetor when more charge flow is required than the flow provided by a single-barrel carburetor. An example of this is shown in Figure 6-2.

Intake manifolds used on most V-type engines are built with runners on two levels. Using this design, relative long runners can be fit between the heads. This can best be shown by the manifold casting cores in Figure 6-3. Successive firing cylinders are fed alternately from the upper and lower runners, so the runner design must match the cylinder firing order. Figure 6-4 shows a V-8 manifold having all of the runners on one level. This allows the hood line of the vehicle to be lowered.

The runners in the intake manifold may be a *log type*. These have the largest possible cross section for maximum air flow. Many have an *H-type* pattern. This design is a compromise between maximum performance and the best use of the space available. It is in use on most of the modern V-type engines. Some runners are called *tuned runners*.

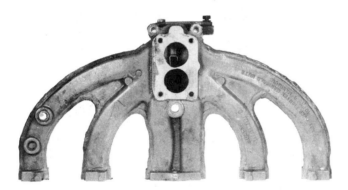

Figure 6-2. A welded aluminum intake manifold for a six cylinder engine using a two barrel carburetor.

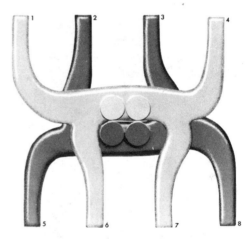

Figure 6-3. Casting cores for an intake manifold that fits a V-eight engine (Courtesy of Ford Motor Company).

6-2 Intake Manifold Features

The intake manifold's primary function is to carry the intake charge from the carburetor to the intake port in the head. As the charge flows, it picks up heat. The heat evaporates the liquid fuel droplets, gradually changing the entire charge into a gaseous air/fuel mixture during part-load operation.

Runners. Siamese intake runners have one intake passage between the carburetor and two side-by-side cylinders. They are used on some economy engines. Most modern engines have a separate runner going to each intake port. This design makes all intake runners in the engine nearly equal in length and in cross section. This runner design helps give an equal air/fuel charge to each cylinder.

Figure 6-4. An intake manifold with all of the runners at the same level for use on a V-type engine.

These runner designs are shown in Figure 6-5. In the tuned runners, the length is designed to take advantage of the natural pressure wave that occurs in a gas column. The pressure wave reaches the cylinder at the exact instant that the intake valve is open. This allows the charge to enter the cylinder with a supercharging or ram effect. The effect of intake manifold tuning is illustrated in Figure 6-6. On V-8 engines using four-barrel carburetors, the primary barrels are often placed approximately on center of the runners. One example is shown in Figure 6-7. This helps to give good low- and midrange performance. The carburetor primary barrels are in use most of the time. The secondary barrels of the carburetor operate only at full throttle. Full throttle is used during a very small percentage of the operating time. Because of this, unequal charge distribution from the secondary barrels is not as critical as it is from the primary barrels.

The way the charge flows through the intake

Figure 6-7. Primary barrels of the carburetor fit on the intake manifold bores near the center of the manifold.

manifold somewhat depends upon the number of sharp runner bends, the smoothness of the runner interior wall, and the cross-sectional runner shape. Sharp bends tend to increase fuel separation. This is illustrated in Figure 6-8. The air, having less mass, is able to make turns much more quickly than the heavy fuel droplets. Rough interior runner surfaces add a drag and turbulence to the charge. This can upset the charge distribution. A round runner shape has the greatest cross-sectional area for its wall surface area; however, a round section is not always the most desirable. Passenger car engine manifold runner floors are flat. Any liquid fuel that drops out of the charge will spread in a thin layer over the manifold floor and rapidly evaporate. The rear of the engine is lower than the front for better drive line positioning. Manifolds are designed so that the manifold floor is level when the engine is mounted in the chassis. The flat manifold floor prevents low spots where liquid fuel can collect. This can be seen

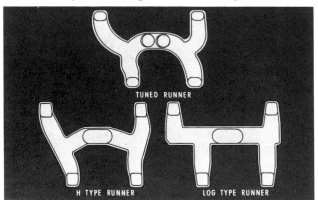

Figure 6-5. Typical intake runner shapes (Courtesy of Chevrolet Motor Division, General Motors Corporation).

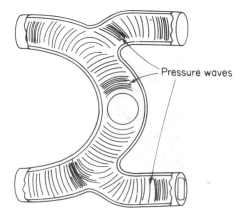

Figure 6-6. Effect of compression waves in a tuned intake manifold.

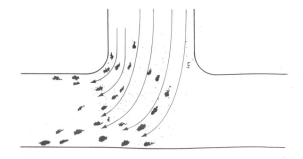

Figure 6-8. Heavy fuel droplets separate as they flow around an abrupt bend in an intake manifold.

FLAT MANIFOLD
FLOOR

Figure 6-9. Flat intake manifold floor (Courtesy of Cadillac Motor Car Division, General Motors Corporation).

in Figure 6-9. Large rectangular and oval shaped cross sections are used on performance engines. They take advantage of the available space for more cross-sectional area at the expense of relatively more wall surface area when compared to round runners. The corners of rectangular runners cause turbulence, called *eddy currents,* that help lift any liquid fuel from the surface. Figure 6-10 illustrates the eddy currents in a cross-sectional view of an intake runner.

Main intake runners have cross-sectional areas of approximately 0.008 in.^{2} per engine cubic inch displacement (CID). Branch runners have cross-sectional areas of approximately 0.006 in.^{2} per CID. Ribs and guide vanes, such as those that can be seen in Figure 6-11, are often positioned in the floor of the manifold runners. They aid in equal distribution of the intake gases to the cylinders, even when some of the fuel remains as a liquid. It is just as important for the fuel to have equal distribution as it is for the air to have equal distribution.

The use of a small third valve in the combustion

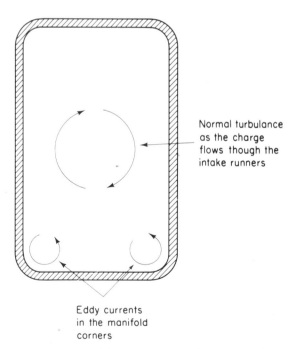

Normal turbulance as the charge flows though the intake runners

Eddy currents in the manifold corners

Figure 6-10. Eddy currents as the charge flows through the intake manifold lift liquid fuel from the flat floor of the manifold.

Figure 6-11. Guide vanes in the floor of the intake manifold.

chamber is relatively new. It has come about in an attempt to reduce exhaust emissions while maintaining reasonably good fuel mileage. The third valve is an intake valve. It supplies a rich air/fuel mixture charge to a small prechamber that also contains the spark plug in the Honda CVCC (Compound Vortex Controlled Combustion) engine. In the MCA (Mitsubishi Clean Air) engine, the third valve supplies fresh air at low speeds and a lean air/fuel mixture at cruising speeds. In both of these engines, a separate runner is required in the intake manifold for the third valve. It runs between the carburetor and the valve port opening in the head.

Manifold Heat. Heat is required in the manifold so that liquid fuel in the charge will evaporate during the time it travels from the carburetor to the combustion chamber. When heat is taken from the air in the intake charge, the charge temperature is lowered. The cooled charge will not evaporate additional fuel as rapidly as a warm charge. In current production engines, additional heat is supplied to the charge. The added heat gives good fuel evaporation for smooth engine operation when the engine is cold. An intake charge temperature range from about 100 to 130 °F (38 to 55 °C) is necessary to give good fuel evaporation. Heat is supplied to the intake manifold during low-temperature operation in most current engines by using *air preheat.* Heat is

picked up from around the exhaust manifold and routed to the air cleaner inlet. A thermostatically controlled bimetal switch adjusts a vacuum motor. It controls the amount of heated air used. Parts of this system are shown in Figure 6-12. Another thermostatic valve, called a *heat riser,* directs exhaust gases against the intake manifold directly below the carburetor. On V-type engines, exhaust gas is

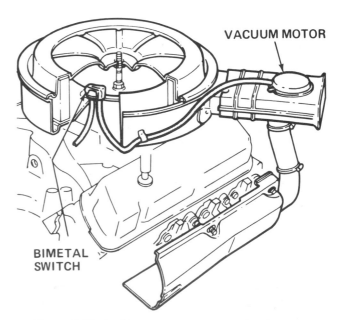

Figure 6-12. A phantom view of a typical carburetor inlet air preheat temperature regulator (Courtesy of AC Division, General Motors Corporation).

72

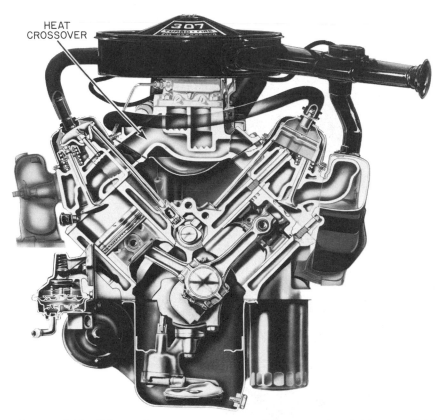

Figure 6-13. Exhaust heat crossover on a V-type engine intake manifold (Courtesy of Chevrolet Motor Division, General Motors Corporation).

routed through a passage, called an exhaust *heat crossover*. Part of the exhaust gas passes against the intake manifold directly under the carburetor. This can be seen in Figure 6-13. On some emission controlled engines, the heat riser valve is operated by a vacuum diaphragm actuator assembly controlled by a temperature-sensitive valve. This system is called *early fuel evaporation* (EFE). A typical EFE valve is shown in Figure 6-14.

When the engine gets fully warmed up, the heat riser valve bypasses the exhaust gas away from the intake manifold and crossover. The exhaust gas is sent directly out through the exhaust system. Excessive heating of the intake manifold will cause the charge to expand. Expansion reduces the mass of the charge that is available to the combustion chamber, thus reducing the engine power. An overheated charge can also lead to abnormal combustion. It is therefore important that the intake manifold is not overheated.

Some engines use the coolant to supply heat to the charge mixture. A passage is provided for warm coolant to flow through a passage below the intake runners, as shown in Figure 6-15 (see following page). Heat from the engine coolant is not available until the engine begins to warm up. Therefore, coolant heat is

Figure 6-14. Typical early fuel evaporation (EFE) system.

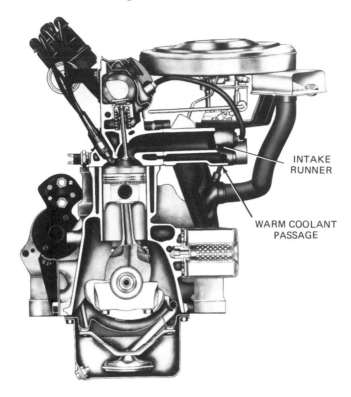

Figure 6-15. Warm coolant passage below the intake manifold runner (Courtesy of Chevrolet Motor Division, General Motors Corporation).

used where the mechanical design of the engine makes it difficult to use exhaust heat. It is also used where a uniform temperature is desired. Engine coolant is always used to provide intake manifold heat on in-line engines when the intake and exhaust manifolds are on opposite sides of the head. An example of these manifolds on a head is shown in Figure 6-16. Manifolds on V-type engines often contain a coolant passage. It connects the cooling system between the V-heads. This passage provides a common cooling outlet for the engine cooling system at the thermostat.

Choke Heat. The carburetor is designed with a choke to provide an excessively rich charge mixture for starting. This is necessary to provide enough of the volatile part of the fuel to make a combustible

Figure 6-16. The head of an engine having the intake and exhaust manifolds on opposite sides of the head. This is called a cross flow head.

mixture. There are two additional problems. First, the intake charge velocity is low during cranking, allowing the fuel to separate from the air. Second, no extra heat is available for fuel evaporation before the engine starts. Most chokes are automatic. They are closed by a temperature-sensing thermostatic spring. In some applications, heat is carried to the thermostatic spring through a tube from a heat chamber, called a *stove*. The choke heat stove is often located in the exhaust manifold where it can pick up exhaust heat. An insulated tube carries warm air from the choke heat stove to the sensing spring on the carburetor. The intake manifold shown in Figure 6-17 has the choke heat stove. In this location the stove is heated by exhaust temperature at the exhaust heat crossover in the intake manifold. Some applications place the heat-sensing choke spring directly in the stove well. A link connects the choke spring with the carburetor choke plate linkage. Figure 6-18 on the following page shows this type of automatic choke.

Exhaust Gas Recirculation. To reduce the emission of oxides of nitrogen, engines have been equipped with *exhaust gas recirculation* (EGR) valves. The EGR valves were first used on California automobiles in 1972. In 1973, they were used on all automobiles. The EGR valve opens during highway speeds on a warm engine. When open, the valve allows a small portion of the exhaust gas to enter the intake manifold. Here, it mixes with and takes the place of some of the intake charge. This leaves less room for the intake charge to enter the combustion chamber. The recirculated exhaust gas does *not* enter into the combustion process. The result is a lower peak combustion temperature. As the combustion temperature is lowered, the production of oxides of nitrogen are also reduced.

The EGR system has some means of interconnecting the exhaust and intake manifolds. The interconnecting passage is controlled by the EGR valve. On V-type engines, the intake manifold crossover is used as a source of exhaust gas for the

Figure 6-17. A typical choke heat stove that supplies heat to a choke spring located on the carburetor.

Figure 6-18. A typical choke heat stove with the heat sensing choke spring located directly in the stove well.

EGR system. A cast passage connects the exhaust crossover to the EGR valve. This was shown in Figure 4-20. The gas is sent from the EGR valve to openings below the carburetor. On inline-type engines, an external tube is generally used to carry exhaust gas to the EGR valve. This tube is often made long so that the exhaust gas is cooled before entering the EGR valve. Figure 6-19 shows a typical long EGR tube. The EGR valve is usually attached to an adapter between the carburetor and intake manifold. Here, it can release the exhaust gas directly into the intake manifold runner.

6-3 Intake Manifold Configuration

There are two general intake manifold designs used on modern V-type engines. The first design is an *open-type manifold*. Runners go through the open-type branches. Other passages provide for exhaust crossover, exhaust gas recirculation, and coolant flow. This design allows good control of manifold runner tuning, is lightweight, and is low cost. Lifter valley covers are needed on engines using open-type manifolds. In some engines, the lifter valley covers are an extension of the intake manifold gasket.

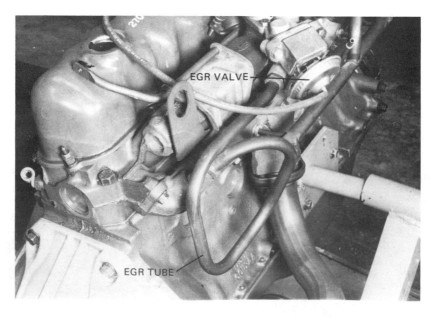

Figure 6-19. EGR tube used to cool the exhaust gas being fed into the intake charge.

Figure 6-20 shows a typical open-type intake manifold.

The second intake manifold design is a *closed-type manifold*. It is used on V-type engines as illustrated in Figure 6-21. This manifold has cast metal between the runners. It is used as a lifter valley cover as well as a manifold. The closed-type intake manifold is heavier and more expensive to make than the open type. The heavy weight is a disadvantage on lightweight high fuel mileage automobiles. Because of the many joints involved, more care is required when installing it. It is difficult to correctly place the gaskets and seals so that they will not leak. Using the closed-type manifold, it is possible to lower the runners, the carburetor, and the automobile hood line. The large mass of metal used in closed-type manifolds tends to muffle engine noise and this results in quieting the noise of engine operation.

Closed-type manifolds on V-type engines have the exhaust crossover located just above the lifter valley, where engine oil could contact the hot surface. Hot exhaust in the crossover would heat the oil that lands on the surface of the crossover. This would cause coking and oil burning. Eventually, the coking would loosen and contaminate the oil, which, in turn, would lead to frequent oil changes. Therefore, shields are put in engines to keep the oil from hitting these hot surfaces. A sheet metal deflector may be fastened to the lifter valley side of the intake manifold to keep the oil from the hot cross over. This is shown in the section view in

Figure 6-21. A closed type intake manifold on a V-type engine. The runners on this manifold are on two planes.

Figure 6-22. Some engines have a large single-piece manifold gasket to serve this function (Figure 6-23). Some closed-type manifolds are designed with a cast-in passage that provides an insulating chamber between the exhaust crossover passage and the lifter valley. This results in good insulation and noise reduction. The insulating chamber is shown in a section view in Figure 6-24.

On inline engines, the intake manifold is less complicated than on V-type engines. The intake manifold can be a simple log-type that directs the charge from the carburetor to the head intake ports in the most convenient manner. A log-type manifold used on a four-cylinder engine is shown in

Figure 6-20. An open type intake manifold on a V-type engine. The runners on this manifold are on a single plane.

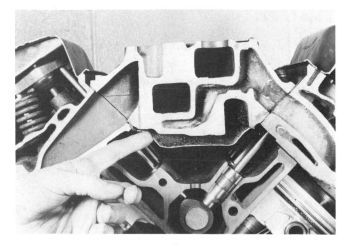

Figure 6-22. The finger points to a sheetmetal deflector oil shield fastened to the lifter valley side of the intake manifold.

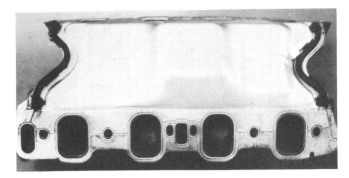

Figure 6-23. A single-piece large intake manifold gasket that serves as an oil shield.

Figure 6-25. A log-type intake manifold used on a four cylinder engine.

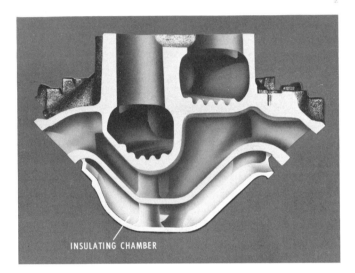

Figure 6-24. Cast-in insulating chambers that forms an oil shield in the intake manifold (Courtesy of Chevrolet Motor Division, General Motors Corporation).

Figure 6-26. A riser tube directly under the carburetor in the intake manifold.

Figure 6-25. When one runner feeds two adjacent cylinders, it is said to be a *siamese* runner. They are often slightly offset to equalize the charge going to each cylinder. Some intake manifolds are designed to minimize restrictions by using separate runners to each intake port. When an inline engine is equipped with a carburetor having one primary and one secondary barrel, both barrels feed a common passage below the carburetor. This passage is sometimes called a *plenum*. These manifolds may have *riser tubes* in the plenum directly under the carburetor as pictured in Figure 6-26. They aid in mixing and vaporizing the fuel by directing the liquid fuel away from the wall of the manifold. When the intake manifold on an inline engine is equipped with a heat riser, the exhaust manifold is bolted to the intake manifold immediately below the carburetor. The heat riser is located at this junction, as

illustrated in Figure 6-27. When an exhaust heat riser is not used, the manifold has a passage through which warm coolant flows. This may be a passage that also leads to the vehicle heater.

Historically, with a few exceptions, intake manifolds have been made of cast iron. These manifolds were easy to cast and easy to machine. This kept their cost down. Weight has become more critical as fuel economy has improved. Lightweight aluminum is more frequently being used for intake manifolds. In some cases, an upper and lower aluminum manifold casting are welded together to form the manifold. One of these was pictured in Figure 6-2. Aluminum manifolds are more expensive and their rate of expansion is greater than cast iron. By careful design and specialized manufacturing techniques, their weight reduction compensates for their added cost.

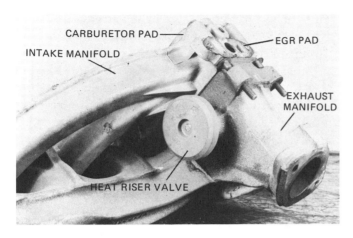

Figure 6-27. Heat riser at the junction of the intake and exhaust manifold.

6-4 Exhaust Manifold Criteria

The exhaust manifold is designed to collect high-temperature spent gases from the head exhaust ports. The hot gases are sent to an exhaust pipe, catalytic converter, exhaust silencer or muffler, resonator, and on to the tailpipe, where they are vented to the atmosphere. This must be done with the least possible restriction or back pressure while keeping the exhaust noise at a minimum.

Exhaust gas temperature will vary according to the power produced by the engine. The manifold must be designed to operate at both engine idle and continuous full power. Under full-power conditions, the exhaust manifold will become red hot, causing a great deal of expansion. At idle, the exhaust manifold is just warm, causing little expansion. After casting, the manifold may be annealed. Annealing is a heat treating process that takes out the brittle hardening of the casting to reduce the chance of cracking from the temperature changes. In passenger car operation, the engine will normally be run under light-load, part-throttle conditions, where the manifold temperatures will not reach the high-temperature extremes. Generally, the exhaust manifold is made from cast iron that can withstand extreme and rapid temperature changes. Rapid temperature changes are called *thermal shock*. It is bolted to the head in a way that will allow expansion and contraction. In some cases, hollow-headed bolts are used to maintain a gastight seal while still allowing normal expansion and contraction. Some manifolds are designed so that no parting surface gasket is required, whereas others require a gasket.

The condition of the exhaust system is more critical now that exhaust-emission laws have been passed. The exhaust system was first used to carry smelly fumes away from the occupants. Soon, mufflers were added to silence the exhaust noise. The exhaust systems have been refined to the point where the wind and tire noises are greater than the exhaust noise.

On some racing engines, the exhaust manifold is replaced by a welded steel tubing *header*. The header allows a smooth, nearly ideal exhaust manifold design that will handle the large volumes of exhaust gas produced when the engine is operating at high speeds. Headers do not improve performance at legal road speeds, they cost more to make, and they have a relatively short useful life. Therefore, they are used only where high engine speeds require their use. Sometimes, owners install headers for the appearance of speed and power, regardless of their cost.

6-5 Exhaust Manifold Compromises

The exhaust manifold is designed to allow the free flow of exhaust gas. Some manifolds use internal cast-rib deflectors or dividers to guide the exhaust gases toward the outlet as smoothly as possible. The exhaust passage cross section has well-proportioned flow areas, smooth flow paths, and maximum branch separation within the limits of the chassis space. The chassis front suspension, steering gear box, and fender skirts limit the space that is available for the exhaust manifold. In general, the chassis and engine are designed first and then the manifold is designed to fit into the remaining space. This is not as much of a compromise as it might seem. Severe bends have no measurable effect on the flow of exhaust gases as long as the required manifold cross section is maintained. Figure 6-28 shows two types of exhaust manifolds that fit inline engines.

Some exhaust manifolds are designed to go above the spark plug, whereas others are designed to go below. The spark plug and carefully routed ignition wires are usually shielded from the exhaust heat with sheet metal deflectors. Typical deflectors can be seen in Figure 6-29.

Often, a stove for the automatic choke is also incorporated in the exhaust manifold. Some

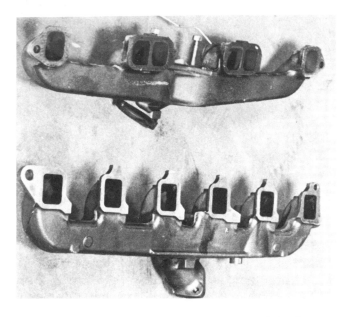

Figure 6-28. The upper exhaust manifold has siamese runners. The lower exhaust manifold has separate runners.

Figure 6-29. An example of heat deflector shields placed between the exhaust manifold and the spark plugs.

manifolds have even been designed with brackets on which are mounted such components as generators and air conditioning compressors.

Exhaust systems are especially designed for the engine–chassis combination. The exhaust system length, pipe size, and silencer are designed, where possible, to make use of the tuning effect of the gas column resonating within the exhaust system. Tuning occurs when the exhaust pulses from the cylinders are emptied into the manifold between the pulses of other cylinders. Here, the pressure is less.

This helps scavenge the exhaust gas from the cylinder. With more exhaust removed, there is more useful space for the fresh intake charge. This leads to more useful engine power at the same throttle position. Exhaust tuning is most effective at one engine speed. Manifolds are therefore tuned to the most desirable engine rpm for the particular vehicle involved. Short pipe lengths favor high-rpm power peaks. Longer pipe lengths tend to increase engine torque in the midrange speed used in passenger car applications. Figure 6-30 compares an exhaust manifold used on a tuned exhaust system to a standard design.

Heat Riser. On some inline engines, the intake manifold is attached to the exhaust manifold. The heat riser valve is located at this attachment point. On V-type engines, the heat riser valve partially blocks one exhaust manifold exit, increasing the exhaust pressure in the exhaust manifold on that side of the engine. This forces the exhaust gases to flow through the intake manifold exhaust heat crossover

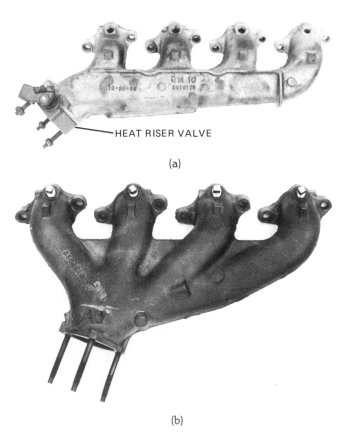

HEAT RISER VALVE

(a)

(b)

Figure 6-30. Exhaust manifold designs. (a) Standard engine and (b) performance engine.

passage to the opposite exhaust manifold. The heat riser valve for a V-type engine can be seen in Figure 6-30.

Thermal Reactor. Engines designed to operate on very lean air/fuel mixtures may use a *thermal reactor*. A thermal reactor is used in place of the exhaust manifold. It is a double-walled insulated chamber made of high-temperature metal. Its purpose is to keep the exhaust gas at high temperature after it leaves the combustion chamber. The high temperature in the reactor gives the remaining hydrocarbons more time to combine with oxygen. This will reduce HC in the exhaust gas to help meet emission standards. A thermal reactor is pictured in Figure 6-31.

Converter. An exhaust pipe is connected to the manifold or header to carry the gases through a catalytic converter, then to the muffler or silencer. In single-exhaust systems used on V-type engines, the exhaust pipe is designed to collect the exhaust gases from both manifolds using a Y-shaped pipe. Dual exhaust systems have a complete exhaust system coming from each of the manifolds. In most cases, the exhaust pipe must be made of several

Figure 6-31. A thermal reactor that takes the place of the exhaust manifold.

parts in order to assemble it into the space available under the automobile. A typical exhaust system is illustrated in the phantom view in Figure 6-32.

The catalytic converter is installed between the manifold and muffler to help reduce exhaust emissions. The converter has a heat-resistant metal housing. A bed of catalyst-coated pellets or a catalyst-coated honeycomb grid is inside the housing. As the exhaust gas passes through the catalyst, most of the hydrocarbons and carbon monoxide remaining in the exhaust gas are oxidized. An air injection system or pulse air system is used on the engine to supply additional air that is needed in the

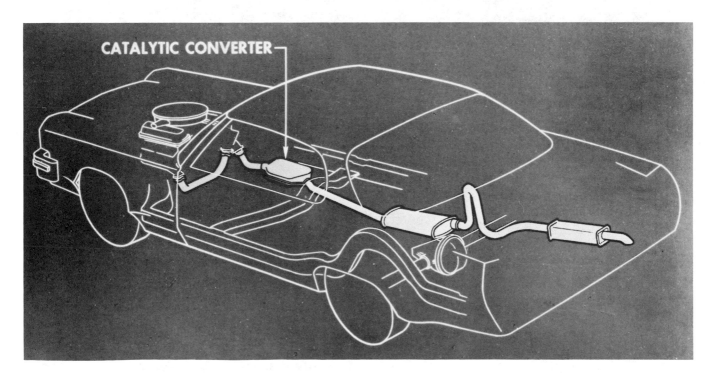

Figure 6-32. A phantom view of a complete catalytic converter exhaust system (Courtesy of General Motors Research).

oxidation process. These air systems inject air into the exhaust stream as close to the exhaust valve as possible.

Automobiles using catalytic converters require the use of unleaded gasoline. Lead in gasoline will contaminate the catalyst so that it will no longer oxidize the exhaust gas. Converter pellets can be removed and replaced using special equipment. When the honeycomb-type catalyst becomes ineffective, the entire converter must be replaced.

Muffler. When the exhaust valve opens, it rapidly releases high-pressure gas. This sends a strong air pressure wave through the atmosphere which produces a sound we call an explosion. It is the same sound produced when the high-pressure gases from burned gunpowder are released from a gun. In an engine, the pulses are released, one after another. The explosions come so fast they blend together in a steady roar.

Sound is air vibration. When the vibrations are large, the sound is loud. The muffler catches the large bursts of high-pressure exhaust gas from the cylinder, smoothing out the pressure pulses and allowing them to be released at an even and constant rate. It does this through the use of perforated tubes within the muffler chamber. The smooth-flowing gases are released to the tail pipe. In this way, the muffler silences engine exhaust noise. Sometimes resonators are used in the exhaust system. They provide additional expansion space at critical points in the exhaust system to smooth out the exhaust gas flow. A sectioned muffler is pictured in Figure 6-33.

The tail pipe carries the exhaust gases from the muffler to the air, away from the automobile. In most cases, the tail pipe exit is at the rear of the automobile below the rear bumper. In some cases, it is released at the side of the automobile just ahead of or just behind the rear wheel.

The muffler and tail pipe are supported with brackets called hangers. The hangers are rubberized fabric with metal ends that hold the muffler and tail pipe in position so that they do not touch any metal part. This helps to isolate the exhaust noise from the rest of the automobile.

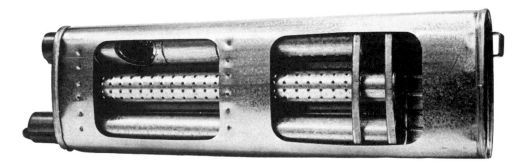

Figure 6-33. A muffler cutaway to show the interior.

REVIEW QUESTIONS

1. What is required for an engine to run smoothly? [INTRODUCTION]
2. Why is it necessary to have intake charge velocities above 50 ft/sec? [6-1]
3. How does a racing intake manifold effect idle? [6-1]
4. Why can fuel-injected engines use large intake manifolds and still operate at low engine speeds? [6-1]
5. Why does a single barrel carburetor work as well on four cylinders as it does on one cylinder? [6-1]
6. What causes the intake charge to become a gaseous air/fuel mixture during part-load operation? [6-2]
7. What is the advantage of having the intake runners on two levels? [6-2]

8. What is a tuned runner in an intake manifold? [6-2]

9. Why are the primary barrels of a four barrel carburetor often centered on the manifold? [6-2]

10. What is the advantage of having a flat floor and guide ribs in the intake manifold? [6-2]

11. Why is it necessary to add heat to the intake charge? [6-2]

12. Name three ways in which heat is added to the charge? [6-2]

13. Why do intake manifolds often contain cooling passages? [6-2]

14. Why is a choke required for engine starting? [6-2]

15. When does the EGR system add exhaust gas to the intake charge? [6-2]

16. What are the advantages and disadvantages of an open-type intake manifold? [6-3]

17. Why is it necessary to keep the engine oil off of the surface of the exhaust crossover passage in the intake manifold? [6-3]

18. What exhaust manifold design features allow for expansion and contraction from heat? [6-4]

19. What are the advantages and disadvantages of exhaust headers? [6-4]

20. What is the purpose of the heat riser in an exhaust system? [6-5]

21. Name an advantage of a tuned exhaust system? [6-5]

22. How does a thermal reactor reduce emissions? [6-5]

23. How does a catalytic converter operate? [6-5]

24. Why must unleaded gasoline be used in engines that have catalytic converters? [6-5]

25. How does a muffler reduce the noise of the engine exhaust? [6-5]

CHAPTER
7

Cylinder Heads and Valves

The passenger car engine is designed to operate smoothly at all engine loads and speeds. At the same time, the engine is expected to develop high power and high efficiencies and to produce low emissions. Smoothness is defined as the lack of objectionable or disagreeable engine vibration that can be felt inside the passenger compartment. Engine vibration can be caused by mechanical unbalance or by abnormal combustion. An unbalanced rotating crankshaft and reciprocating pistons cause vibration and roughness. Abnormal combustion causes very rapid pressure buildup in the combustion chamber. This leads to engine roughness.

Combustion (discussed in detail in Chapter 4) is a very complex chemical process. It is caused by fuel reactions within the combustion chamber. These reactions will differ according to the fuel type, combustion chamber shape, cooling system efficiency, location of the spark plugs and valves, compression ratio, and the quantity of the intake charge. One of the most important of these is the combustion chamber shape.

The bottom of the combustion chamber is shaped by the top of the piston. The side is shaped by the cylinder wall. The top is shaped by the cylinder head. The piston is nearly at the top of the stroke when the combustion takes place. At this position, very little of the cylinder wall is exposed to combustion. The combustion chamber is the space between the top of the piston and the pocket in the cylinder head. The shape of these parts have a great deal to do with the control of combustion.

7-1 Head Removal

It is necessary to remove the spark plugs and manifolds and to disconnect the valve train before the head is removed. On pushrod engines, the rocker arm covers are removed. Engines with rocker shafts will have to have the entire shaft, including the rocker arms, removed as an assembly. The shaft hold-down bolts or cap screws should be loosened alternately. This is done so that there is little bend in the shaft as the valve spring tension is released. Rocker arm pivots that are attached with individual studs with nuts may only require loosening to free the pushrods. Other pivot types will have to be completely removed. The pushrods can then be removed. On overhead cam engines, the belt or chain drive will have to be disconnected before the head can be removed. The cam and followers usually remain attached to the head until the head has been removed. After the valve train is disconnected, the head cap screws can be removed and the head lifted from the block. The combustion chambers can then be seen.

The overhead camshaft will either have one-piece bearings in a solid bearing support or it will have split bearings and a bearing cap. When a one-piece bearing is used, the valve springs will have to be compressed with a fixture before the camshaft can be pulled out endwise. One type of fixture to hold the followers down is shown in Figure 7-1.

When bearing caps are used, they should be loosened alternately so that bending loads are not placed on either the cam or bearing caps (Figure 7-2).

The valves are removed by compressing the valve springs with a valve spring compressor as shown in Figure 7-3. The valve locks are removed and the spring compressor released. The valve spring, retainers, and valve can then be removed from the head.

7-2 Combustion Chamber Types

Combustion chambers of modern automotive overhead valve engines have come from two basic

Figure 7-2. Overhead camshaft with bearing caps on split bearings.

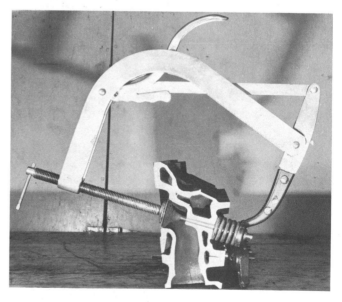

Figure 7-3. Compressor in use to compress the valve spring so that the valve locks can be removed.

Figure 7-1. A fixture to hold the valves compressed so the overhead camshaft can be removed.

types. One is the nonturbulent hemispherical chamber and the other is the turbulent wedge chamber. Each has advantages. Engine designers try to combine these advantages to form the best possible compromise combustion chamber design.

Hemispherical Combustion Chamber. In nonturbulent hemispherical combustion chambers, the charge is inducted through widely slanted valves. The charge is compressed and then ignited from a centrally located spark plug (Figure 7-4). The spark plug is as close as possible to all edges of the combustion chamber. Combustion radiates out from the spark plug, completely burning in the shortest possible time. This tends to reduce the formation of NO_x. The end gases ahead of the flame front have little time to react, so that knock is reduced. The rapidly burning charge in the hemispherical combustion chamber causes pressure to rise very rapidly. The rapid rise of combustion pressure makes the engine rough and noisy, when it is run under medium and heavy loads at low engine speeds. If this type of operation is encountered in automobiles, it may be objectionable to the operator. Laboratory and road tests have shown that the hemispherical combustion chamber is the best type for use in race-car application, where low-speed conditions are seldom encountered.

Hemispherical combustion chambers are usually fully machined to form the hemispherical shape. This is an expensive operation that increases the cost of the engine.

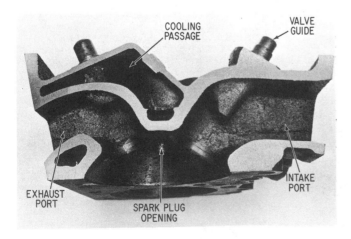

Figure 7-4. A section view of a hemispherical combustion chamber.

Wedge Combustion Chamber. The wedge-shaped combustion chamber is designed to produce smooth uniform burning by controlling the rate of combustion. A section view of a wedge-shaped combustion chamber is shown in Figure 7-5. In wedge-shaped combustion chambers, the charge is inducted through parallel valves. As the piston nears the top of the compression stroke, the piston approaches close to a low or flat portion of the head. The gases are squeezed from between the piston and head surface area. This area is called a squish or *quench area.* The gases squeezed from the quench area produce turbulence within the charge. This was shown in Figure 4-8. **The turbulence thoroughly mixes the air and fuel in the charge.** The spark plug is positioned in

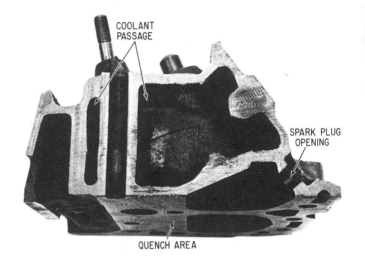

Figure 7-5. A section view of a wedge-shaped combustion chamber.

the highly turbulent part of the charge. Ignition is followed by smooth and rapid burning of the turbulent charge. The combustion flame front radiates out from the spark plug. The end gases which would burn abnormally remain in the quench area. Here the end gases are cooled and they do not react, because this area is squeezed very thin, less than 0.100 in. (2.54 mm) when the piston is at the top center. The designer controls the combustion by changes in the combustion chamber shape. Seemingly unimportant modifications in the combustion chamber shape result in large changes in the amount and the timing of the combustion pressure. The combustion chamber is designed to specific engine and fuel requirements. Turbulent combustion chambers usually remain as cast in the head, with no machining being done. This helps keep the production cost down.

Surface Quenching. Unburned hydrocarbon emission from engines is critical. The charge adjacent to the combustion chamber surface, from 0.002 to 0.020 in. (0.005 to 0.050 mm) thick, does not burn. The temperature of the combustion chamber surface is less than the temperature required for combustion. This cools the part of the charge that is next to the surface to a temperature below its burning temperature. The combustion flame goes out and this leaves unburned hydrocarbons. These unburned hydrocarbons are expelled with the burned gases on the exhaust stroke. Combustion chamber surface quenching is one of the major causes of unburned hydrocarbons in the exhaust gas. Combustion chambers having a low surface area for their volume, such as the hemispherical combustion chamber, produce low unburned hydrocarbons. The wedge combustion chamber, which has a relatively high surface area-to-volume ratio, produces high unburned hydrocarbons.

The trend in combustion chamber design is to take advantage of the best features of each basic type of combustion chamber. At the same time, the less desirable features are eliminated. This is done by changing the valves of the wedge head to different angles, repositioning the spark plug, reducing the quench area, reducing the combustion chamber surface area-to-volume ratio, and inducing a stratified charge. The resulting combustion is very efficient and smooth burning while running on available pump grades of gasoline. They are used as cast rather than having expensive machined chambers. These chambers are called by names such as polyspherical, hemiwedge, kidney shapes, and pent roof. Two of these shapes are illustrated in Figure 7-6 and 7-7.

There is an ideal location for the spark plug for best combustion. It is generally impossible to get a spark plug in this location on production engines. The spark plug requires space in the head and the service technician must be able to reach it for replacement with standard tools. Two sizes of spark plug threads are used: 14 mm and 18 mm. The larger spark plug improves combustion and reduces cold fouling tendencies. The

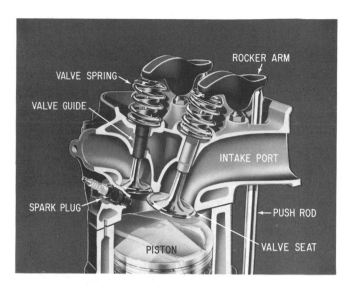

Figure 7-6. A phantom view of a modified wedge combustion chamber (Courtesy of Chevrolet Motor Division, General Motors Corporation).

smaller spark plug, which has become the industry standard, requires less space in the cylinder head.

7-3 Stratified Charge Combustion Chamber

The combustion chambers on some of the import engines are quite different from basic domestic engines. In the broadest sense, these different heads

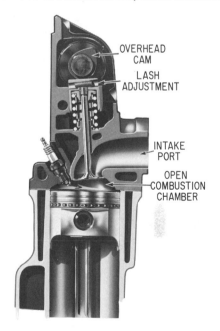

Figure 7-7. A section view of an open combustion chamber (Courtesy of Chevrolet Motor Division, General Motors Corporation).

have stratified charge combustion chambers. Basically, the shape or the induction process of a stratified charge combustion chamber causes a *swirling* of the air/fuel mixture charge. Swirling causes the charge to rotate around in the combustion chamber. The stratified engine gets its name from the layers or strata of different air/fuel mixtures that are formed within the swirling charge. Some of the layers have a rich air/fuel mixture, whereas others have a lean mixture. The overall stratified charge has a very lean air/fuel mixture.

A rich air/fuel mixture is easier to ignite with a spark plug than is a lean air/fuel mixture. For this reason, a rich strata surrounds the spark plug. When a hot flame develops in this rich strata after ignition, it puts both heat and pressure on the lean remaining charge. The lean part of the charge will then burn.

Two types of stratified charge engines are being studied. One with a carburetor uses two combustion chambers, a main chamber and a prechamber. The other injects fuel into the swirling air near the end of the compression stroke. The spark plug is located near the injector where it is surrounded by a rich fuel mixture.

A great deal of engineering research is being placed on the stratified charge principle. This has been brought about because there are three important advantages to the stratified charge combustion chamber. First, it produces good part-load fuel economy. Second, it can operate with lower octane fuels. Third, it appears to produce lower exhaust emissions.

Imported Stratified Charge Engines. A number of imported production stratified charge combustion chamber engines use two combustion chambers. The small prechamber contains the rich charge and the main chamber contains the lean charge. Others do not use a prechamber. Several examples of these combustion chambers are shown in Figure 7-8.

The first production stratified charge engine

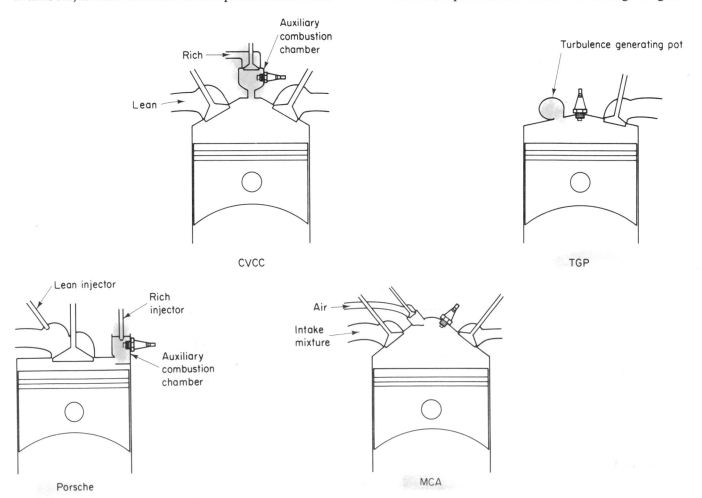

Figure 7-8. Line drawings showing the principle of import stratified charge combustion chambers.

was the Honda CVCC (Compound Vortex Controlled Combustion) engine announced in 1972. In this engine, a small second intake valve and a spark plug are located in the prechamber. Honda calls the prechamber an auxiliary combustion chamber. A rich air/fuel charge is drawn into the prechamber at the same time that a lean charge is being drawn into the main chamber. The ignited rich charge rushes from the prechamber into the main chamber like a flame from a blowtorch. This hot torch flame ignites the lean charge in the main chamber.

A second type of stratified combustion chamber engine is the Toyota TGP (Turbulence Generating Pot). The TGP is a prechamber designed to generate turbulence in the main combustion chamber. A carburetor supplies the intake charge to the single intake valve. During compression, some of the charge is forced into the relatively hot TGP. The spark plug, located at the opening of the TGP, ignites the charge. The charge in the TGP burns rapidly. This causes the flame to jet out into the main combustion chamber to produce turbulence. The resulting flame and turbulence causes the lean charge mixture to burn within the main combustion chamber.

A third type of stratifed charge combustion chamber has been developed by Porsche. Fuel is injected into a prechamber that contains the spark plug. This forms a rich charge mixture. A lean fuel charge is injected into the main chamber when the intake valve opens. After combustion occurs, the burning and turbulence are very similar to the CVCC engine.

A fourth type of stratified charge combustion chamber engine is the MCA (Mitsubishi Clean Air) system. It does not use a prechamber. Instead, it has a small second intake valve. This valve is operated by the same cam follower that operates the main intake valve. Air from above the throttle plate comes through the small intake valve. At low engine speeds, a large flow of air enters the combustion chamber through the small valve. The air flow develops swirl and turbulence for complete mixing of the charge. Because of the location of the air passage intake in the carburetor, very little air will flow into the combustion chamber at high engine speeds. Only normal turbulence occurs during high engine speeds.

Domestic Stratified Charge Engines. Two domestic stratified charge engines are nearing limited production at the time of this writing. The oldest of these designs is the Texaco Controlled Combustion System (TCCS). This engine has no throttle, so a full charge of air is taken in on each intake stroke. Only enough fuel for the power required is injected directly into swirling compressed air within the combustion chamber. The spark plug ignites the rich part of the charge in the swirling air directly downstream from the injector. The other domestic stratified charge engine is the Ford PROCO (Programmed Combustion) engine. This engine uses air throttling as well as fuel injection. Both air and fuel are controlled to form a combustible mixture. The injector and the two spark plugs in the PROCO engine are located close to the center of the combustion chamber. The combustion chambers of both TCCS and PROCO engines are located in the *head* of the piston. This can be seen in Figure 7-9.

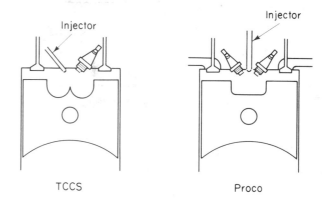

Figure 7-9. Line drawings showing the principles of domestic stratified charge combustion chambers.

7-4 Intake and Exhaust Ports

The part of the intake and exhaust system passages that is cast in the cylinder head is called a *port*. Ports lead from the manifolds to the valves. The most desirable port shape is not always possible because of space requirements in the head. Space is required for the head bolt bosses, valve guides, cooling passages, and for the pushrod openings. Inline engines may have both intake and exhaust ports located on the same side of the engine. Often, two cylinders share the same port because of the restricted space available. Shared ports are called *siamesed ports.* Examples are shown in Figures 7-10

Figure 7-10. Typical siamesed ports in a cylinder head.

and 7-11. Each cylinder uses the port at a different time. Some inline engines have a separate port for each cylinder on one side of the head, as shown in Figure 7-12. This arrangement allows equal cylinder filling. Larger ports and better breathing is possible in engines that have the intake port on one side of the head and the exhaust port on the opposite side. This type of head is a *cross-flow head* design. The hemispherical head shown in Figure 7-4 has this type of port. Another cross-flow head shown in Figure 7-13 allows the valve to be located and angled for most efficient engine breathing. It also allows the spark plug to be placed near the center of the combustion chamber. V-type engines have the cross-flow head design.

One design objective of the intake and exhaust system is to meet the engine's maximum power needs with the minimum restriction. At the same time, the induction system must provide satisfactory charge distribution at part throttle and at idle speeds. To do this the air velocity must not become too slow. The engine designer makes use of air flow measuring equipment to develop a satisfactory

Figure 7-11. Close up view of a siamesed exhaust port.

compromise of the port shape that will meet the engine requirements.

The flow of gases is often different than one might think. At times a restricting hump (Figure 7-14) within a port may actually increase the air flow capacity of the port. It does this by redirecting

Figure 7-12. Separate ports for each valve in a cylinder head.

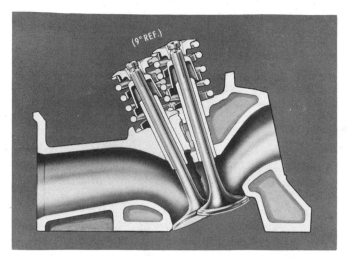

Figure 7-13. A cross flow cylinder head having a long intake ports on the left and a short exhaust port on the right.

Figure 7-14. A hump in the intake port that actually increases the air flow capacity of the port.

the flow to an area of the port that is large enough to handle the flow. Modifications in the field, such as *porting* or *relieving,* would result in restricting the flow of such a carefully designed port.

The intake port in the head is relatively long, whereas the exhaust port is short. The typical length of the ports in a cross-flowhead can be seen in Figure 7-12. The long intake port wall is heated by coolant flowing through the head. The heat aids in vaporizing the fuel in the intake charge. The exhaust port is short so that there is the least amount of exhaust heat transferred to the engine coolant. Sometimes, there is a cast opening around the crossover port. The opening can be seen in Figure 7-15. This opening keeps the exhaust heat from transferring to the coolant. Keeping the heat in the exhaust has two advantages. First, the cooling

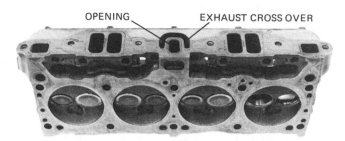

Figure 7-15. A heat insulating cast opening can be seen going partly around the exhaust cross over passage in the head.

system can be made smaller and this reduces the cost. Second, the exhaust will remain hot for a longer time and this reduces HC and CO emissions in the exhaust. Both the thermal reactor and the catalytic converter require exhaust gas heat to operate. They are not as effective when heat is removed from the exhaust gases. Some engines are designed with an insulating exhaust port liner. This more effectively conserves exhaust heat to reduce exhaust emission.

7-5 Coolant Passages In The Head

Coolant flow within the engine is designed to flow from the coolest portion of the engine to the warmest portion. The coolant pump takes the coolant from the bottom of the radiator. It is pumped into the block, where it is directed all around the cylinders. The coolant then flows upward through the gasket to the cooling passages cast into the cylinder head. The heated coolant is collected at a common point and returned to the radiator to be cooled and recycled. Typical coolant passages in a head are shown in Figure 7-16.

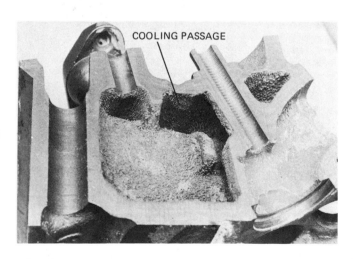

Figure 7-16. Coolant passages can be seen in the sectioned head.

There are relatively large holes in the gasket surface of the head leading to the head cooling passages. The large holes are necessary to support the cooling passage core through these openings while the head is being cast. After casting, the core is broken up and removed through these same openings. Core support openings to the outside of the engine are closed with expansion plugs or soft plugs. These plugs are often mistakenly called freeze plugs. The openings between the head and the block are usually too large for the correct coolant flow. When the openings are too large the head gasket performs an important coolant flow function. Special sized holes are made in the gasket. These holes will correct the coolant flow rate at each opening. Therefore, it is important that the head gasket is installed correctly for proper engine cooling. A head gasket with special sized holes to cover the head openings is shown in Figure 7-17.

Special cooling nozzles, carefully located openings, or deflectors may be designed into the head. They direct the coolant toward a portion of the head where localized heat must be removed. Usually, this is in the area of the exhaust valve. Some of the deflectors are cast in the cooling passages. Others are pressed-in sheet-metal nozzles. Pressed-in nozzles are replaceable.

7-6 Lubricating Overhead Valves

Lubricating oil is delivered to the overhead valve mechanism, either through the valve pushrods or through drilled passages in the head and block casting. There are special openings in the head gasket to allow the oil to pass between the block and head without leaking. After the oil passes through the valve mechanisms, it returns to the oil pan through oil return passages. Some engines have drilled oil return holes, but most of the engines have large cast holes that allow the oil to return freely to the engine oil pan. The cast holes are large and do not become easily plugged. They also tend to lighten the head casting, thus reducing cost and total engine weight.

7-7 Valve Mechanisms

Automotive engine valves are a *poppet valve* design. The valve is opened through a valve train that is operated by a cam. The cam is timed to the piston position and crankshaft cycle. The valve is closed by one or more springs.

The cam is driven by timing gears, chains, or belts, located at the front of the engine. The gear or sprocket on the camshaft has twice as many teeth, or notches, as the one on the crankshaft. This results in two crankshaft turns for each turn of the camshaft. The camshaft turns at one-half the crankshaft speed in all four-stroke cycle engines.

Most valve design features are the same for all valves. Typical valves are shown in Figure 7-18. Intake valves control the inlet of cool, low-pressure induction charges. Exhaust valves handle hot, high-pressure exhaust gases. This means that exhaust valves are exposed to more severe operating conditions. They are, therefore, made from much higher-quality materials than the intake valves. This makes them more expensive.

The valve stems are supported by a valve *guide*. The valve guide may be a machined hole in the head or it may be a replaceable insert guide. The guide is centered over the valve *seat* so that the valve face and seat make a gastight fit. The face and seat will have either of two angles; 30° or 45°. These are the

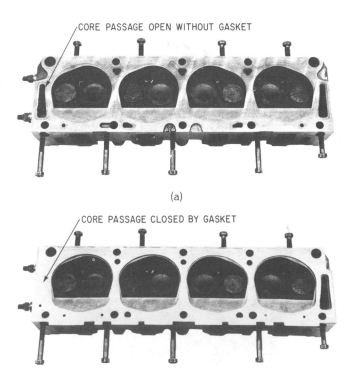

(a)

(b)

Figure 7-17. Coolant flow control. (a) Head core passages open without a gasket and (b) gasket covering the left-hand core passage opening.

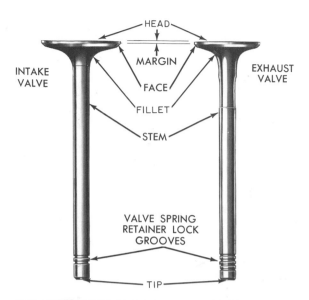

Figure 7-18. Identification of valve parts (Courtesy of Chrysler Motors Corporation).

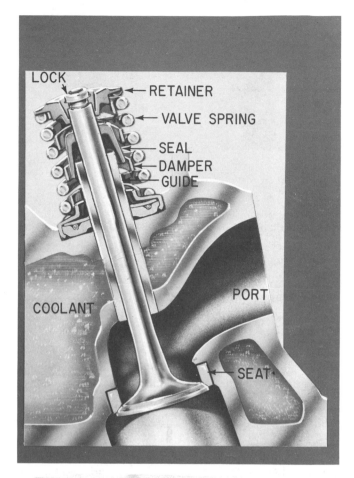

Figure 7-19. Identification of the parts of a typical valve assembly (Courtesy of General Motors Corporation).

nominal angles. Actual service angles might be a degree or two different from these. Most engines use a nominal 45° valve and seat angle. A *valve spring* holds the valve against the seat. The valve *lock* secures the spring *retainer* to the stem of the valve. For removal, it is necessary to compress the spring and remove the valve lock. Then the spring, valve seals, and valve can be removed from the head. A typical valve assembly is shown in Figure 7-19.

Valve Design. Extensive testing has shown that there is a normal relationship between the different dimensions of valves. Engines with cylinder bores from 3 to 8 in. (80 to 200 mm) will have intake valves approximately 45% of the bore size. The exhaust valve is approximately 38% of the cylinder bore size. The intake valve must be larger than the exhaust valve to handle the same mass of gas. The larger intake valve controls low-velocity, low-density gases. The exhaust valve, on the other hand, controls high-velocity, high-pressure, denser gases. These gases can be handled by a smaller valve. Exhaust valve heads are, therefore, approximately 85% of the intake valve head size. For satisfactory operation, valve head diameter is nearly 115% of the valve port diameter. The valve must be large enough to close over the port. The amount the valve opens, called *valve lift,* is close to 25% of the valve diameter.

Poppet valve heads may be designed from a

rigid valve to an *elastic* valve, as shown in Figure 7-20. The rigid valve is strong, holds its shape, and conducts heat readily. It also causes less valve recession. Unfortunately, it is more likely to leak and burn than other valve head types. The elastic valve, on the other hand, is able to conform to valve seat shape. This allows it to seal easily, but it runs hot and the flexing to conform may cause it to break. A popular shape is one with a small cup in the top of the valve head. It offers a reasonable weight, good strength, and good heat transfer at a slight cost penalty. Elastic valve heads are more likely to be found on intake valves and rigid on exhaust valves.

The valve face angles have been carefully selected to give the best compromise. With an equal valve lift, the opening space around the valve face increases as the face angle is reduced. This opening can be calculated by using the trigonometric function, cosine, of the valve face angle. A flat valve

(a)   (b)   (c)   (d)

Figure 7-20. Valve head types from rigid (a) to elastic (d).

with a face angle of 0° provides the maximum valve opening for this lift. A flat face angle is very difficult to seal when it is closed. Poor sealing will lead to valve burning and a short useful valve life. The relationship of the seating angles to the opening size when each valve has the same lift is illustrated in Figure 7-21.

The sealing force on the valve seat is increased as the valve angle is increased. Forty-five-degree face angles are used on exhaust valves and on intake valves where high seating pressures are needed. The high pressure will either crush or wipe off deposits to prevent valve leakage. A 30° valve face angle is used where high gas flow rates are desirable and where durability is no problem.

Valve Materials. Valve shape and gas flow considerations have been well established. They provide satisfactory performance in the modern engine. Increasing durability and cost will continue to be subjects of valve development studies. These problems are most apparent in valve metallurgy and manufacturing techniques. New manufacturing techniques must be developed as new valve materials are used to maintain valve economy. Most of the recent valve development work has been done on exhaust valves. These valves are being required to operate in an increasingly severe environment.

Alloys used in exhaust valve materials are chromium for oxidation resistance, with small amounts of nickel, manganese, and nitrogen added. Heat treating is used whenever it is necessary to produce special valve properties. Some exhaust valves are manufactured from two different materials where a one-piece design cannot meet the hardness and corrosion-resistance specifications desired. This can be seen on a new valve in Figure 7-16. The joint cannot be seen after valves have been used. The valve heads are made from special alloys that can operate at high temperature, have physical strength, resist lead-oxide corrosion, and have indentation resistance. These heads are welded to stems that have good wear-resistance properties. Figure 7-22 shows an inertia welded valve before final machining. In severe applications, facing alloys such as stellite, are welded to the valve face

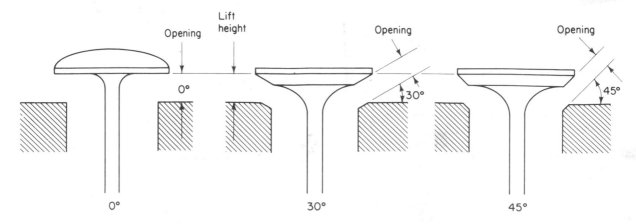

Figure 7-21. The relationship of the valve seating angles to the opening size with the same amount of valve lift.

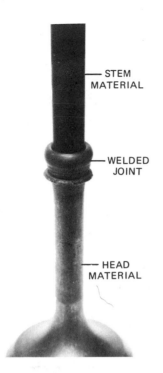

Figure 7-22. Inertia welded valve stem and head before machining.

Figure 7-23. A hollow valve stem (Courtesy of Sealed Power Corporation).

and valve tip. The valve is aluminized where corrosion may be a problem. Aluminized valve facing reduce valve recession when unleaded gasoline is used. Aluminum oxide forms to separate the valve steel from the cast-iron seat to keep the face metal from sticking.

Some heavy-duty applications use hollow stem exhaust valves that are partially filled with metallic sodium. An unfilled hollow valve stem is shown in Figure 7-23. The sodium in the valve becomes a liquid at operating temperatures. As it splashes back and forth in the valve stem, the sodium transfers heat from the valve head to the valve stem. The heat goes through the valve guide into the coolant. In general, a one-piece valve design using properly selected materials will provide satisfactory service for automotive engines.

Valve Guides and Seats. The valve face closes against a valve seat to seal the combustion chamber. The seat is generally formed as part of the cast-iron head of automotive engines. This is called an integral seat. The seats are usually induction hardened so that unleaded gasoline can be used. This minimizes valve recession as the engine operates. Valve recession is the wearing away of the seat so

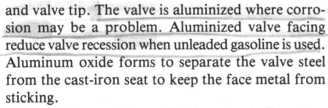

the valve seats further into the head. Insert seats are used in applications where corrosion and wear resistance are critical. Insert seats and guides are always required in aluminum heads. It should be noted that the exhaust valve seat runs as much as 180°F (100°C) cooler in aluminum heads than in cast-iron heads. Insert seats are also used as a salvage procedure when reconditioning integral automotive engine valve seats that have been badly damaged. Typical insert valve seats and guides can be seen in Figure 7-18.

Valve seat distortion is one of the major causes of premature valve failure. **Distortion is twisting and warping.** Valve seat distortion may be temporary as the result of pressure and thermal stress. It may become permanent as the result of mechanical stress. Stress is a force put on a part that trys to change its shape. Valve seat distortion must be kept at a minimum for maximum valve life. This means that a properly designed engine must be correctly and carefully assembled using proper parts.

The valve guide supports the valve stem so that the valve face will remain perfectly centered or *concentric* with the valve seat. The valve guide is generally integral with the head casting for better heat transfer and for lower manufacturing costs.

Insert valve guides are always used where the valve stem and head materials are not compatible.

Valve Springs and Locks. A valve spring holds the valve against the seat when the valve is not being operated. One end of the valve spring is seated against the head. The other end of the spring is attached under compression to the valve *stem* through a valve spring *retainer* and a valve spring *keeper* or *lock* as shown in Figure 7-24.

Valves usually have a single inexpensive valve spring. The springs are generally made of chromium–vanadium alloy steel. When one spring cannot control the valve, additional devices are added. Variable-rate springs add spring force when the valve is in its open position. This is accomplish-

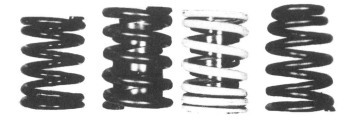

Figure 7-25. Valve spring types. On the left the spring has equally spaced coils. The next spring has a damper inside the spring coil. This is followed by a spring that has closely spaced coils and a damper. On the right is a taper wound coil spring.

springs generally have their coils wound in opposite directions. This is done to control valve spring surge and excessive valve rotation. **Valve spring surge is the tendency of a valve spring to vibrate.**

A large number of valve lock types have been used on the end of the valve stem to retain the spring. They have evolved from simple, low-cost lock pins and horseshoes to the current high-quality, split-cone valve lock or valve keeper. The inside surface of the split lock uses a variety of grooves or beads. The design depends upon their holding requirments. The outside of the split lock fits into a cone-shaped seat in the center of the valve spring retainer (see Figure 7-26).

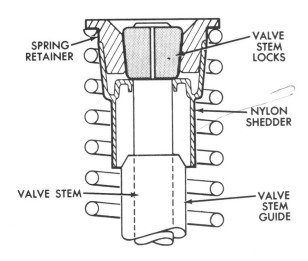

Figure 7-24. Parts of a valve lock and retainer assembly (Courtesy of Cadillac Motor Car Division, General Motors Corporation).

ed by using *closely spaced coils* on the cylinder head end of the spring. The closely spaced coils also tend to dampen vibrations that may exist in an equally wound coil spring. The damper helps to reduce valve seat wear. Some valve springs use a *flat coiled damper* inside the spring. This eliminates spring surge and adds some valve spring tension. The normal valve spring winds up as it is compressed. This causes a small but important turning motion as the valve closes on the seat. The turning motion helps to keep the wear even around the valve face. Figure 7-25 illustrates typical valve springs.

Multiple valve springs are used where large lifts are required and a single spring does not have enough strength to control the valve. Multiple valve

(a)

(b)

Figure 7-26. Valve split lock types (a) and stem grooves (b).

One-piece valve spring retainers are forged from high-quality steel. They will hold their shape under the pounding they receive in operation. Some retainers have built-in devices called valve rotaters. They cause the valve to rotate in a controlled manner as it is opened. One type uses small steel balls and slight ramps. Each ball moves down its ramp to turn the rotor sections as the valve opens. A second type uses a coil spring. The spring lays down as the valve opens. This action turns the rotor body in relation to the collar. The rotor can be seen in Figure 7-27. Valve rotors are only used where it is desirable to increase the valve service life because rotors cost more than plain retainers.

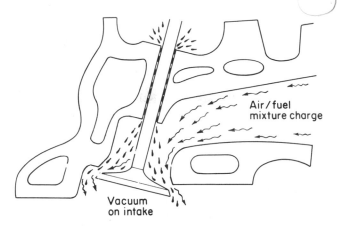

Figure 7-28. Oil pulled through an intake valve guide by the intake vacuum in the manifold.

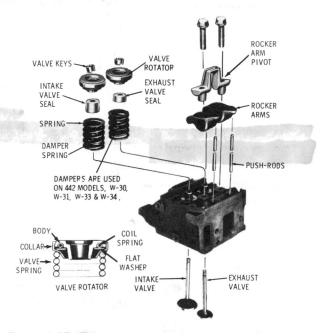

Figure 7-27. Parts of a valve assembly showing the location of the valve rotator (Courtesy of Oldsmobile Division, General Motors Corporation).

Valve Oil Seals. Leakage past the valve guides is a major oil consumption problem in the overhead-valve engine. This is especially true of leakage around the intake valve stem. A high vacuum exists in the port, as shown in Figure 7-28. This pulls the oil from the rocker arm area of the engine. A lot of design effort has gone into the development of deflectors and valve guide seals. Some early designers used umbrellas above the spring retainers. They were designed to deflect the lubricating oil to the area outside the valve spring. Later designs used synthetic rubber seals between the valve stem and

retainer so that the retainer acted in the same manner as the umbrella. These can be seen in Figure 7-29. As more control was required, synthetic rubber umbrella cups were placed on the valve stem. These extended over the valve guide. Some float and some are fastened. This umbrella-type seal is shown in Figure 7-30. Positive-type valve seals have come into some use. The positive seal jacket is attached to the valve guide. The valve slides through the seal, as shown in Figure 7-31.

Care must be used when installing valve seals to be sure they are correctly installed and not damaged. Careless installation will result in oil leakage that will lead to excessive oil consumption.

Rocker Arms. Rocker arms reverse the upward movement of the pushrod to produce a downward movement on the tip of the valve. Engine designers make good use of the rocker arm. It is designed to

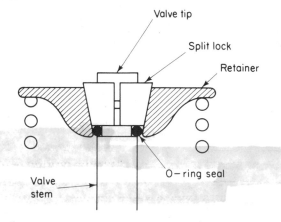

Figure 7-29. O-ring synthetic rubber seal between the retainer and valve stem.

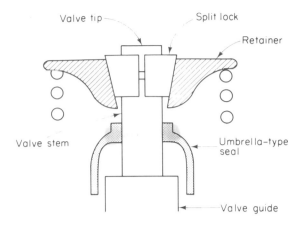

Figure 7-30. The location of umbrella-type valve stem oil seals.

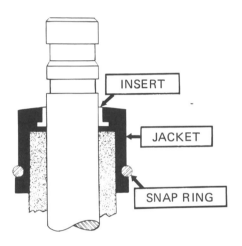

Figure 7-31. The location of a positive-type valve stem oil seal (Courtesy of Dana Corporation).

reduce the travel of the cam follower or lifter and pushrod while maintaining the required valve lift. This is done by using a rocker arm ratio of approximately 1.5:1, as shown in Figure 7-32. For a given amount of lift on the pushrod, the valve will open 1.5 times the pushrod lift distance. This ratio allows the camshaft to be small so the engine can be smaller. It also results in lower lobe to lifter rubbing speeds.

Rocker arm design has undergone change in new engines. This was brought on by cost-reduction programs. These changes also allow for divergent valve angles that are required in some advanced combustion chamber designs. Rocker arms may be cast, forged, or stamped. Forged rocker arms are the strongest, but require expensive manufacturing operations. Rocker arms may have bushings or bearings installed to reduce friction and increase durability. Cast rocker arms cost less to make and do not usually use bushings, but they do require

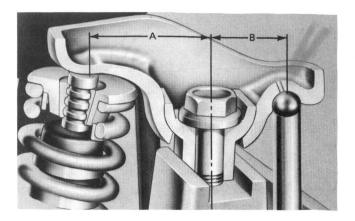

Figure 7-32. Method of measuring the rocker arm to determine the rocker arm ratio.

several machining operations. They are not as strong as forged rocker arms, but are satisfactory for passenger car service. Typical cast rocker arms are shown in Figure 7-33.

Several types of stamped rocker arms have been developed (Figure 7-34). These are the least expensive type to manufacture. They are lightweight and very strong. Two general types are in use—those that operate on a ball or cylindrical pivot, and those that operate on a shaft. The ball and cylindrical pivot types are lubricated through hollow pushrods.

Figure 7-33. Typical cast rocker arm types.

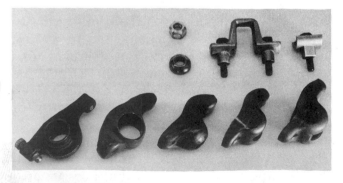

Figure 7-34. Typical stamped rocker arm types.

The shaft type is lubricated through oil passages that come from the block, through the head and into the shaft, then to the rocker arms.

Overhead camshaft engines use several methods of opening the valves. One type opens the valves directly with a cam follower or bucket (Figure 7-35). This can also be seen in the section view of an engine in Figure 6-15. The second type uses a finger follower that provides an opening ratio similar to a rocker arm (Figure 7-36). The valve opens approximately one-and-one-half times the cam lift. The pivot point of the finger follower may have a mechanical adjustment or it may have an automatic hydraulic adjustment. A third type moves the rocker arm directly through a hydraulic lifter (Figure 7-37).

All engines have some method to keep the rocker arm correctly positioned over the valve tip. Rocker arms are held in position on rocker shafts with springs and spacers. This can be seen in Figure 7-38. The rocker shaft keeps the rocker arms from twisting. Cylindrical pivots hold the rocker arm

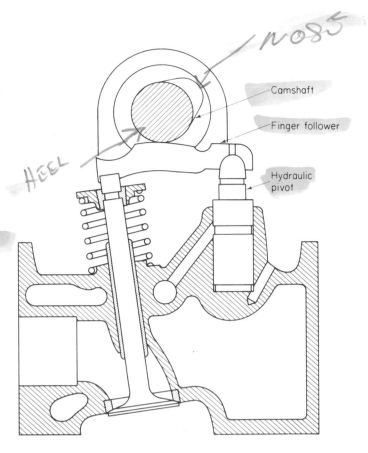

Figure 7-36. The overhead cam operating on a finger follower. A hydraulic pivot is on the right end of the finger follower. The finger follower operates the valve on the left.

Figure 7-35. The overhead cam operating directly on top of the bucket-type cam follower.

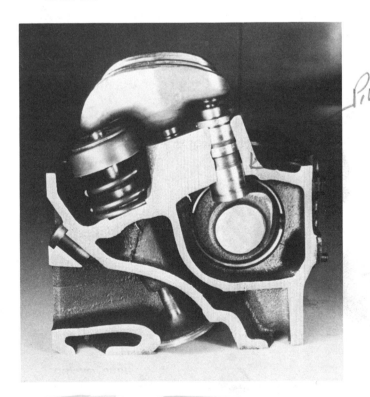

Figure 7-37. The overhead cam operates the rocker arm through a hydraulic lifter.

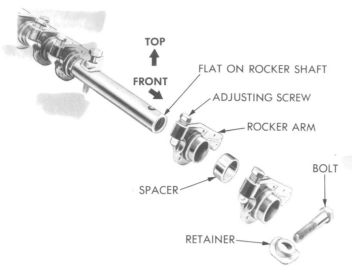

Figure 7-38. The rocker arms are held in position on this rocker shaft with spacers (Courtesy of Chrysler Corporation).

Figure 7-40. Finger follower held in place over the valve tip with a slot.

position and keep it from twisting (Figure 7-39). Rocker arms with ball pivots and finger followers are held with a slot over the valve tip (Figure 7-40) or a guide that holds the pushrod in position (Figure 7-41).

Pushrods. Pushrods are designed to be as light as possible and still maintain their strength. They may be either solid or hollow. If they are to be used as passages for oil to lubricate rocker arms, they *must* be hollow. Pushrods use a convex ball on the lower

(a)

Figure 7-39. A typical rocker arm cylindrical pivot being positioned for assembly.

(b)

Figure 7-41. Rocker arms held in place over the valve tip with a push rod guide (a) case in the head and (b) a stamped plate held in place under the rocker pivot stud.

end that seats in the lifter. The rocker arm end is also a convex ball, unless there is an adjustment screw in the pushrod end of the rocker arm. In this case, the rocker arm end of the pushrod has a concave socket. It mates with the convex ball on the adjustment screw in the rocker arm. This can be seen in Figure 7-42. Pushrod end types are shown in Figure 7-43.

Lifters and Tappets. Valve lifters or tappets follow the contour or shape of the camshaft lobe. This arrangement changes the cam geometry to a reciprocating motion in the valve train. The majority of lifters have a relatively flat surface that slides on the cam. Some lifters, however, are designed with a roller to follow the cam contour, rather than a flat surface. An example of a roller lifter is shown in Figure 7-44. Because of the expense of making the roller lifters, flat lifters are used wherever possible. When the flat lifters are not able to provide satisfactory performance, roller lifters are used.

Figure 7-42. Convex rocker arm clearance adjustment ball fitting in a socket on the upper end of the push rod.

Figure 7-43. Types of push rod ends.

Figure 7-44. Typical roller lifters.

The valve train, like other manufactured parts, is made with a tolerance. Tolerance is the maximum and minimum specified limit. This tolerance makes it necessary to have some means of clearance adjustment in the valve train system so that the valves will positively seat. This valve train clearance is sometimes called valve lash. Valve train clearance must not be excessive or it will cause noise or result in premature failure. Two methods are commonly used to make the necessary valve clearance adjustments. One is a solid valve lifter with a mechanical adjustment and the other is a lifter with an automatic hydraulic adjustment built into the lifter body.

In L-head engines, the solid valve lifter usually has an adjusting screw in the lifter. Overhead valve engines with mechanical lifters have an adjustment screw at the pushrod end of the rocker arm or an adjustment nut at the ball pivot. Adjustable pushrods are available for some specific applications.

Valve trains using solid lifters require the valve train to run with some clearance to ensure positive valve closure, regardless of the engine temperature. This clearance is matched by a gradual rise in the cam contour, called a *ramp*. The ramp will take up the clearance before the valve begins to open. It also has a closing ramp to ensure quiet operation. Valve trains using hydraulic lifters run with no clearance because the hydraulic unit is designed to take up the lash. Because of this they do not use a ramp on the cam. Some models of engines have mechanical adjustments as well as hydraulic adjustment. The mechanical adjustments are used to place the hydraulic unit in its midrange position for normal operation. Typical solid lifters are shown in Figure 7-45.

A lifter is solid in the sense that it transfers motion as a solid piece from the cam to the pushrod or valve. Its physical construction is a lightweight cylinder, either hollow or with a small-diameter center section and full diameter ends. In some types that transfer oil through the pushrod, the external appearance is the same as hydraulic lifters.

The major parts of a hydraulic lifter are a hollow cylinder body enclosing a closely fit hollow plunger, a check valve, and a pushrod cup. Lifters that feed oil up the push rod have a metering disc or restrictor valve located under the push rod cup. Engine oil pressure is fed by an engine passage to the exterior lifter body. An undercut portion allows the oil under pressure to surround the lifter body. Oil under pressure goes through holes in the undercut into the center of the plunger. From there, it goes down through the check valve to a clearance space between the bottom of the plunger and interior bottom of the lifter body base. It fills this space with oil at engine pressure. Slight leakage designed into the lifter allows the air to bleed out and allows the lifter to leak down if it should have become overfilled. The operating principle of a hydraulic lifter is shown in Figure 7-46.

The pushrod fits into a cup in the top open end of the lifter plunger. A hole in the pushrod cup, pushrod end, and hollow pushrod allows oil to transfer from the lifter piston center past a metering disc or restrictor valve, up through the pushrod to the rocker arm. Oil leaving the rocker arm lubricates the rocker arm assembly.

As the cam starts to push the lifter against the valve train, the oil below the lifter plunger is squeezed and so it tries to return to the lifter plunger center. A lifter check valve, either ball or disc type, traps the oil below the lifter plunger. This hydraulically locks the operating length of the lifter. The hydraulic lifter then opens the engine valve like a solid lifter. When the lifter returns to the base circle of the cam, engine oil pressure again replaces any oil that may have leaked out of the lifter.

The hydraulic lifter's job is to take up all clearance in the valve train. Occasionally, engines are run at excessive speeds. This tends to throw the valve open, causing *valve float*. During valve float, clearance exists in the valve train. The hydraulic lifter will take up this clearance as it is designed to do. When this occurs, it will keep the valve from closing on the seat. This is called *pump up*. Pump up will not occur when the engine is operated in its designed speed range.

Hydraulic lash adjusters are used with finger followers on overhead-cam engines. Their internal parts and operating principles are the same as hydraulic lifters. They are, however, designed with more clearance between the plunger and body. The extra clearance increases the leakdown rate of the adjuster. Since the overhead-cam valve train has almost no deflection, the high leakdown rate is necessary to compensate for small variations in shape of the cam base circle.

Figure 7-45. Typical solid valve lifters. The external appearance of the two lifters on the right is the same as a hydraulic lifter. The lifter on the far right is disassembled to show the internal parts required to control oil flow to the push rod.

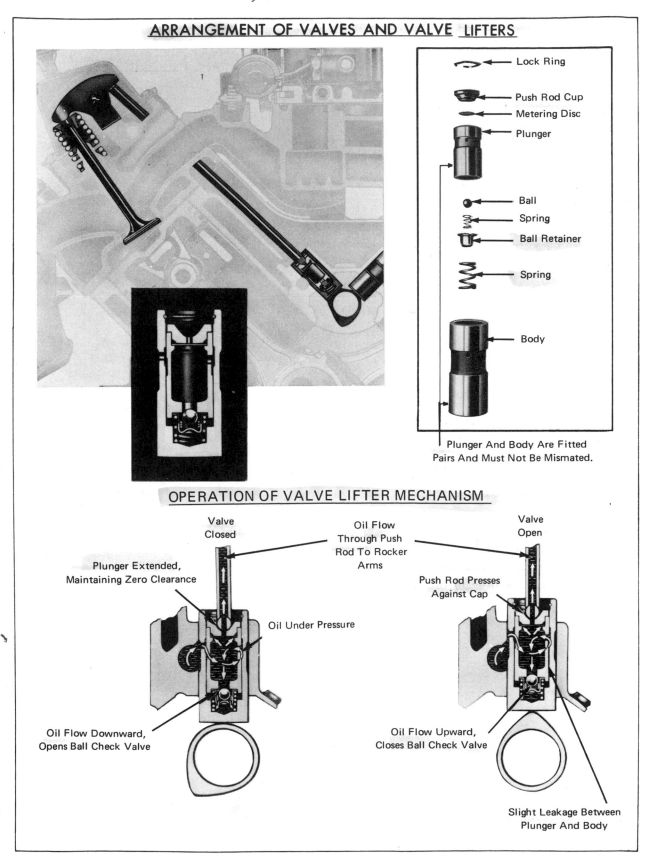

ARRANGEMENT OF VALVES AND VALVE LIFTERS

Lock Ring

Push Rod Cup

Metering Disc

Plunger

Ball

Spring

Ball Retainer

Spring

Body

Plunger And Body Are Fitted
Pairs And Must Not Be Mismated.

OPERATION OF VALVE LIFTER MECHANISM

Valve
Closed

Oil Flow
Through Push
Rod To Rocker
Arms

Valve
Open

Plunger Extended,
Maintaining Zero Clearance

Push Rod Presses
Against Cap

Oil Under Pressure

Oil Flow Downward,
Opens Ball Check Valve

Oil Flow Upward,
Closes Ball Check Valve

Slight Leakage Between
Plunger And Body

Figure 7-46. Operating principle of hydraulic lifters (Courtesy of Cadillac Motor
Car Division, General Motors Corporation).

REVIEW QUESTIONS

1. What forms the top and bottom of the combustion chamber? [INTRODUCTION]
2. Why are the fasteners that hold a rocker shaft loosened alternately? [7-1]
3. What must be removed before the head can be taken off the block? [7-1]
4. What are the advantages and disadvantages of a hemispherical combustion chamber? [7-2]
5. State the advantages and disadvantages of a wedge combustion chamber [7-2]
6. What results if there is combustion chamber surface quenching? [7-2]
7. Why do modern engines use 14 mm spark plugs? [7-2]
8. What happens to the intake charge in a stratified charge combustion chamber? [7-3]
9. How does the air/fuel ratio affect ignition of the charge? [7-3]
10. What are the advantages of a stratified charge combustion chamber? [7-3]
11. Why does the siamesed port work as well as separate ports in passenger car engines? [7-4]
12. What is a cross-flow head design? [7-4]
13. Why are some ports long and others short? [7-4]
14. What is the advantage of keeping exhaust heat from the cooling system? [7-4]
15. Why is it necessary to have large openings from the gasket surface into the head coolant passages? [7-5]
16. What coolant flow devices are used in and on the head? [7-5]
17. How does oil get through the head to the overhead valve mechanism? How does it return from the head? [7-6]
18. How many times does the crankshaft rotate when the camshaft rotates once in a four-stroke cycle engine? [7-7]
19. Why are intake valves larger than exhaust valves? [7-7]
20. Why are exhaust valves more expensive than intake valves? [7-7]
21. What are the normal valve and seat angles used in automobile engines? [7-7]
22. What are the advantages and disadvantages of a rigid valve head compared to an elastic valve head? [7-7]
23. Where are 30° and 45° valve face and seat angles used? [7-7]
24. When is it necessary to make a valve in two pieces? [7-7]
25. When are sodium-filled valve stems used? [7-7]
26. What is valve recession? [7-7]
27. When are insert valve seats used? [7-7]
28. When are insert valve guides used? [7-7]
29. Why do some valve springs use closely wound coils at one end? [7-7]
30. Why do some valve springs use a damper? [7-7]
31. When are multiple valve springs used? [7-7]
32. When are valve rotators used? [7-7]
33. Name two methods used to operate valve rotators. [7-7]
34. List the types of valve stem seals described in this book. [7-7]
35. Why is the rocker arm ratio important? [7-7]
36. When are the rocker arms lubricated through the push rods? [7-7]

37. How are the valves opened on overhead cam engines? [7-7]
38. When is it necessary to have hollow push rods? [7-7]
39. Why is it necessary to have an adjustment for clearance in the valve train? [7-7]
40. When is a ramp used on a cam? Why is it necessary? [7-7]
41. What is meant by the term "solid lifter"? [7-7]
42. Describe the operation of a hydraulic lifter. [7-7] _Take up the lash_
43. What causes the hydraulic lifter to pump up? [7-7] _High Speed_

CHAPTER
8

Pistons, Rings, and Rods

All of an engine's power is developed by burning fuel in the presence of air in the combustion chamber. Heat from the combustion causes the burned gas to increase its pressure. The force of this pressure is converted into useful work through the piston, connecting rod, and crankshaft.

The *piston* forms a movable bottom to the combustion chamber. It is attached to the connecting rod with a piston pin or wrist pin. The piston pin allows for a rocking movement called a swivel joint at the piston end of the connecting rod. The connecting rod is connected to the part of the crankshaft, called a *crank throw, crank pin,* or *connecting rod bearing journal.* This provides another swivel joint. The crank throw is the amount that the large end of the connecting rod is offset from the center of the crankshaft main bearing centerline. The original crankshafts were assembled with a metal rod or pin that the connecting rod is attached to. This term is carried over as the crankpin. The crankpin has been smoothed to operate with the connecting rod bearing and so the surface is called the bearing journal.

Piston rings seal the small space between the piston and cylinder wall, keeping the pressure above the piston. When the pressure builds up in the combustion chamber, it pushes on the piston. The piston, in turn, pushes on the piston pin and upper end of the connecting rod. The lower end of the connecting rod pushes on the crank throw. This provides the force to turn the crankshaft. The turning force of the crankshaft turns the drive wheels through a drive train. **This turning force is torque.**

As the crankshaft turns, it develops inertia. **Inertia is the force that causes the crankshaft to continue turning.** This action will bring the piston back to its original position, where it will be ready for the next power stroke. While the engine is running, the combustion cycle keeps repeating as the piston reciprocates (moves up and down) and the crankshaft rotates. These motions put mechanical forces on the engine parts. The combustion heat and mechanical forces are a major consideration in the design of the parts.

8-1 Piston and Rod Removal

Many schools have nearly new engines donated to them by automobile manufacturers. These can be readily disassembled using the following procedure. If a high-mileage engine is being studied, the disassembly procedure in Chapter 13 should be followed.

It is necessary to remove the head and oil pan before the piston and rod assembly can be removed from the block. The head removal procedure was discussed in Chapter 7. The oil should be drained before the pan is removed. The rod and caps should be checked for markings that identify their location. Typical number markings are shown in Figure 8-1. If the rod and caps are not marked, they should be marked *before* disassembly. If number stamps

are not available, punch marks, as shown in Figure 8-2, can be used. The crankshaft is turned until the piston is at the *bottom* of its stroke. This places the connecting rod nuts or cap screws where they are easily accessible. They are removed and the rod cap is taken off. This may require *light* tapping on the connecting rod bolts with a *soft-faced* hammer. Protectors are placed over the rod bolt threads. They will protect the threads and the surface of the crankshaft journal. This can be seen in the cutaway

Figure 8-2. Punch marks used on connecting rod, rod cap, and main bearing cap to identify their location in the engine.

Figure 8-1. Typical connecting rod, rod cap, and main bearing cap numbers that identify their location in the engine.

engine pictured in Figure 8-3. The piston and rod assembly is pushed out, taking care to avoid hitting the bottom edge of the cylinder with the rod. Figure 8-4 shows how the rod will hit the edge when it is carelessly removed. If the cylinder is hit, it will raise a burr. If the burr is not removed, it will score the piston after the engine is reassembled and run, as shown in Figure 8-5. The rod caps should be reattached to the rod after the assembly has been removed from the cylinder. The rod caps are not interchangeable between rods. The assembly must be handled carefully. Neither the piston nor the rod should be clamped in a vise. Careless clamping will

Figure 8-3. Protectors placed over the rod bolt threads to keep from damaging the crankshaft journal.

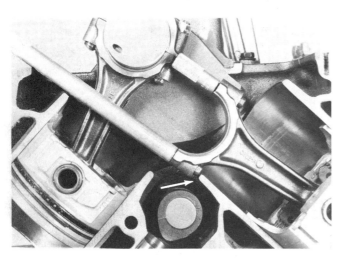

Figure 8-4. Careless removal will allow the connecting rod to hit the bottom of the cylinder skirt, raising a burr.

Figure 8-5. A piston skirt that was scored by a burr raised by hitting the bottom of the cylinder skirt with the connecting rod as it was being removed.

cause them to warp or even crack. They should be placed on a parts stand so that they do not strike each other. The aluminum piston can be easily scratched or nicked.

The rings are carefully removed from the piston to avoid damage to either the piston or the ring. The best way to remove them is to use a piston ring expanding tool. A good type of ring expanding tool is shown in Figure 8-6.

Most piston pins have a press-fit in the rod. Removal requires special support and pressing fixtures to prevent piston damage. Removal should not be attempted without these special tools. It is not usually necessary to remove the piston pins on a study engine.

8-2 Piston Criteria

When the engine is running, the piston starts at the top of the cylinder. It accelerates downward to a maximum velocity slightly before it is halfway down. The piston comes to a stop at the bottom of the cylinder at 180° of crankshaft rotation. During the next 180° of crankshaft rotation, the piston moves upward. It accelerates to a maximum velocity slightly above the halfway point, and them comes to a stop at the top of the stroke. Thus, the piston starts, accelerates, and stops twice in each

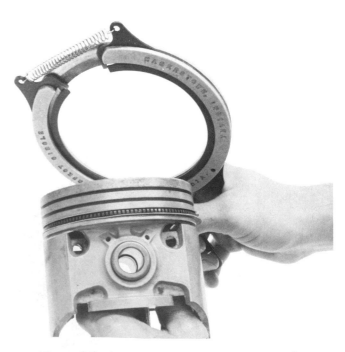

Figure 8-6. A good type of piston ring expanding tool.

crankshaft revolution. This reciprocating action of the piston produces large *inertia forces*. **Inertia is the force that causes a part that is stopped to stay stopped or in motion to stay in motion.** The lighter the piston can be made, the less inertia force is developed. Less inertia will allow higher engine operating speeds. For this reason, pistons are made as light as possible while still having the strength that is needed.

The piston operates with its head exposed to the hot combustion gases while the skirt contacts the relatively cool cylinder wall. This results in a temperature difference of about 275 °F (147 °C) between the top and bottom of the piston. The temperature difference between cast and forged pistons is shown in Figure 8-7.

The automotive engine piston is more than a cylinder plug to convert the combustion pressure to a force on the crankshaft. It is a fine compromise between strength, weight, and thermal expansion control. It must also support piston sealing rings. The piston must have satisfactory durability to **operate under these conditions while sliding against a cylinder wall.**

Aluminum alloy has proved to be the best material from which to make pistons. It is lightweight and it provides adequate strength. It does, however, increase the expansion problem. The reasons for this is that aluminum alloys expand approximately twice as much as does cast iron when both are heated the same number of degrees. Cast iron is used in the blocks and cylinder bores in which the aluminum pistons operate. The aluminum piston expands faster than the cylinder. This tends to reduce the operating clearance as the engine warms up.

To further complicate piston design problems, modern automobile styling and size limits the space available for the engine. At the same time, customers expect good engine performance. The easiest way to provide enough power is to have a large engine displacement. Displacement is controlled by the cylinder bore size and lengthening the piston stroke. Each of these requires space. In down-sized automobiles, space is not available. The passenger car engine designer has been able to provide enough engine displacement with a small external engine size. This is done by keeping the height of the piston to a bare minimum and bring it close to the crankshaft at the bottom of the stroke as shown in Figure 8-8. This piston must still have enough

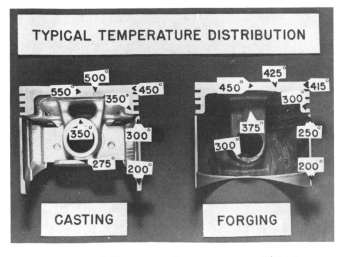

Figure 8-7. Differences in the temperature within pistons operated under the same conditions. (Courtesy of Michigan Division of TRW).

Figure 8-8. The piston is very close to the crankshaft counterweight when the piston is on the bottom of the stroke. The piston has a slipper skirt when the connecting rod is short as shown here.

strength to support combustion pressure and reciprocating loads. It must also have enough piston skirt to guide it straight in the bore. In addition, the piston must have heat expansion control for quiet, long-life operation. Finally, it holds the piston rings perpendicular to the cylinder wall so they can seal properly.

Piston Heads. Because the piston head forms a portion of the combustion chamber, its shape is very important to the combustion process. Generally, low-cost, low-performance engines have *flat-top* pistons. Some of these flat-top pistons come so close to the cylinder head that *recesses* are cut in the piston top for valve clearance. Pistons used in high-powered engines may have raised domes or a *pop-up* on the piston heads. These are used to increase the compression ratio. Pistons used in other engines

may be provided with a depression or a *dish*. The depth of the dish provides different compression ratios required by different engine models. A number of piston head shapes are shown in Figure 8-9.

The piston head must have enough strength to support combustion pressures. Ribs are often used on the underside of the head to maintain strength while at the same time reducing material to lighten the piston. These ribs are also used as cooling fins to transfer some of the piston heat to the engine oil. Typical ribs on the underside of the piston can be seen in Figure 8-10.

Piston Ring Grooves. Piston ring *grooves* are located between the piston head and skirt (Figure 8-11). The width of the grooves, the width of the *lands,* and the number of rings are a major factor in

(a) (b) (c)

(d) (e) (f)

Figure 8-9. Piston head shapes. (a) Flat, (b) recessed, (c and d) pop-up, and (e and f) dished.

Figure 8-10. Ribs on the underside of the piston.

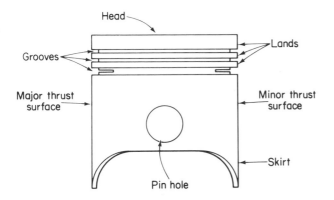

Figure 8-11. Names of piston parts.

determining minimum piston height. The outside diameter of the lands is about 0.020 in. (0.5 mm) smaller than the *skirt* diameter. Some pistons for heavy-duty engines have oil ring grooves located on the piston skirt below the piston pin. Passenger car engines use two compression rings and one oil control ring. They are all located above the piston pin. Some experimental engines have used two ring pistons; one compression ring and one oil ring. These have not been able to satisfactorily seal the combustion chamber with present piston ring technology.

The piston ring groove must be deep enough to prevent the ring from hitting the base of the groove when the ring is pressed in so that it is flat with the land face. Its groove depth becomes critical for

some piston ring expander designs. These expanders wedge between the back of the ring and the base of the groove. The sides of the groove must be square and flat so that the side of the piston ring will seal on the side of the groove. Oil ring grooves are vented in the base so that oil scraped from the cylinder wall can flow through the vents to the crankcase. This venting is done through drilled holes or slots, as shown in Figure 8-12.

Piston Skirt. Piston expansion was a minor problem in old engines with cast-iron pistons. Owners of these engines would usually accept piston slap noise that resulted from large piston-to-cylinder wall clearances on cold engines, because the slap noise would usually stop when the engine warmed up. Some means of piston expansion control was required, however as the owners demanded quiet engine operation.

Piston expansion was first controlled through a *slot* on the minor thrust surface of the piston skirts. These pistons were fitted in the cylinder with very little piston-to-cylinder clearance. The piston skirt would expand into the slot as the piston was heated during operation. The most popular slot types were the *U-slot* and the *T-slot*. The U-slot design had two slots on the piston skirt that were connected together near the top of the piston skirt, forming an inverted U shape. The T-slot had one slot down the piston skirt with a cross slot at the upper edge of the piston skirt to form the T-slot design. This

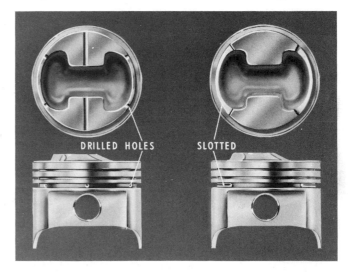

Figure 8-12. Oil ring groove venting using drilled holes and slots (Courtesy of Chevrolet Motor Division, General Motors Corporation).

method of expansion control carried over into the early aluminum pistons.

Aluminum pistons expand approximately twice as much as cast-iron pistons with the same increase in temperature. With this much expansion, the expansion slot that was required made the pistons too weak. A better method of expansion control was devised using a *cam ground* piston skirt. The piston thrust surfaces closely fit the cylinder, while the piston pin boss diamter is fitted loosely. As the cam ground piston is heated, it will expand along the piston pin so that it will become nearly round at its normal operating temperatures. A cam ground piston skirt is illustrated in Figure 8-13.

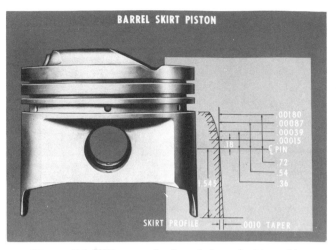

Figure 8-14. Design of a barrel shaped piston skirt (Courtesy of Chevrolet Motor Division, General Motors Corporation).

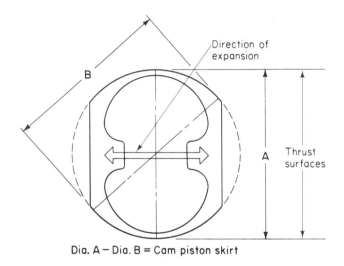

Figure 8-13. Piston cam shape. The largest diameter is across the thrust surfaces and perpendicular to the piston pin (lettered A).

Some later model engines have modified the cam ground piston skirts. This was done by adding a progressively round skirt cam drop or *barrel shape* (Figure 8-14). The lowest part of the piston skirt is the greatest distance from the combustion chamber. It does not run as hot, so it expands less than the upper part. This allows the lower part of the piston skirt to be made larger than the upper part when measured across the thrust surface diameter. The lower part of the skirt will have a close cold fit in the cylinder for quiet operation and a satisfactory piston service life. When the piston gets hot the upper part will expand the most so the piston skirt is straight. It then matches the cylinder wall.

Most pistons have horizontal separation *slots* that act as *heat dams.* These slots reduce heat

transfer from the hot piston head to the lower skirt. This, in turn, will keep the skirt temperature lower and there will be less skirt expansion. By placing the slot in the oil ring groove, the slot is used for oil drain back as well as for expansion control. Engines built by the Buick Motor Division of General Motors place a slot below the piston pin. This isolates the lower skirt from piston pin boss deflections caused by stress that occurs on the power stroke. The lower skirt can better maintain its size. These heat dam slots can be seen in Figures 8-15 and 8-16.

Figure 8-15. Heat dam cast slots just below the oil ring groove.

Figure 8-16. Heat dam saw slot in the bottom of the oil ring groove and a cast slot below the piston pin.

A major development in expansion control was accomplished by casting the piston aluminum around two stiff steel *struts*. The struts are not chemically bonded to the aluminum, nor do they add any strength to the piston. There is only a mechanical bond between the steel and aluminum. The bimetallic action of this strut in the aluminum forces the piston to bow outward along the piston pin. This keeps the piston skirt thrust surfaces from expanding more than the cast-iron cylinder in which the piston operates. Pistons with steel strut inserts allow good piston-to-cylinder wall clearance at normal temperatures. At the same time, they allow the cold operating clearance to be as small as 0.0005 in. (0.0127 mm). This small clearance will prevent cold piston slap and noise. A typical piston expansion control strut is visible in Figure 8-17.

Heavy-duty pistons are cylindrical castings with ring grooves at the top, using a *trunk-type* skirt. The piston shown in Figure 8-16 has a trunk-type skirt. As automotive passenger car requirements have increased, the number and thickness of the piston rings have decreased and the cast-aluminum piston skirt has been reduced to a minimum by using an open-type *slipper* skirt. Examples of the slipper skirt piston are shown in Figures 8-15 and 8-17. High-performance engines need pistons with added strength. They use impact-extruded forged pistons whose design falls between these two extremes of heavy-duty and automotive pistons. Figure 8-18 shows a forged aluminum piston with a trunk skirt and Figure 8-19 shows a forged aluminum piston with a slipper skirt.

Skirt Finish. For maximum life, the piston skirt surface finish is important. Turned grooves or waves 0.0005 in. (0.0125 mm) deep on the surface of some piston skirts produce a finish that will carry oil for lubrication. Other piston skirts are relatively smooth. Figure 8-20 shows typical surfaces of piston skirts. A thin tin-plated surface (approx-

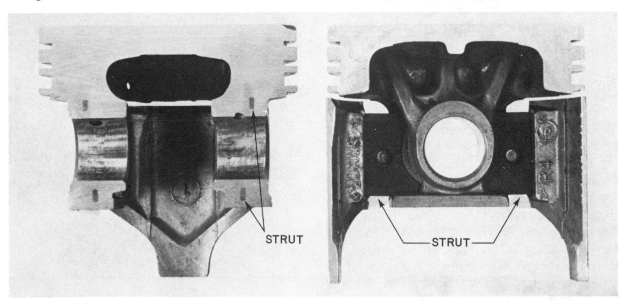

Figure 8-17. Two section views of a piston showing the expansion control strut.

Figure 8-18. Grain flow lines can be seen in this forged aluminum piston with a trunk skirt.

Figure 8-19. A forged aluminum piston with a slipper skirt.

imately 0.00005 in. or 0.00125 mm thick) is also used on some aluminum pistons to help reduce scuffing and scoring during occasional periods of minimum lubrication. The piston skirt normally rides on a film of lubricating oil. Any time the oil film is lacking, metal-to-metal contact will occur, and this starts piston scuffing. This condition often occurs when a faulty cooling system causes the engine to overheat. Overheating causes the oil to thin and the heat overexpands the piston. Piston scuffing leads to poor oil control, short piston life, roughened cylinder bores, and scuffed rings.

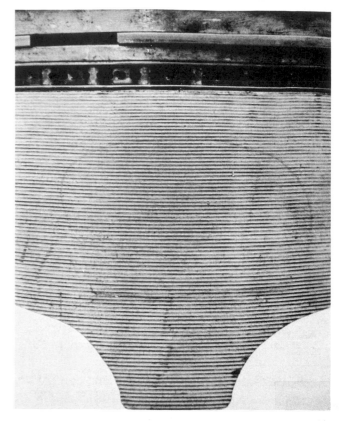

Figure 8-20. Typical piston skirt surfaces in common use.

Piston Balance. Pistons are provided with enlarged *pads* or skirt flanges (Figure 8-21) that are used for controlling piston weight. Material is removed from the surface of these pads by the manufacturer as the last machining operation to bring the piston within the correct weight tolerances.

8-3 Piston Pins

Piston pins are used to attach the piston to the connecting rod. The piston pin transfers the force produced by combustion chamber pressures and piston inertia forces to the connecting rod. The piston pin

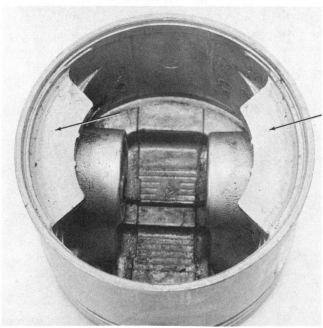

Figure 8-21. Typical location of pads used to balance pistons.

is made from high-quality steel in the shape of a tube to make it both strong and light. Sometimes, the interior hole of the piston pin is tapered, large at the ends and small in the middle of the pin. This gives the pin strength that is proportional to the location of the load placed on it. A double-taper hole such as this is more expensive to manufacture, so it is used only where its weight advantage merits the extra cost.

Piston Pin Offset. The piston pin holes are not centered in the piston. They are located toward the *major thrust surface* approximately 0.062 in. (1.57 mm) from the piston center line, as shown in Figure 8-22. Pin offset is designed to reduce piston slap and noise that could result as the large end of the connecting rod crosses over upper dead center.

The minor thrust side of the piston head has a greater area than the area on the major side. This is caused by the pin offset. As the piston moves up in the cylinder on the compression stroke it will be

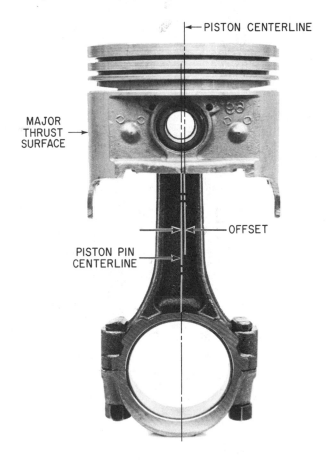

Figure 8-22. Piston pin is offset toward the major thrust surface.

riding against the minor thrust surface. When compression pressure becomes high enough, the greater head area on the minor side will cause the piston to cock slightly in the cylinder. This keeps the *top* of the minor thrust surface on the cylinder. It forces the *bottom* of the major thrust surface to contact the cylinder wall. As the piston approaches top center, both thrust surfaces are in contact with the cylinder wall. When the crankshaft crosses over top center, the force on the connecting rod moves the entire piston toward the major thrust surface. The lower surface of the major thrust surface has already been in contact with the cylinder wall. The rest of the piston skirt wipes into full wall contact just after the crossover point, thereby controlling piston slap. This action is illustrated in Figure 8-23.

Locating the piston offset toward the minor thrust surface would provide a better mechanical advantage. It also would cause less piston-to-cylinder friction. For these reasons, offset is often placed toward the minor thrust surface in racing engines. Noise and durability are not as important as maximum performance in racing engines.

Piston Pin Fits. The finish and size of piston pins are very closely controlled. Piston pins have a smooth finish like a mirror. Their size is held to tenths-of-thousandths of an inch so that exact fits can be maintained. If the piston pin is loose in the piston or in the connecting rod, it will make a rattling sound while the engine is running. If the piston pin is too tight in the piston, it will restrict piston expansion along the pin diameter. This will lead to piston scuffing. Normal piston pin clearances range from 0.0005 to 0.0007 in. (0.0126 to 0.0180 mm).

Piston Pin Rotaining Methods. It is necessary to retain or hold piston pins so that they stay centered in the piston. If piston pins were not retained, they would move endwise and groove the cylinder wall. Piston pins are retained by one of three general methods. The piston pin may be full floating, with some type of stop located at each end. It may be fastened to the connecting rod. In a few engines, the pin is fastened to the piston.

Full-floating piston pins in automotive engines are retained by lock rings located in grooves in the piston pin hole at the ends of the piston pin (Figure 8-24). Some engines use aluminum or plastic plugs in both ends of the piston pin. These plugs will touch the cylinder wall without scoring, to hold the piston pin centered in the piston.

Piston pins have been retained in connecting rods by a clamp bolt located in the piston end of the connecting rod (Figure 8-25). The piston pin used in the clamp has an undercut through which the edge of the clamp bolt fits. The clamp bolt locates the pin in the piston center and clamps the rod around the pin to hold it securely. The modern method of retaining the piston pin in the connecting rod is to make the connecting rod hole slightly smaller than the piston pin. The pin is installed by heating the

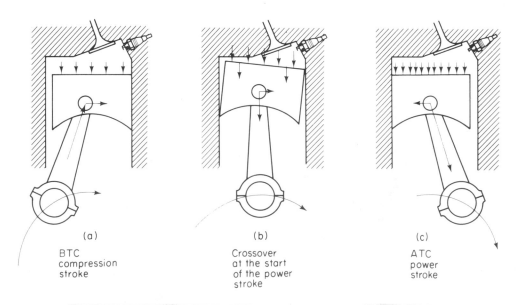

(a)
BTC
compression
stroke

(b)
Crossover
at the start
of the power
stroke

(c)
ATC
power
stroke

Figure 8-23. The effect of piston pin offset as it controls piston slap.

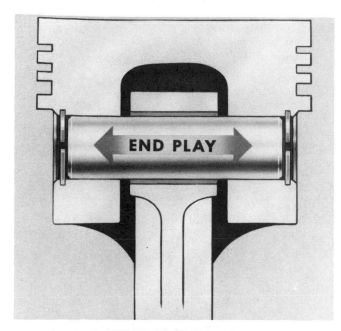

Figure 8-24. Full floating piston pin retained by a lock ring on each end of the piston pin.

Figure 8-25. Clamp bolt used to retain a piston pin.

rod to expand the hole or by pressing it into the rod. This retaining method will securely hold the pin. This press or shrink fit is called an *interference fit.* Care must be taken to have the correct hole sizes and the pin must be centered in the connecting rod. The interference fit method is the least expensive to produce. It is, therefore, found in the majority of passenger car engines.

On automobile engines, the piston is free to move on the piston pin. On some heavy-duty engines, a cap screw on one side of the piston boss enters a hole or contacts a flat on the piston pin to retain the pin. The cap screw is only placed on *one side* of the piston. In this way, clamping does not interfere with the normal piston expansion that takes place along the pin.

8-4 Piston Rings

Piston rings provide two major functions in engines. They form a sliding combustion chamber seal that prevents the high-pressure combustion gases from leaking past the piston. They also keep engine oil from getting into the combustion chamber. In addition, the rings transfer some of the piston heat to the cylinder wall, where it is removed from the engine through the cooling system.

Piston rings are classified as two types—*compression rings,* located toward the top of the piston, and *oil rings,* located below the compression rings. The first piston rings were made with a simple rectangular cross section. This cross section was modified by tapers, chamfers, counterbores, slots, rails, and expanders. Piston ring materials have also changed from plain cast iron to materials such as pearlitic and nodular iron as well as steel. The use of ductile-iron as a piston ring material has been limited to heavy-duty applications. It is beginning to be used in some automotive engines. Piston rings may be faced with chromium or molybdenum materials.

8-5 Compression Rings

A compression ring is designed to form a seal between the moving piston and cylinder wall. This is necessary to get maximum power from the combustion pressure. At the same time, the compression ring must keep friction at a minimum. This is done by providing only enough static or built-in mechanical tension to hold the ring in contact with the cylinder wall during the intake stroke. Combustion chamber pressure during the compression, power, and exhaust strokes is applied to the top and back of the ring. This pressure will add the force on the ring that is required to seal the combustion chamber during these strokes. Figure 8-26 illustrates how the combustion chamber pressure adds force to the ring.

Ring Forces. Mechanical *static tension* of the ring results from the ring shape, material characteristics, and expanders used. Rings are manufactured so that they have a cam shape in their free state. When the piston ring is compressed to the cylinder size, it becomes round and develops the required static tension. Additional piston ring control causes the ring

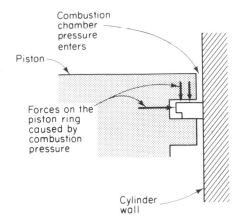

Figure 8-26. Combustion pressure forces the top piston ring downward and outward against the cylinder wall.

to *twist* toward the chamfers and counterbores when it is compressed to the size of the cylinder (Figure 8-27). Twist is used to provide *line contact* sealing on the cylinder wall and in the piston ring groove. The line contact can be seen on the slightly used rings pictured in Figure 8-28. Line contact provides a relatively high unit pressure for sealing. At the same time, it allows low total ring force against

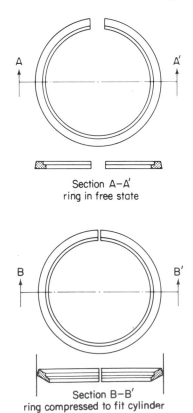

Section A–A'
ring in free state

Section B–B'
ring compressed to fit cylinder

Figure 8-27. A piston ring counterbore causes the ring to twist when it is compressed to fit the cylinder.

Figure 8-28. Piston rings are slightly used so only the line contact shows. The upper barrel faced ring has line contact in the center. The second taper faced ring has line contact along the lower edge of the ring.

the cylinder. This results in low ring friction. Expanders are sometimes used under compression rings when additional static tension is required. Pressure in the combustion chamber acts on the top piston ring. The pressure forces the ring to flatten on the bottom side of the piston ring groove. This action will seal the ring-to-piston joint. The ring groove must have a flat and square side for this seal. **Pressure behind the ring will also force it against the** cylinder wall to seal ring-to-cylinder wall contact surface. This action (Figure 8-26) produces a *dynamic sealing force* that makes an effective moving combustion chamber seal.

Ring Gap. The piston *ring gap* will allow some leakage past the top compression ring. This leakage is useful in providing pressure on the second ring to develop a dynamic sealing force. The amount of piston ring gap is critical. Too much gap will allow excessive *blow-by*. **Blow-by is the leakage of combustion gases past the rings.** Blow-by will blow oil from the cylinder wall. This oil loss is followed by piston ring scuffing. Too little gap, on the other hand, would allow the piston ring ends to butt when the engine is hot. Ring end butting increases the mechanical force against the cylinder wall, causing excessive wear and possible engine failure. In general, stronger piston rings require larger ring gaps.

A butt-type piston ring gap is the most common type used in automotive engines. This is the least expensive to manufacture. Some *low-speed* industrial engines and some diesel engines use a more expensive tapered or seal-cut ring gap. These gaps are necessary to reduce losses of the high-pressure combustion gases. At low speeds, the gases have more time to leak through the gap. Typical ring gaps are illustrated in Figure 8-29.

Ring Cross Section. As engine speeds have increased, inertia forces on the piston rings have also increased. As a result, engine manufacturers have found it desirable to reduce inertia forces of the rings by reducing their weight. This has been done by narrowing the piston ring in fractional steps from ¼ in. to as low as ¹⁄₁₆ in. Narrow rings require less material, they take less space, and minimize scuffing. As the rings get narrower, the piston ring groove is also made narrow. Narrow ring grooves are difficult to manufacture. Ring grooves of ¹⁄₁₆ in. (1.6 mm) appear to be the practical minimum that can be made within the state of the manufacturing art.

Typical compression ring cross sections are illustrated in Figure 8-30. A discussion of piston ring cross sections must start with a rectangular shape. This was first modified with a *taper face* that would contact the cylinder wall at the lower edge of the piston ring. When either a chamfer or counterbore relief is made on the *upper inside* corner of the piston ring, the ring cross section is unbalanced. This will cause the ring to twist in the groove in a positive direction. *Positive twist* will give the same wall contact as the taper face ring. It will also provide a line contact seal on the bottom side of the groove. Sometimes, twist and a taper face are used on the same compression ring.

Some second rings are notched on the *outer lower* corner. This, too, provides a positive ring twist. The sharp lower-outer corner becomes a scraper that helps oil control, but it has less compression control than the preceding types.

By chamfering the ring lower-inner corner, a *reverse twist* is produced. This seals the lower outer section of the ring and piston ring groove, thus improving oil control. Reverse twist rings require a greater taper face or barrel face to maintain the desired ring face-to-cylinder wall contact.

Some rings replace the outer ring taper with a barrel face. The barrel is 0.0003 in. per 0.100 in. (0.0076 mm per 0.254 mm) of piston ring width. Barrel faces are found on rectangular rings as well as on torsionally twisted rings. The upper ring in Figure 8-28 shows the line contact that occurs with a

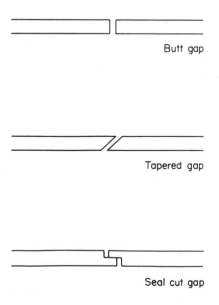

Figure 8-29. Typical ring gaps.

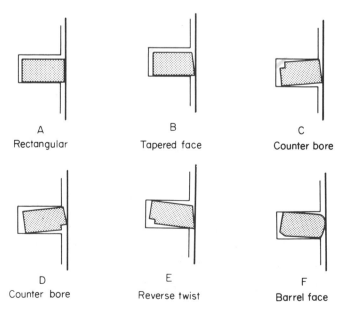

Figure 8-30. Typical compression ring cross sections.

barrel-faced compression ring and the lower ring shows a taper face contact.

Ring Facings. Piston ring facing materials are very important to provide maximum service life. Prior to World War II, rings were made entirely of cast iron. They were given a phosphorus coating to minimize rusting and initial scuffing during startup. Other coatings had names such as Ferrox and Graphitox. These coatings are ferrous oxide and graphite, respectively. This practice was modified during World War II as a result of aircraft engine piston ring development. Techniques to put hard *chromium* on the piston ring face were developed. Chromium greatly increases piston ring life, especially where *abrasive materials* are present in the air. During manufacture, the chromium-plated ring is slightly chamfered at the outer corners. About 0.0004 in. (0.010 mm) of chrome is then plated on the ring face. Chromium-faced rings are prelapped or honed before being packaged and shipped to the customer. The finished chromium facing is shown in a section view in Figure 8-31.

Early in the 1960s, *molybdenum* piston ring faces were introduced. These rings also proved to have good service life, especially under *scuffing conditions*. Most molybdenum-faced piston rings have a 0.004 to 0.008 in. (0.1 to 0.2 mm) deep groove cut on the ring face. This groove is filled with molybdenum, using a metallic spray method, so that there is a cast-iron edge above and below the molybdenum. This edge may be chamfered on some

applications. A section view of a molybdenum-faced ring is shown in Figure 8-32.

New ring coatings have developed. In one design, a thin layer of fine-grained chromium with microscopic cracks is placed over the hard chromium. These rings can be used without being prelapped. A super molybdenum face coating has also been developed. It does not have to be placed in a groove, so it does a better job of resisting both mechanical shock and flaking.

Molybdenum-faced piston rings will survive under heat and scuffing conditions better than chromium-faced rings. Under abrasive wear conditions, chromium-faced rings will have a better service life. There is little measurable difference between these two facing materials with respect to blow-by, oil control, break-in, and horsepower. Piston rings with either of these two types of facings are far better than plain cast-iron rings with phosphorus coatings. When used, a molybdenum-faced ring will be found in the top groove and a chromium-faced ring in the second groove.

8-6 Oil Control Rings

Originally, piston rings were not divided into compression and oil rings. All piston rings were plain rectangular rings. The first rings to be called oil rings were tapered rings. The lower scraping edge removed a large part of the oil from the cylinder wall on the downstroke of the piston. In the next development, the oil rings were vented by machining slots through the ring. This allowed oil to return

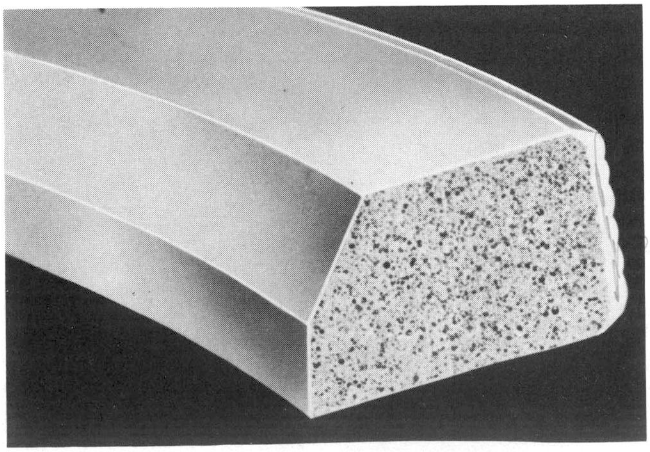

Figure 8-31. Chromium facing can be seen on the right side of the section view of the ring (Courtesy of Sealed Power Corporation).

Figure 8-32. Molybdenum facing can be seen on the right side of the section view of the ring (Courtesy of Sealed Power Corporation).

through the ring and openings in the piston. This machining, as shown in Figure 8-33, produced two scraping edges that performed better than the single edge. Steel spring expanders were placed in the ring groove back of the ring to improve static radial tension. They forced the ring to conform to the cylinder wall. **Conform means to change their shape to match the shape of the part they are in contact with.** Many expander designs are used. One type of expander (Figure 8-34) acts as a spring between the ring groove base and the ring. The force of another type of expander (Figure 8-35) results from radial force when the two ends of the expander butt together. This forms static tension as the ring is forced into the piston ring groove by the cylinder.

As the oil ring requirements became greater, cast iron was no longer satisfactory. Steel *rails* with chromium or other type of facings replaced the cast-iron scraping edges. The rails are backed with *expanders* and separated with a *spacer*. This can be seen in Figure 8-36.

Some expander designs provide the spacing function as well as the expansion function. An oil ring with this type of construction is lightweight, having desirable low inertia. It is well ventilated, so the oil can easily flow through it to the crankcase. It provides excellent oil control and it has a long service life. This type of oil ring is shown in Figure 8-37.

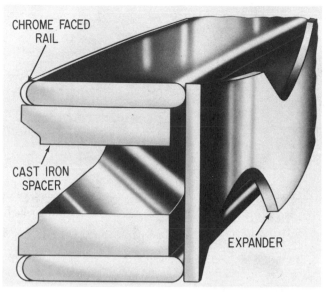

Figure 8-34. An oil ring with a cast iron spacer, two chrome faced rails, and an expander (Courtesy of Dana Corporation).

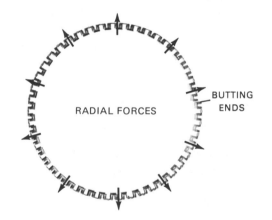

Figure 8-35. An oil ring expander type that provides radial force as it is compressed with the ends butting together.

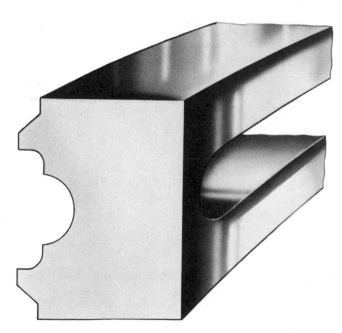

Figure 8-33. Cast iron oil ring with a machined slot for oil venting (Courtesy of Dana Corporation).

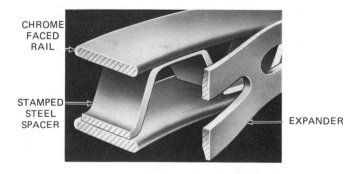

Figure 8-36. An oil ring with a stamped steel spacer, two chrome faced rails and an expander (Courtesy of Sealed Power Corporation).

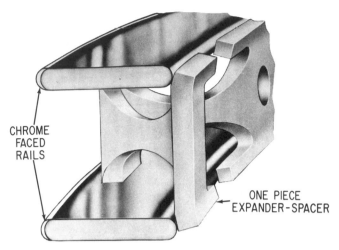

Figure 8-37. An oil ring with a one piece expander-spacer and two chrome faced rails (Courtesy of the Dana Corporation).

The latest design is a two-piece oil ring. One complex piece is an expander-spacer-rail. The other is a single rail. Two- and three-piece oil rings are compared in Figure 8-38.

The piston ring cannot do its job unless it can seal against the cylinder wall through its entire normal service life. The cylinder wall is honed with a *crosshatch* pattern (Figure 8-39). The crosshatch must be smooth enough for use with the piston ring facing materials. In general, it should have a satin finish of about 25 microinches. **Microinches are a measure of the average surface smoothness from the highest ridge to the lowest depression.**

Figure 8-38. On the left is an oil ring having an expander-spacer and two chrome faced rails. On the right is an oil ring having one chrome faced rail and a combination expander-spacer-rail.

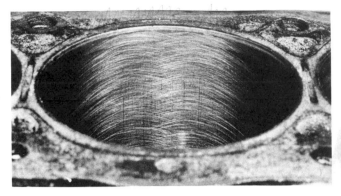

Figure 8-39. Typical cylinder wall finish before use.

8-7 Connecting Rods

The connecting rod transfers the force and reciprocating motion of the piston to the crankshaft. The small end of the connecting rod reciprocates with the piston. The large end rotates with the crank pin. These dynamic motions make it desirable to keep the connecting rod as light as possible and still have a rigid beam section. Lightweight rods also reduce the total connecting rod material cost.

Connecting rods are manufactured by both casting and forging processes. Forged connecting rods have been used for years. They are always used in high-performance engines. They are generally used in heavy-duty engines. Casting materials and processes have been improved so that they are used in most high-production-standard passenger car engines. Cost of cast rods is lower than forged rods, both in the initial casting cost and in the machining cost. A typical rough connecting rod casting is shown in Figure 8-40. Generally speaking, the forging method produces lighter weight, stronger, but more expensive connecting rods.

The connecting rod design, as shown in Figure 8-41, is basically two ring forms. One encircles the piston pin and the other encircles the crank pin. Each of these ring forms is blended into a tapered

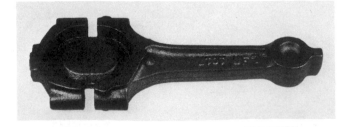

Figure 8-40. A rough casting for a connecting rod.

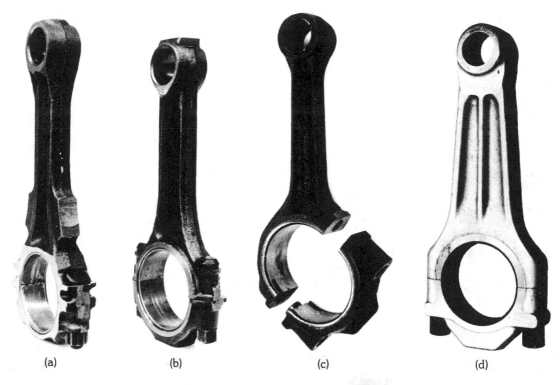

Figure 8-41. Connecting rod types. (a) Cast iron, (b) forged steel, (c) cap separated at an angle on a forged rod, and (d) forged aluminum racing rod.

I-beam section. The large split-ring form for the crankshaft end is machined *after* the cap is assembled on the rod. The hole will be a perfect circle. Therefore, the rod caps must not be interchanged. Assembly bolt holes are closely reamed in both the cap and connecting rod to assure alignment. The connecting rod bolt diameters have *piloting surfaces* which closely fit these reamed holes. The parts of a typical connecting rod are shown in Figure 8-42. The connecting rod bolt heads are formed so that they have two or three sides which hold against the rod bolt bosses. The fourth side is left off so that there is enough cylinder skirt clearance as the crankshaft turns. Some connecting rod bolt heads are made on an angle to give this clearance. These bolt head shapes can be seen in Figure 8-43.

In some engines, offset connecting rods provide the most economical distribution of main bearing space and crankshaft cheek clearance. V-6 engines commonly have the connecting rods offset approximately 0.100 in. (2.54 mm). An example of this is pictured in Figure 8-44. The amount of offset of the piston pin end and the crankshaft journal end is measured in the lengthwise direction of the engine. The offset may be divided equally between each end, keeping the connecting rod column perpen-

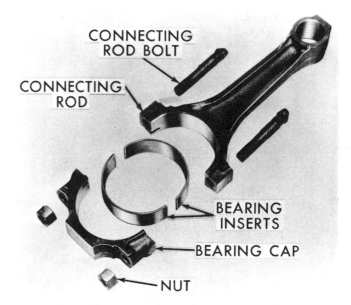

Figure 8-42. Parts of a typical connecting rod assembly (Courtesy of Cadillac Motor Car Division, General Motors Corporation).

dicular. Rods with a large offset do not have as good a bearing endurance quality as rods that are not offset. Bearing failure occurs in the bearing edge halves nearest the center of the rod where the loads are greatest.

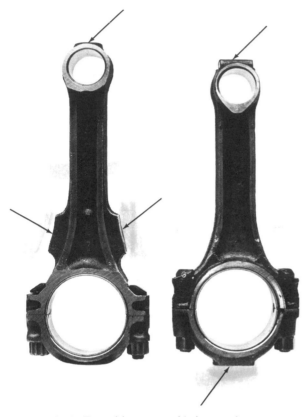

Figure 8-43. Typical locations of balancing bosses on connecting rods.

Figure 8-44. An offset connecting rod used in a V-6 engine.

The sweep or path of the connecting rod must clear all engine parts as the crankshaft rotates. This requires minimum bolt-center-line distance and minimum bolt head and nut size. The large end of the connecting rod, however, needs to be large enough to carry a connecting rod bearing that is designed to support the dynamic loads. The length of the bearing (from edge to edge) usually determines the high-speed endurance capacity of the bearing.

Connecting rods are made with balancing bosses, so that their weight can be adjusted to specifications. Some have balancing bosses only on the rod cap. Others have a balancing boss above the piston pin as well. Some manufacturers put balancing bosses on the side of the rod near the center of gravity of the connecting rod. Typical balancing bosses can be seen in Figure 8-43. Balancing is done on automatic balancing machines as the final machining operation before the rod is installed in an engine.

Most connecting rods have a *spit hole* that bleeds some of the oil from the connecting rod journal. Typical examples are shown in Figure 8-45. The hole may be drilled or it may be a chamfer on the cap parting surface. On inline engines, oil is thrown up from the spit hole into the cylinder that the rod is in. On V-type engines, it is thrown into a cylinder in the opposite bank. The oil that is spit from the rod is aimed so that it will splash into the interior of the piston. This helps to lubricate the piston pin. Occa-

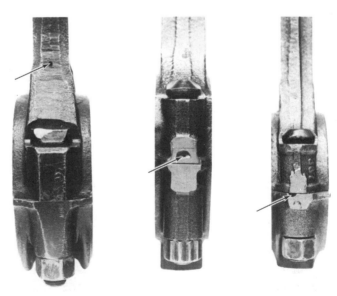

Figure 8-45. Connecting rod spit and bleed holes.

sionally, adequate piston pin and cylinder wall lubrication is obtained without a spit hole. A hole similar to the spit holes may be used. It is called a *bleed hole*. Its only purpose is to control the oil flow through the bearing. Some heavy-duty engine con-

necting rods are drilled lengthwise. Oil flows through the drilled passage from the crank pin to the piston pin. This is an expensive process and is used only where the spit-hole method will not supply enough piston pin lubrication.

REVIEW QUESTIONS

1. List the order of part removal to take the piston and rod assemblies from the engine you are studying. [8-1]

2. Why must the piston be placed at the bottom of the stroke during removal of the piston and rod assembly? [8-1] *For easy removal*

3. Why are protectors placed on the connecting rod bolts before removal of the piston and rod assembly? [8-1]

4. How are piston and rod assemblies cared for after they have been removed from the cylinder? [8-1] *Put one a Bench - separate from each other so no damage occurs*

5. What are the advantages and disadvantages of aluminum as a piston material? [8-2]

6. List the physical requirements that must be met by the piston. [8-2]

7. At what part of the four-stroke cycle would it be necessary to have recesses in the piston head for valve clearance? [8-2] *Compression*

8. List the compression and oil ring groove requirements. [8-2]

9. What methods are used to control piston heat expansion? [8-2] *U slots — T slot on side - pin*

10. How does a cam-shaped piston skirt differ from a barrel-shaped piston skirt? [8-2] *Forces the piston to expand along the piston pin*

11. How do the two steel struts in a piston help control piston expansion? [8-2]

12. How does a slipper piston skirt differ from a trunk-type piston skirt? [8-2] *Heavy duty uses trunk-type. Car - uses slipper type*

13. Why are some piston skirts tin plated? [8-2] *To controll piston weight*

14. What is the purpose of piston balance pads? [8-2]

15. When does the piston pin have a double tapered hole? [8-3] *This gives strength proportioned to the load*

16. Describe the action of the piston pin offset as it controls the piston slap. [8-3] *Offset is design to reduce piston slap and noises*

17. What will happen inside the engine if the piston pin does not fit correctly? [8-3] *If loose it will rattle - if to tight it restrict expantion of piston - causin scoffin.*

18. Name three general methods used to retain the piston pin? [8-3] *Clamp, press-fit, cap screw*

19. Why is the interference fit piston pin retaining method used on most passenger car engines? [8-3] *Because it is cheaper and good enough for cars*

20. Where does piston and rod movement take place when the piston pin has an interference fit? [8-3] *On the piston pin*

21. What are two purposes of the piston rings? [8-4] *Pressure pushes ring against the cylinder walls*

22. How does combustion pressure help seal the combustion chamber? [8-5]

23. What causes line contact on the cylinder wall when new piston rings are first installed? [8-5] *Twist is used to provide line contact sealing*

24. What would be the result of having either too little or too much ring gap? [8-5] *To much friction — to much blow-by*

25. When is a tapered or seal-cut ring gap used? [8-5] *Engine industrial and diesel*

26. What limits how narrow the piston ring grooves can be made? [8-5]

27. During which strokes does ring twist aid the piston ring? [8-5] *Power - intake*

28. What are the advantages and disadvantages of piston ring negative twist? [8-5] IT IMPROVES OIL CONTROL

29. How does the ring face-to-cylinder wall contact differ between taper face and barrel face rings? [8-5] DIFFERENT PARTS - WILL MATE CONTACT SE

where ABRASIVE MATERIAL IS PRESENT IN AIR

30. Chromium face rings operate best under what operating conditions. [8-5]

31. Under what operating conditions do molybdenum face rings operate best? [8-5] SCUFFING CONDITION

SEPARATE THE TWO STEEL RAIL

32. What is the purpose of an oil ring expander? [8-6] FORCES THE RING TO CONFORM-

33. What is the purpose of an oil ring spacer? [8-6] THE CYLINDER WALL

34. What are the advantages of an oil ring having a combination expander-spacer? [8-6] IT PROVIDE EXCELENT OIL CONTROL - and LAST ALONG Ti.

CAST IS Cheaper Fo.

35. What are the cylinder wall requirements for good piston ring service life? [8-6] IT HAS TO BO HONED CROSS HATCH PATTERN, AND HAS TO BE SMOOTH TO A SATIN FINISH

36. What are the advantages and disadvantages of cast connecting rods? [8-7]

37. Why is it important to keep the connecting rod cap with the same rod and installed in the correct way? [8-7] THERE ARE INDIVIDUALLY FIT

38. What aligns the cap correctly on the connecting rod? [8-7] THE CRANK PIN

39. Why are some connecting rods offset? [8-7] On V8 IT IS more economical

40. What is the difference between a spit hole and a bleed hole in a connecting rod? [8-7] SPIT-Hole HELP LUBRICATE THE Piston-PIN Bleed Hole - Controls the flow or oil THROUGH THE BEARING

CHAPTER
9

Shafts and Bearings

Automotive engines have only two major rotating parts—the *crankshaft* and the *camshaft*. Power from expanding gases in the combustion chamber is delivered to the crankshaft through the piston, piston pin, and connecting rod. The connecting rods and their bearings are attached to a bearing journal on the crank throw. **The crank throw is offset from the crankshaft center line.** The combustion force is applied to the crank throw after the crankshaft has moved past top center. This produces the turning effort or *torque,* which rotates the crankshaft. The camshaft is rotated by the crankshaft through gears, with chain- or belt-driven sprockets. The camshaft drive is timed so that the valves will be opened in relation to the position of the piston.

The crankshaft rotates in main bearings. These bearings are split in half so that they can be assembled around the crankshaft main bearing journals. The camshaft, in pushrod engines, rotates in sleeve bearings that are pressed into bearing bores within the engine block. Overhead camshaft bearings may be either sleeve-type bushings or split-type bearings, depending on the design of the bearing supports.

Both shafts must support the intermittent variable loads applied on them. In addition, they must have the necessary metal properties to function as good bearing journals. **The bearing journal is the surface of the crankshaft that operates in a bearing.**

If the bearings and cap are not marked, they should be marked bcfore disassembly, as discussed in Chapter 8. The vibration damper is removed with a threaded puller similar to the puller shown in Figure 9-1. A hook puller must not be used, because it could separate the damper. The coolant pump and timing cover (Figure 9-2) have to be removed next. Then the crankshaft and camshaft can be removed. It is assumed that the pan, head, and valve train have already been removed, as discussed in Chapter 8. When used, the timing chain and at least one sprocket are removed together. In some engines, both sprockets have to be removed along with the timing chain.

The main bearing cap bolts are loosened and the bearing caps are removed. Make sure that the caps do not become interchanged or turned around. They only fit in one position. The crankshaft can then be lifted from the block. The camshaft on pushrod engines can be carefully slid out through the front of the block. It can be reached and supported through the block as it is removed. Take every precaution to avoid scratching the cam bearings with the camshaft as it is being removed. It is helpful to place the block on end so that the camshaft can be lifted straight up to remove it, as illustrated in Figure 9-3.

9-1 Crankshaft

All of the engine power is delivered through the crankshaft. The shaft must have the necessary shape and must be made from proper materials to meet these power demands.

Crankshaft Requirements. Each time combustion occurs, the force deflects the crankshaft as it transfers torque to the output shaft. This deflection occurs in two ways, to *bend* the shaft sideways and

Figure 9-1. Typical puller used to remove the crankshaft vibration damper.

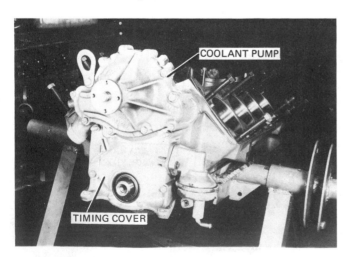

Figure 9-2. The coolant pump and timing cover must be removed before the crankshaft and camshaft can be removed from a pushrod engine.

Figure 9-3. Removing the camshaft by lifting it straight up.

to *twist* the shaft in torsion. The crankshaft must be rigid enough to keep the deflection low.

Crankshaft deflections are directly related to engine roughness. When back and forth deflections occur at the same vibrational *frequency* (vibrations per second) as another engine part, the parts will vibrate together. When this happens, the parts are said to *resonate*. These vibrations may become great enough to reach the audible level, producing a "thumping" sound. If this type of vibration continues, the part may fail.

Harmful crankshaft twisting vibrations are dampened with a torsional *vibration damper*. It is also called a *harmonic balancer*. This damper or balancer usually consists of a cast-iron *inertia ring* mounted to a cast iron *hub* with an *elastomer sleeve*. Two examples are shown in Figure 9-4. **Elastomers are synthetic-rubber-like materials.** The inertia ring size is selected to control the *amplitude* of the crankshaft vibrations for each specific engine model. **Amplitude is the strength of the vibration.**

Crankshaft Material and Manufacturing. Crankshafts used in high-production automotive engines may be either *forged* or *cast*. Forged crankshafts are stronger than the cast crankshaft, but they are more expensive. Casting materials and techniques have improved cast crankshaft quality so that they are used in most production automotive engines. Forged crankshafts have a wide separation line where the flashings have been ground off. Cast crankshafts have a fine line where the mold parted. The flashing and parting lines can be used for identification.

Forged Crankshaft. Forged crankshafts are made from SAE 1045 or a similar type of steel. The crankshaft is formed from a hot steel billet through a series of forging dies. Each die changes the shape of the billet slightly. The crankshaft blank is finally formed with the last die. The blanks are then machined to finish the crankshaft. Forging makes a very dense, tough crankshaft with the metal's grain structure running parallel to the principal direction of stress. Figure 9-5 shows a typical forged crankshaft with wide separation lines where the flashings have been removed.

Two methods are used to forge crankshafts. One method is to forge the crankshaft in place. This is followed by straightening. The *forging in-place* method is usually used with forged four- and six-cylinder crankshafts. A second method is to forge the crankshaft in a *single plane*. It is then twisted in the main bearing journal to index the throws at the

(a)

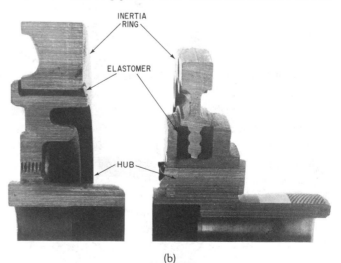

(b)

Figure 9-4. Crankshaft torsional vibration damper. (a) Front view and (b) section view of two different types of dampers.

Figure 9-5. Wide separation lines where the flashings have been removed from this forged crankshaft show that it has been twisted to index the crank throws.

desired angles. **Throws are the offset part of the crankshaft.** The amount of throw offset determines the piston stroke. The throw is one-half the stroke. Crankshaft blanks are straightened as part of the twisting process. The forged crankshaft design shape is limited by the die geometry and the forging parting lines. This means that forged crankshaft design must be a compromise between crankshaft requirements and manufacturing capabilities.

Cast Crankshaft. Cast automotive crankshafts may be cast in steel, nodular iron, or malleable iron. The major advantage of the casting process is that crankshaft material and machining cost are less than forging. The reason for this is that the crankshaft can be made close to the required shape and size, including all complicated counterweights. The only machining required on a carefully designed cast crankshaft is grinding bearing journal surfaces and finishing front and rear drive ends. Metal grain structure in the cast crankshaft is uniform and random throughout. Because of this, the shaft is able to handle loads from all directions. Counterweights on cast crankshafts are slightly larger than

counterweights on a forged crankshaft because the cast shaft metal is less dense and therefore somewhat lighter. The narrow mold parting surface lines can be seen on the cast crankshaft pictured in Figure 9-6.

Engines with forged crankshafts are usually balanced internally. When a cast crankshaft is used, some of the balancing is done on the damper, flywheel, and converter drive plate. Heavy vibration will occur if the wrong damper, flywheel, or converter drive plate are used. This vibration can lead to a broken crankshaft. The broken cast crankshaft shown in Figure 9-7 had the incorrect damper installed.

9-2 Crankshaft Design

The angle of the crankshaft throws in relation to each other are selected to provide smooth power output. The engine firing order is determined by the angle selected for the crankshaft throws. Counterweights are used to balance static and dynamic forces that occur during engine operation.

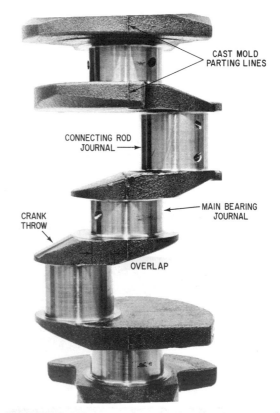

Figure 9-6. Cast crankshaft showing the bearing journal overlap and a straight narrow cast mold parting line.

Figure 9-7. A crankshaft broken as a result of using the wrong torsional vibration damper.

Six-Cylinder Crankshaft. Before the fuel shortage of 1973, two domestic engine designs were in common use; the inline six and the V-8. Crankshafts for these are shown in Figure 9-8. The inline six-cylinder engine has six crank throws in three matched pairs. **The throw is ground to make a crank pin. The smooth surface of the crank pin is called**

Figure 9-8. Crankshafts for a seven main bearing inline six engine (left) and a five main bearing V-8 engine (right).

the bearing journal. Each pair of throws on a six-cylinder engine is 120° from the other pairs. This causes one pair of pistons to reach top center each **120° of crankshaft rotation. Pistons in cylinders 1 and 6, 2 and 5, and 3 and 4 move together as pairs.** Each pair of pistons is 360° out of phase with its mate in the 720° four-stroke cycle. This arrangement gives smooth, low-vibration operation. There are even power strokes, one at each 120° of crankshaft rotation. This is illustrated in Figure 9-9. The crankshafts for these engines usually have one main bearing journal between each throw, making seven main bearings. There are some that have two throws between each main bearing, with four main bearings. The slant-six engine built by Chrysler is one of these.

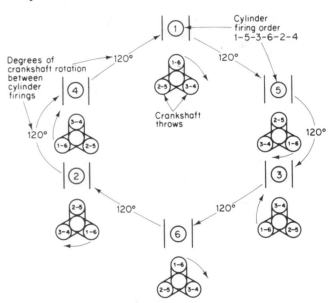

Figure 9-9. Even firing inline six cylinder crankshaft.

Eight-Cylinder Crankshaft. The V-8 engine has four inline cylinders in each of the two blocks placed at 90°. Each group of four inline cylinders is called a *bank*. The crankshaft for the V-8 engine has four throws. The connecting rods from two cylinders are connected to each throw, one from each bank. This can be seen in the cut away V-type engine pictured in Figure 9-10. This arrangement results in a minimum unbalanced condition. The V-8 crankshaft has two *planes,* so there is one throw every 90°. **A plane is a flat surface that cuts through the part.** These could be seen if the crankshaft were cut lengthwise through the center of the main bearing and crank pin journals. Looking at

Figure 9-10. Connecting rods from two opposite cylinders connect on the same crankshaft throws in V-type engines.

the front of the crankshaft with the first throw at 360° (up), the second throw is at 90° (to the right), **the third throw is at 270° (to the left), and the** fourth throw is at 180° (down). There is one main bearing journal between each throw, so there are five main bearings in a V-8 engine. In operation with this arrangement, one piston reaches top center each 90° of crankshaft rotation so that the engine operates smoothly with even firing at each 90° of crankshaft rotation. This can be seen in Figure 9-11.

Four-Cylinder Crankshaft. The majority of import engines have four cylinders inline. The crank-

shaft on these engines has four throws on a single plane. There is usually a main bearing journal between each throw, making a five-main-bearing crankshaft (Figure 9-12). Pistons move as pairs in this engine, too. Pistons in 1 and 4 cylinders move together and pistons 2 and 3 move together. Each piston in the pair is 360° out of phase with the other piston in the 720° four-stroke cycle. With this arrangement, the four-cylinder inline engines fires one cylinder each 180° of crankshaft rotation. This is illustrated in Figure 9-13. A four-cylinder opposed engine and a 90° V-4 engine have crankshafts that look like the four-cylinder inline crankshaft.

One of the ways the power of an engine can be changed is to change the number of cylinders. In 1961, Pontiac had need for a small engine for their compact automobile. They made this engine by casting an engine block having only one four-cylinder bank of their V-8 engine. This was called a

Figure 9-12. A five main bearing crankshaft for an inline four cylinder engine.

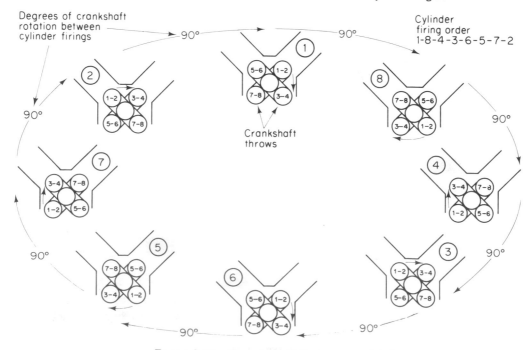

Figure 9-11. Even firing V-8 engine crankshaft.

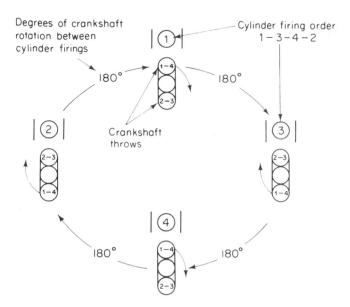

Figure 9-13. Even firing inline four cylinder engine crankshaft.

slant-four engine. The crankshaft was like other four-cylinder engines. It had a single plane with the four crank throws 180° apart. Five main bearings were used. In this way, they were able to build a four-cylinder engine using much of the machining on the V-8 engine production line. This engine also used the same pistons, rod, bearings, head, and valve train as the V-8 engine. In this way Pontiac made the four-cylinder engine at much less cost than a completely redesigned engine.

Five- and Three-Cylinder Engines. In Europe, some manufacturers had need for a more powerful engine than the power produced by the inline four. They found that the most economical way to make the more powerful engine was to add cylinders to the four-cylinder engine. A six-cylinder engine was too large, especially for front-wheel-drive automobiles, so they made a five-cylinder inline engine. This engine was made on the same production line as the four-cylinder engine. The engine used the same pistons, rods, bearings, and valve train as the four-cylinder engine. The inline five-cylinder engine has a five-throw crankshaft with one throw at each 72°. There are six main bearings used on this crankshaft. The piston in one cylinder reaches top center each 144° of crankshaft rotation. The throws were arranged to give a firing order of 1–2–4–5–3. Dynamic balancing was one of the major problems with this engine design. The vibration was satisfactorily balanced and isolated so that it was acceptable to the operator.

In Japan, Daihatsu developed a three-cylinder inline four-stroke cycle engine. It uses a 120° three-throw crankshaft with four main bearings. This engine requires a balancing shaft that turns at crankshaft speed, but in the opposite direction, to reduce the vibration to an acceptable level.

V-Six Cylinder Engine. In 1962, Buick introduced a V-6 engine with two banks of three cylinders. The banks were placed at 90° to each other so that the engine could be machined on the V-8 production line. An engine block of this type can be seen in Figure 9-14. Essentially, this engine was a V-8 with one cylinder removed from each bank. The 1962 90° V-6 engine used a three-throw crankshaft with four main bearings. The throws are 120° apart. Like typical V-type engines, each crank throw had two connecting rods attached, one from each bank. The object of this engine was to produce a smaller displacement engine and still use as much of the parts and the manufacturing equipment from the V-8 engine as possible. This V-6 engine design did not have even firing impulses because the pistons, connected to the 120° crankpins, did not reach top center at even intervals. The engine had a firing pattern of 150°–90°–150°–90°–150°–90°, as illustrated in Figure 9-15. This firing pattern produced unequal pulses that had to be isolated with engine mounts that were carefully designed. Even at that, the engine operated

Figure 9-14. A 90° V-6 engine block.

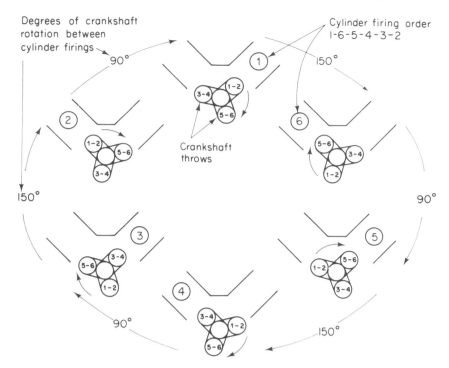

Figure 9-15. A diagram illustrating the even firing pattern produced by a crankshaft with the crank throws 120° apart operating in a 90° V-6 block.

rougher than either the inline six or V-8 engine. By 1965, the performance race was on, so this engine design was dropped and the manufacturing equipment sold.

Interest again turned to the V-6 engine during the fuel crisis in 1973. The original V-6 manufacturing equipment was for sale and Buick bought it back. The engine was updated and reintroduced in 1975. As vehicles were down-sized in the late 1970s, the V-6 engine was adapted to more luxurious automobiles. Because the customers for these automobiles had been accustomed to smoothly operating V-8 engines, Buick decided to modify the V-6 engine to make it even-firing. The even-firing 90° V-6 engines were introduced as a running change during 1977.

The crank throws for the even-firing V-6 engine were split, making separate crank pins for each cylinder. The split throw can be seen in Figure 9-16. The crank pins were made 0.25 in. larger in diameter to give adequate strength. The four main bearings were kept. The crank pin journals were split; one was moved 30° ahead. This angle between the crank pins on the crankshaft throws is called a *splay angle.* Figure 9-17 illustrates how the 30° splay angle allows even firing. A flange of 0.120 to 0.180 in. was left between the split crank pin journals. This provides a continuous fillet or edge for machining and grinding operations. It also provides a normal flange for the rod and bearing. This flange between the splayed crank pin journals is sometimes

Figure 9-16. Split crank pin journals splayed at 30° used in an even firing 90° V-6 block.

134

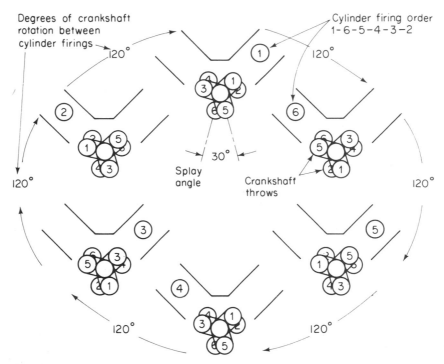

Figure 9-17. A diagram illustrating even firing of a 90° V-6 engine using a crankshaft with 30° splayed crank pin journals.

called a *flying web*. The even-firing, 90° V-6 engine has a firing order of 1–6–5–4–3–2.

In 1978, Chevrolet introduced a 90° V-6 engine. This engine was adapted from the small block Chevrolet V-8 by removing cylinders 3 and 6. They took a different approach than Buick to the crankshaft and vibration. The four main bearing

crankshaft of the 90° V-6 Chevrolet engine has an 18° splay angle between the adjacent crank pins. This design results in a firing pattern of 132°–108°–132°–108°–132°–108°. The firing pattern is illustrated in Figure 9-18. Some torque variation remains between power strokes but it is still 62% less than the vibration of a 90° V-6 engine

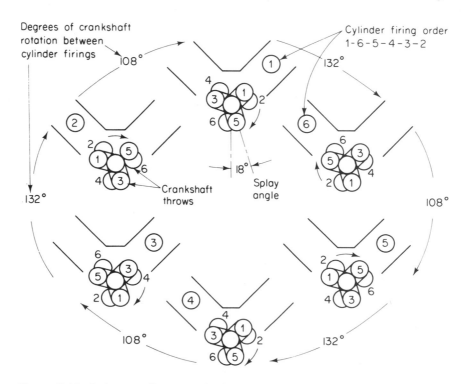

Figure 9-18. A diagram illustrating the firing pattern of a 90° V-6 engine using a crankshaft with 18° splayed crank pin journals.

135

having only three crank pins. The vibration from the Chevrolet 90° V-6 engine is satisfactorily isolated with properly designed engine mounts and accessory brackets.

The 90° V-6 engine is shorter and lighter than an inline six. It is, therefore, the best engine of the two for the lightweight standard front-engine, rear-wheel-drive automobiles. It is wider than it is long. This makes it too wide to be used as a transverse engine (crossways) in front-wheel-drive automobiles. The inline six is also too long to be used for a transverse front wheel drive. A large four-cylinder engine is also unsatisfactory for front wheel drives because its large torque impulses cause rough running operation. The obvious compromise was to make a 60° V-6 engine for transverse front-wheel-drive automobiles. This type of engine has the required power and it is narrow enough to be transverse mounted in a front-wheel-drive automobile. In 1960, General Motors produced a 60° V-6 engine for some of the smaller trucks. Ford imported one from Germany in the 1970s. In 1980, Chevrolet introduced a 60° V-6 engine for their newly designed lightweight front-wheel-drive intermediate-size automobile.

In many ways, the 60° V-6 engine is similar to the even-firing 90° V-6 engine. To be even-firing, the adjacent pairs of crank pins on the crankshaft used in the 60° V-6 engine have a splay angle of 60°. This design allows even firing as shown in Figure 9-19. With this large 60° splay angle, the flange or flying web between the splayed crank pins is made heavier than on crankshafts with smaller splay angles. This is necessary to give strength to the crankshaft. The crankshaft of the 60° V-6 engine also uses four main bearings.

9-3 Crankshaft Strength

Strength or torsional stiffness is one of the most important crankshaft design requirments. The crankshaft is made strong by using materials with the correct physical properties. It is designed to have a large bearing journal and crank cheek size. Stress concentration is minimized through large fillets and properly placed lightening holes. Main and rod crank pins overlap each other as shown in Figure 9-20. The overlap increases crankshaft strength because more of the load is carried through the overlap area rather than through the fillet and

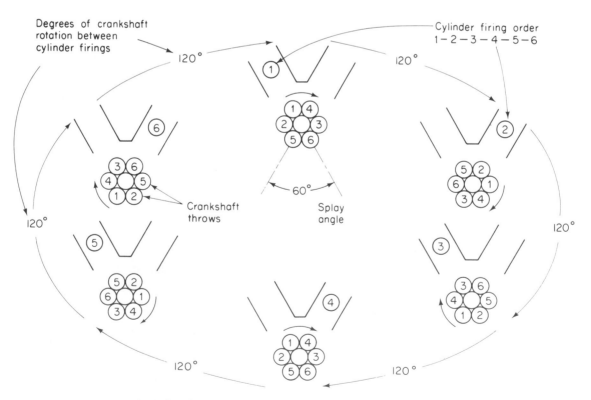

Figure 9-19. A diagram showing the even firing pattern of a 60° V-6 engine using a crankshaft with 60° splayed crank pin journals.

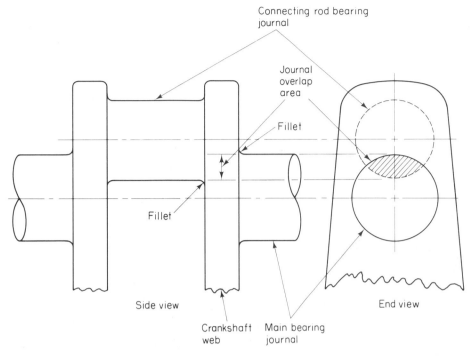

Figure 9-20. A drawing showing the main and rod crank pin bearing journal overlap.

crankshaft web. This is especially true for crankshafts having splayed crank pins.

Stress tends to concentrate at oil holes drilled through the crankshaft journals. A typical oil hole can be seen in the cut crankshaft pictured in Figure 9-21. These holes are usually located where the crankshaft loads and stresses are the lowest. The oil holes lead the top center of the crank pin by approximately 80°. Straight drilling eliminates places where dirt could become trapped. The edges of the oil holes are carefully chamfered to relieve as much stress concentration as possible. Chamfered oil holes are shown in Figure 9-22.

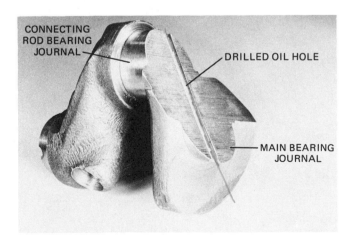

Figure 9-21. Typical oil hole drilling through the crankshaft between the main bearing journal and the connecting rod bearing journal.

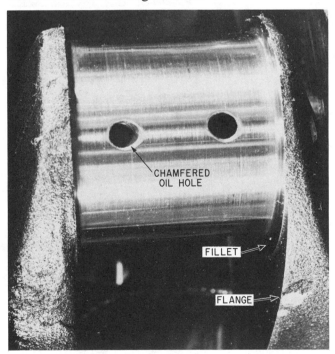

Figure 9-22. A typical chamfered hole in a crankshaft bearing journal.

Lightening holes in the crank pins do not reduce their strength if the hole size is less than half of the crank pin diameter. Lightening holes will often increase crankshaft strength by relieving some of the natural stress in the crankshaft. The hole in the center of the crank throw is used for balancing, by controlling the hole depth (Figure 9-23).

Crankshaft Balance. Most crankshaft balancing is done during manufacture. Holes for balance are drilled in the counterweight to lighten them. Sometimes, these holes are drilled after the crankshaft is installed in the engine. Some manufacturers are able to control their casting quality so closely that counterweight machining for balancing is not necessary. Engines with cast crankshafts usually have some external balancing. External balance of these engines is accomplished by adding weights to the damper hub and to the flywheel or automatic transmission drive plate. Typical methods of adding balance weights are shown in Figure 9-24.

Thrust Surface. Automatic transmission pressure in the torque converter and clutch release forces tend to push the crankshaft toward the front of the engine. Thrust bearings in the engine will support thrust loads as well as maintain the crankshaft position. Smooth thrust-bearing journal surfaces are ground on a small boss located on the crankshaft cheek next to one of the main bearing journals (Figure 9-25). One main bearing has thrust bearing flanges that ride against these thrust bearings. Thrust bearings may be located on any one of the main bearing journals.

Bearing Journals. It has been found through experience that journal polishing direction is very im-

(a)

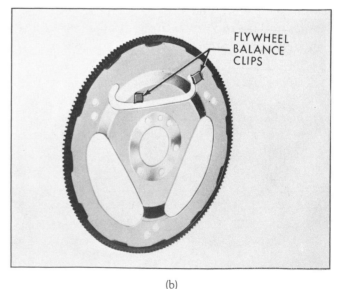

(b)

Figure 9-24. External balance weights added to the vibration damper (a) and flywheel balance clips to the torque converter drive plate (b) (Courtesy of Buick Motor Division, General Motors Corporation).

Figure 9-23. A balance hole drilled in a crank pin.

Figure 9-25. Thrust bearing located on one of the crankshaft main bearings.

Pushrod engines have the cam located in the block. They are smaller and lighter than overhead cam engines. The overhead cam engine, on the other hand, has fewer moving parts in the valve train, so they can operate at high speed. With fewer moving parts, the valve train is more rigid. Because of this, the valve action follows the cam more closely. The valves can open faster than with pushrods, so the engine can operate with less valve overlap. This allows the high-speed engine to idle more smoothly than a pushrod engine that has high valve overlap. The overhead cam engine operates the valves directly, through pivoting finger followers or through an overhead lifter and rocker arm assembly.

Dual overhead cams allow the intake and exhaust valves to have different angles for combustion chambers that are designed for power. At the same time, they allow the valves to be operated directly by the cam to minimize flexibility in the valve train. Dual overhead cams are only used for high-performance engines. They have more complex and noisy valve trains than do single overhead cams. This makes dual overhead cam engines heavy and expensive. They generally have poor low-speed performance.

portant. The bearing will last much longer when the journal is polished against the direction of normal rotation. This can be illustrated by realizing that the surface finish left by grinding has slightly bent whiskers or fuzz like the teeth of a very fine file. It feels smooth when the shaft turns with the direction of the teeth, but it acts like a fine milling cutter when the direction of rotation is toward the teeth. Polishing removes this fuzz.

There are many crankshaft designs. Engineers base their selection on previous crankshaft performance and experience, modified by cost considerations. New understandings in metallurgy and manufacturing techniques, along with a better understanding of the load requirements, will continue to lead to better crankshafts.

9-4 Camshaft

The second rotating shaft is the camshaft. Its major function is to operate the valve train. Cam shape or *contour* is the major factor in the operating characteristics of the engine.

Camshaft Requirements. The camshaft is timed to the crankshaft so that the valves are opened and closed in relation to crankshaft angle and piston position. This is illustrated as a graph in Figure 9-26. The engine will have maximum volumetric efficiency at the engine speed selected with the proper valve shape and timing. Cam lobe shape has more control over engine performance characteristics than does any other single engine part. Engines identical in every way except cam lobe shape may have completely different operating characteristics

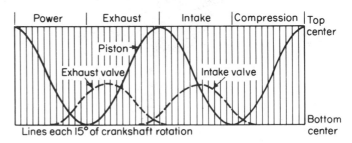

Figure 9-26. Valve opening related to the position of the piston throughout the four stroke cycle. Note that the valve is fully open when the piston is moving most rapidly.

and performance. Two cam shapes for a small block Chevrolet V-8 are shown in Figure 9-27.

The camshaft is driven by the crankshaft through gears, sprockets and chains, or sprockets and timing belts. Timing chains are not as wide as timing belts, so engines with timing chains can be shorter. Timing chains often have dampers pressing on the unloaded side of the chain. The damper pad is a Nylatron molding that is filled with molybdenum disulfide to give it low friction. The damper is held against the chain by either a spring or hydraulic oil pressure (Figure 9-28). The gears or sprockets are keyed to their shafts so that they can be installed in only one position. The gears and sprockets are then indexed together by marks on the gear teeth or chain links. When the crankshaft and camshaft timing marks are properly lined up, the cam lobes are indexed to the crankshaft throws of each cylinder so that the valves will open and close correctly in relation to the piston position.

As the camshaft lobe pushes the lifter upward against the valve spring force, a backward twisting force is developed on the camshaft. After the lobe goes past its high point, the lifter moves down the back side of the lobe. This makes a forward twisting force (Figure 9-29). This action produces an alternating torsion force forward, then backward at each cam lobe. These alternating torsion forces are multiplied by the number of cam lobes on the shaft. the camshaft must have sufficient strength to minimize torsion twist. It must also be tough enough to minimize fatigue from the alternating torsion forces.

Most valve trains use a spherical lifter face, 50

DAMPER HELD WITH A SPRING

DAMPER HELD WITH OIL PRESSURE

Figure 9-28. Timing chain dampers.

Figure 9-27. The shape of two small block Chevrolet V-8 cam lobes. A standard cam is on the left and a performance cam is on the right.

140

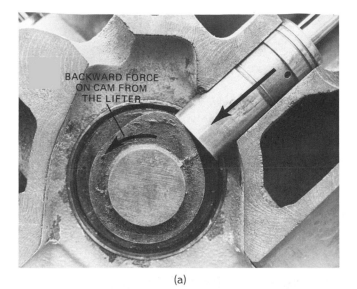

(a)

(b)

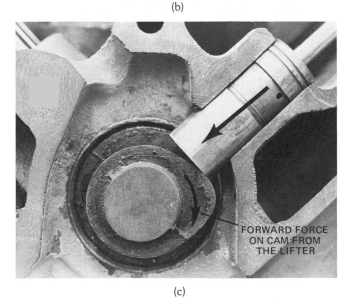

(c)

Figure 9-29. Lifter contact on the cam lobe. (a) The cam lobe is beginning to lift the lifter, (b) the lifter is fully raised, and (c) the lifter is lowering.

to 80 in. (1270 to 1432 mm) in radius that slides against the cam lobe. This produces a slightly convex surface on the lifter face. It contacts the lobe slightly off center. This produces a small turning force on the lifter to cause some lifter rotation for even wear. In operation, there is a wide line contact between the lifter and the high point of the cam lobe. These are the highest loads that are produced in an engine. The lifter contact on the top of the cam lobe can be seen in Figure 9-30. A great amount of design effort has gone into the metallurgy, heat treatment, design, and lubrication of the cam-to-lifter contact surface. This surface is the most critical lubrication point in an engine.

Camshaft Materials. Most automotive camshafts are made from hardenable alloy cast iron. It resists wear and provides the required strength. The very hardness of the camshaft causes them to chip through edge loading or through careless handling. An example of this is shown in Figure 9-31.

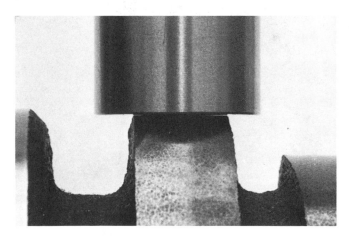

Figure 9-30. Typical lifter contact on the top of the cam lobe.

CHIPPED CAM LOBE

Figure 9-31. A hardened camshaft that was chipped through careless handling by students in an auto shop.

141

Some heavy-duty engine camshafts are made of steel. These must have case-hardened journals and lobes to give them the durability required. Steel camshafts are also required in engines that use roller lifters.

The crankshaft gear or sprocket that drives the camshaft is usually made of sintered iron. When gears are used on the camshaft, the teeth must be made from a soft material to reduce noise. Usually, the whole gear is made of aluminum or fiber. When a chain and sprocket are used, the camshaft sprocket may be made of iron or it may have an aluminum hub with nylon teeth for noise reduction. Figure 9-32 shows each of these sprocket types. The

timing chain is either a silent chain or roller chain. They are pictured in Figure 9-33.

Camshaft Design Features. The camshaft is a one-piece casting with lobes, bearing journals, drive flanges, and accessory gear blanks close to finished size. The drive end is finished first so that the cam lobes will be correctly indexed to the proper angle. The accessory drive gear is finished with a gear cutter. The lobes and journals are ground to the proper shape. The remaining portion of the camshaft surface is not machined.

On pushrod engines, camshaft bearing journals must be larger than the cam lobe so that the camshaft can be installed in the engine through the cam bearings. Some overhead cam engines have bearing caps on the cam bearings. These cams can have large cam lobes with small bearing journals. Cam bearings on some engines are progressively smaller from the front journal to the rear. Other engines use the same-size camshaft bearing on all of the journals.

Some engines transfer lubrication oil from the main oil gallery to the crankshaft around the camshaft journal or around the outside of the camshaft bearing. One example of a passage around the outside of the bearing is shown in Figure 9-34. Cam bearing clearance is critical in these engines. If the clearance is too great, oil will leak out and the crankshaft bearings will not get enough oil. Other engines use drilled holes in the camshaft bearing journals to meter lubricating oil to the overhead rocker arm. Oil goes to the rocker arm each time the

Figure 9-32. Two types of sprockets that can be used on the same engine. A cast iron sprocket is on the left and an aluminum-nylon sprocket is on the right.

Figure 9-33. Timing chain types. A silent chain is on the left and a roller chain is on the right.

Figure 9-34. A groove in the camshaft bearing bore forms a passage for oil to flow around the outside of the cam bearing.

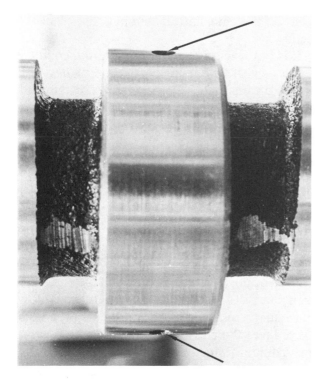

Figure 9-35. A hole through a camshaft bearing journal. The hole meters oil to a rocker shaft when it indexes oil passages in the cam bearing.

Figure 9-36. A typical thrust plate located between the cam gear and a flange on the camshaft.

Figure 9-37. Oil pump turning effort produces end thrust on the camshaft.

holes index between the bearing oil gallery passage and the outlet passage to the rocker arm. Camshaft oil metering holes are shown in Figure 9-35.

Each camshaft must have some means to control the shaft end thrust. Two methods are in common usage. One method is to use a *thrust plate* between the camshaft drive gear or sprocket and a flange on the camshaft (Figure 9-36). This thrust plate is attached to the engine block with cap screws. A second method is to use the thrust developed by the *oil pump turning effort* to hold the camshaft into the block (Figure 9-37). A flange on the back of the camshaft drive gear or sprocket rides against the front of the block. This keeps the camshaft from moving backward into the engine. In a few camshafts, a button, spring, or retainer that contacts the timing cover limits forward motion of the camshaft. It is only useful when the engine is accidentally rotated backward.

An eccentric cam lobe for the fuel pump is often cast as part of the camshaft. The fuel pump is operated by this eccentric with a long pump arm or pushrod. Some engines use a steel cup-type eccentric that is bolted to the front of the cam drive gear. This allows a damaged fuel pump eccentric to be replaced without replacing an entire camshaft. it also places the fuel pump well forward on the engine, away from the exhaust. Here it will be in the cool air coming into the front of the vehicle. This helps reduce the chance of vapor lock in the fuel pump and lines. **Vapor lock is evaporating gasoline from heat on the inlet side of the fuel pump.** When vapor is pumped rather than liquid fuel, the engine will not run. Typical fuel pump eccentrics are identified on a number of camshafts pictured in Figure 9-38.

Camshaft materials and manufacturing methods follow standard industrial practices. The cam lobe shape greatly affect an engine's performance. These shapes are the most critical camshaft design features and require very careful machining.

9-5 Auxiliary Shafts

Pushrod engines operate all of the accessories from either the crankshaft or camshaft. External engine accessories are driven by belts from a crankshaft pulley on the front of the engine. Inside the engine, the oil pump, fuel pump, and distributor are usually driven by the camshaft at one-half the crankshaft speed.

It is not as easy to drive the internal engine accessories with the camshaft on engines using overhead camshafts. These engines often use a small auxiliary shaft. Sometimes, this shaft is called a jack shaft. It is driven by the timing belt or timing chain. Figure 9-39 shows a typical auxiliary shaft in the engine block.

Some engines use balance shafts to dampen normal engine vibrations. **Dampening is reducing the vibration to an acceptable level.** A balance shaft turning at crankshaft speed but in the opposite direction is used on a three-cylinder inline engine. Weights on the ends of the balance shaft move in a direction opposite to the direction of the end piston. When the piston goes up, the weight goes down, and when the piston goes down, the weight goes up. This reduces the end-to-end rocking action on this three-cylinder inline engine.

Another type of balance shaft is designed to

(a)

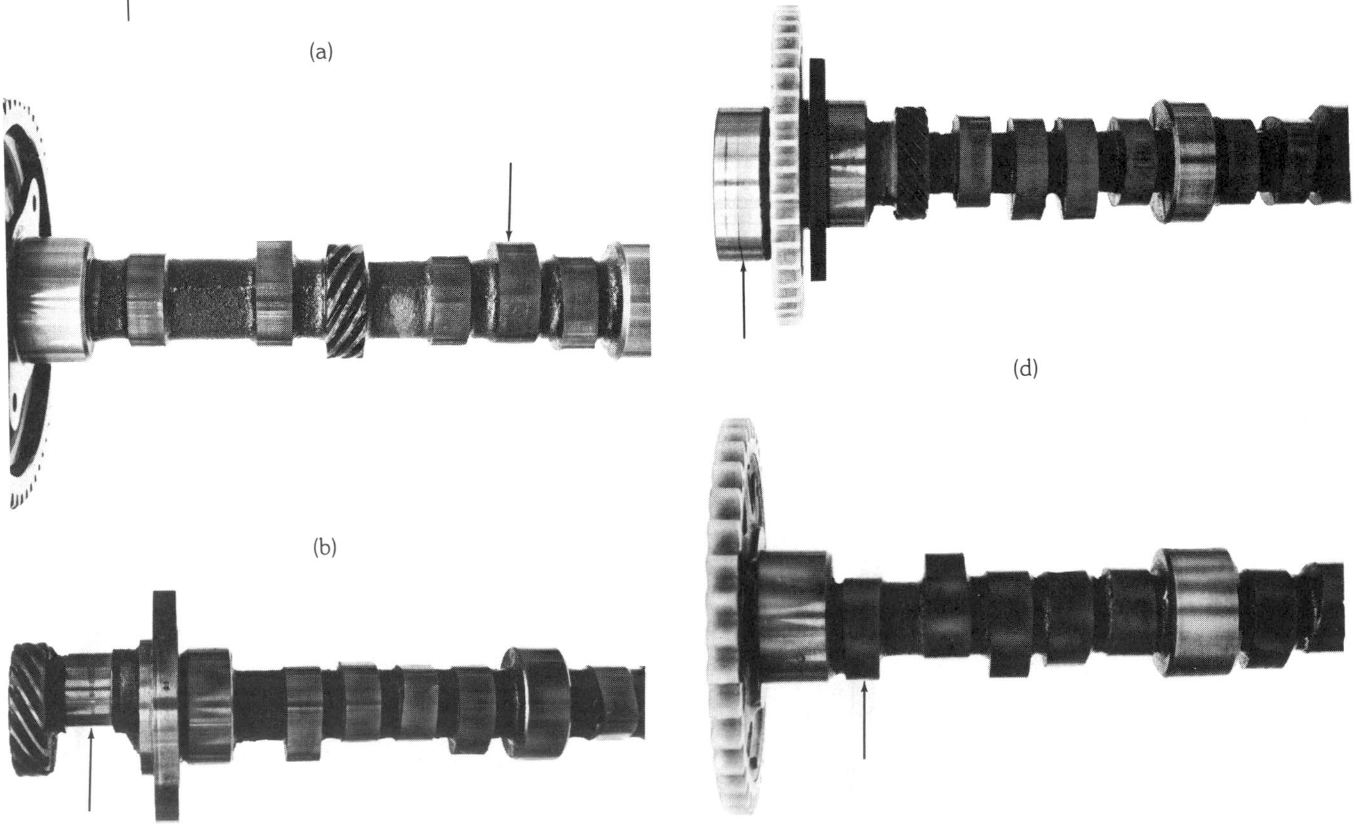

(b)

(c)

(d)

(e)

Figure 9-38. Typical fuel pump eccentric locations on camshafts used in pushrod engines.

Figure 9-39. A typical auxiliary shaft.

counterbalance vibrations on a four-stroke, four-cylinder engine. Two shafts are used and they turn at *twice* the engine speed. One of the shafts turns in the same direction as the crankshaft and the other turns in the opposite direction. The oil pump gears are used to drive the reverse turning shaft. Counterweights on the balance shafts are positioned to oppose the natural rolling action of the engine as well as the secondary vibrations caused by the piston and rod movements. This design is shown in Figure 9-40.

Auxiliary shafts are used only where they are

necessary. They increase the complexity of the shaft drives with a possible increase in noise. They increase the engine weight and manufacturing cost. They also make engine service more complex.

9-6 Engine Bearings

Engine durability relies on bearing life. Bearing failure usually results in immediate engine failure.

Bearing Requirements. Engine bearings are designed to support the operating loads of the engine and, with the lubricant, provide minimum friction. This must be done at all designed engine speeds. The bearings must be able to operate for long periods of time, even when small foreign particles are in the lubricant.

Most engine bearings are *plain* or *sleeve bearing* types. They need a constant flow of lubricating oil. Roller, ball, and needle bearings, which are called *antifriction bearings,* are used where only minimum lubrication is available. Properly lubricated plain bearings cause no more friction than do the antifriction bearings. This results from the fact that the shaft is actually rolling on a film of oil. In automotive engines, the lubricating system supplies oil to each bearing all the time that the engine runs.

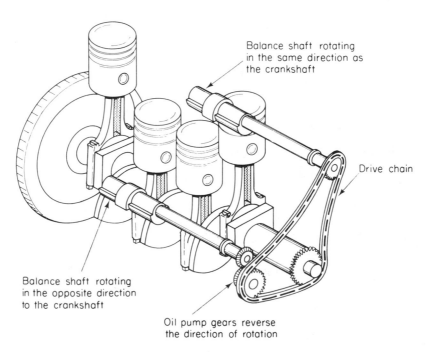

Figure 9-40. Two counter-rotating balance shafts used to counterbalance the vibrations of a four cylinder engine.

Only residual oil remaining from the last run will be on the bearing during engine start before the pressure builds up. During startup, the oil film will be borderline. **The borderline oil film is so thin that the high spots of the shaft and bearing will actually contact each other.** This results in high friction and wear. After the oil film forms, the metal-to-metal contact stops and friction drops, to stop bearing wear. Bearing and journal *only* wear when the parts come in contact with each other or when foreign particles are present.

It is important that the engine have bearings large enough so that the bearing load is within strength limits of the bearing. Bearing load capacity is calculated by dividing the bearing load in pounds by the projected area of the bearing. The projected area is the bearing length multiplied by bearing diameter. The load on engine bearings is determined by developing a polar bearing load diagram which shows the amount and the direction of the instantaneous bearing loads. Bearing load diagrams look similar to drawings shown in Figure 9-41.

Opposing objectives are at work in the selection of bearings. The automobile designers want the engine to develop the greatest power and economy and, at the same time, they want to make the smallest size and the lightest engine possible. This requires the use of small bearings with high bearing loads. As greater bearing loads are applied, bearing life is reduced, unless a higher-quality, more-expensive bearing is installed. To keep costs down, one of the major design objectives of bearing engineers is to select the lowest-cost bearing that will meet the engine's needs.

Bearing Performance Characteristics. Bearings tend to flex or bend slightly under changing loads. This is especially noticeable in reciprocating engine bearings. Bearing metals, like other metals, tend to fatigue and break after being flexed or bent a number of times. Flexing starts fatigue, which shows up as fine cracks in the bearing surface. These cracks gradually deepen almost to the bond between the bearing metal and the backing metal. The cracks then cross over and intersect with each other as illustrated in Figure 9-42. In time, this will allow a piece of bearing material to fall out. The length of time before fatigue will cause failure is called the *fatigue life* of the bearing. Bearings must

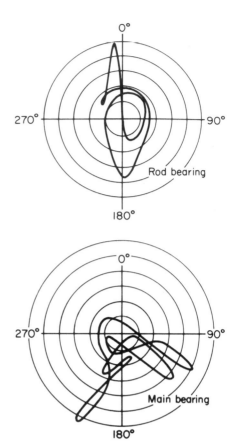

Figure 9-41. Typical rod and main bearing load diagrams. The circles on these polar diagrams indicate the amount of force on the bearing as it rotates.

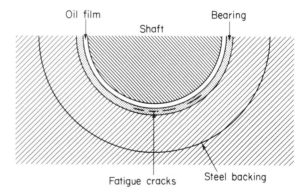

Figure 9-42. The shape of typical fatigue cracks in a bearing.

have a long fatigue life for normal engine service. The harder the bearing material, the longer is its fatigue life. Soft bearings have a low fatigue life and low bearing load strength. They are generally low in cost and can only be used where the bearing requirements are low.

Costs increase as the manufacturing tolerance

becomes closer to the exact design size. To enable manufacturers to build engines with economical manufacturing tolerances, the bearings must have the ability to conform or change their shape slightly. This is necessary to match small variations in the shaft surface. The ability of bearing materials to creep or flow slightly to match shaft variations is called *conformability*. The bearing conforms to the shaft during the engine break-in period. In modern automobile engines, there is little need for bearing conformability or break-in, because automatic processing has held machining tolerances very close to the designed size.

Engine manufacturers have designed engines so as to produce minimum crankcase deposits. This was done by providing them with oil filters, air filters, and closed crankcase ventilation systems that minimize contaminants. Still, some foreign particles get into the bearings. The bearings must be capable of embedding these particles into the bearing surface so that they will not score the shaft. To fully embed the particle, the bearing material gradually works across the particle, completely covering it. This bearing property is called *embedability*. Embedability is illustrated in Figure 9-43.

Under some operating conditions, the bearing will be temporarily overloaded. This will cause the oil film to break down, allowing the shaft metal to come in contact with the bearing metal. As the rotating crankshaft contacts the bearing high spots, the spots become hot from friction. The friction will cause localized hot spots in the bearing material that seize or weld to the crankshaft. The crankshaft then pulls the particles of materials around with it, scratching or scoring the bearing surface. Bearings have a characteristic called *score resistance*. It prevents the

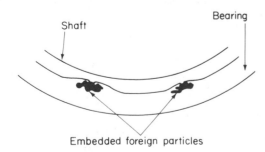

Figure 9-43. Bearing material covers foreign material as it embed in the bearing.

bearing materials from seizing to the shaft during oil film breakdown. This bearing characteristic is the result of relatively low melting temperature bearing material.

Modern motor oils contain a small amount of chemical additives that add characteristics to the oil to satisfy special engine requirements. In time, under high engine temperatures and high bearing loads, the additives break down. They combine with the by-products of combustion and form acids in the oil. The bearings' ability to resist attack from these acids is called *corrosion resistance*. Corrosion can occur over the entire surface of the bearing. This will remove material and increase the oil clearance. It can also leach or eat into the bearing material, dissolving some of the bearing material alloys. Either type of corrosion will reduce bearing life.

Bearing Materials. Three bearing materials are used for automobile engine bearings: babbitt, copper–lead, and aluminum. A 0.010 to 0.020 in. (0.25 to 0.50 mm)-thick layer of the bearing materials is applied over a low-carbon-steel backing. The steel backing with a surface coating of bearing material formed as an engine bearing is called a bearing *shell*. The steel provides support needed for the shaft load. The bearing material meets the rest of the bearing operating requirements.

Babbitt is the oldest automotive bearing material. Its base is either lead or tin. This is alloyed with small quantities of copper and antimony to give it the required strength. Babbitt is still used in applications where soft material is required for soft shafts running under moderate loads and speeds. It will work with occasional borderline lubrication and oil starvation without failure.

Copper–lead is a stronger and more expensive bearing material than babbitt. It is used for intermediate- and high-speed applications. Tin, in small quantities, is often alloyed with the copper–lead bearings. This bearing material is most easily damaged by corrosion from acid accumulation in the engine oil. Corrosion results in bearing journal wear as the bearing is eroded by the acids.

Aluminum is the newest of the three bearing materials to be used for automotive bearings. Automotive-bearing aluminum has small quantities of tin and copper alloyed with it. This makes a stronger but more expensive bearing than either babbitt or copper-lead.

Aluminum, with a small percentage of lead, is used for high-quality intermediate-strength bearings. Most of its bearing characteristics are equal to or better than babbitt and copper-lead. Aluminum bearings are well suited to high-speed, high-load conditions.

Because of its expense, aluminum is often used along with bearings made from other bearing materials. For example, aluminum bearings may be used for the highly loaded lower shell of the main bearing, with babbitt being used for the lightly loaded upper shell of the main bearing on a single main bearing journal.

Bearing Manufacturing. Modern automotive engines are *precision insert-type* bearing shells. The bearing is manufactured to very close tolerances so that it will fit correctly in each application. The bearing, therefore, must be made from accurate materials under closely controlled manufacturing processes. Figure 9-44 shows the typical bearing shell types found in modern engines.

Most of the precision insert bearing shells are manufactured in a continuous-strip process. The low-carbon-steel backing is delivered to the bearing manufacturer in a roll. This steel must be within 0.001 in. (0.025 mm) of the thickness required. In processing, it is cleaned, flattened, and heated to the required bonding temperature.

The bearing material is applied to the steel strip in either of two ways, casting or sintering. In the *casting* process, melted bearing alloy is poured on the backing strip, where it bonds to the steel as it cools. The *sintering* process is similar. Fine particles of the bearing materials are mixed as a powder. The powdered bearing material is spread evenly on the continuous steel backing strip. It is then pressed and heated until it fuses together and bonds to the steel. Bonds of both processes are chemical rather than mechanical. The finished strip is cut into bearing blanks as it leaves this production line. The blanks are formed and coined to size in presses, then punched and machined to the final shape of the bearing shell.

Many of the copper-lead and aluminum bearings have an *overlay* or third layer of metal. This overlay is usually babbitt. Babbitt overlayed bearings have high fatigue strength, good conformity, good embedability, and good corrosion resistance. The overplated bearing is a premium bearing. It is also the most expensive because the overplating layer, from 0.0005 to 0.001 in. (0.0125 to 0.025 mm) thick, is put on the bearing with an *electroplating* process. The layers of bearing material on a bearing shell are illustrated in Figure 9-45.

Overplate reduces bearing damage by cushioning the journal during the first few break-in hours of running. Once the bearing has conformed to the bearing journal, it will have a satisfactory life even when the overplate is gone.

Bearing Design. The physical design of the bearing must consider the loads being applied to the journal. In automotive engines, the load varies in strength and direction. Maximum bearing areas must be located where the forces or loads are the greatest. Oil holes and grooves are located on the lightly loaded areas of the bearing.

Oil enters the bearing through the oil holes and grooves. It spreads into a smooth wedge-shaped oil film that supports the bearing load by hydro-

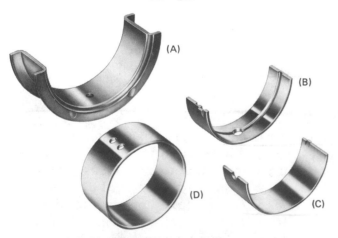

Figure 9-44. Typical bearing shell types found in modern engines. (Courtesy of Sealed Power Corporation).

Figure 9-45. Layers of bearing material on different types of bearing shells (Courtesy of Sealed Power Corporation).

dynamic action of the oil. Under high journal speeds, during high-rpm engine operation, the oil film may no longer be able to maintain its laminar or smooth layer flow. When laminar flow breaks down, turbulent flow occurs and breaks up the bearing oil film. Turbulent flow lubrication leads to bearing failure.

Many bearings have *oil bleed holes* so that the bearing will continue to be supplied with fresh oil. This oil is used for hydrodynamic bearing lubrication and to cool the bearing. Often, the oil bleed hole is made to aim oil at the cylinder wall for lubrication. When it does, the hole is called a spit hole. This was discussed in Section 8-7.

A bearing design has been developed that does not require an oil bleed hole. The bearing is of an *eccentric* design, illustrated in Figure 9-46. **Eccentric means that the inside and outside of the bearing do not have the same circle center.** This bearing has close clearances on the highly loaded bearing areas. The lightly loaded areas have larger clearances for the oil film to develop. At the same time, the larger clearance allows oil to flow from the bearing edges for proper cooling.

The bearing-to-journal clearance may be from 0.0005 to 0.0025 in. (0.025 to 0.060 mm), depending on the engine. Doubling the journal clearance will allow more than *four* times as much oil to flow from the edges of the bearing. The oil clearance must be large enough to allow an oil film to build up, but small enough to prevent excess oil leakage that would cause loss of oil pressure. A large oil leakage at one of the bearings would starve other bearings farther along in the oil system. This would result in failure of the oil-starved bearings.

The bearing design also includes bearing spread and crush. They are illustrated in Figure 9-47. The bearing shell has a slightly larger arc than does the bearing housing. This is called *bearing spread* and is from 0.005 to 0.020 in. (0.125 to 0.500 mm) wider than the housing bore. A lip or *tang* locates the bearing endwise in the housing. The tang can be identified in Figure 9-48. Spread holds the bearing shell in the housing while the engine is being

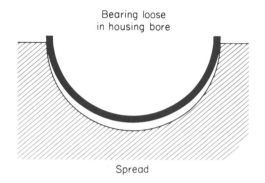
Bearing loose in housing bore

Spread

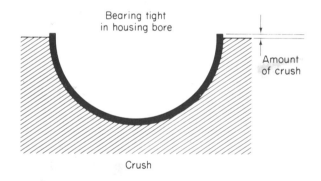
Bearing tight in housing bore

Amount of crush

Crush

Figure 9-47. Bearing spread and crush.

Figure 9-46. Eccentric bearing design (exaggerated).

Figure 9-48. Tang on a bearing used to properly locate the bearing during assembly.

149

assembled. When installed, each end of the bearing shell is slightly above the parting surface. When the bearing cap is tightened, the ends of the two bearing shells touch and are forced together. This force is called *bearing crush*. Crush holds the bearing in place and keeps the bearing from turning when the engine runs. Crush must exert a force of at least 12,000 lb/in.^{2} (82,740 kPa) stress at 250 °F (121 °C) to hold the bearing securely in place. A stress of 40,000 lb/in.^{2} (275,790 kPa) is considered max-

imum without damaging the bearing or housing.

The engineer will select the least expensive bearing that will perform satisfactorily in operation. Replacement bearings should be as good or of better quality than the original bearings. The replacement bearings must also have the same oil holes and grooves. Modified engines have different bearing requirements and therefore usually require a higher-quality bearing to provide satisfactory service.

REVIEW QUESTIONS

1. List the order in which parts are removed to take the crankshaft and the camshaft from the engine. [INTRODUCTION] *DAMPER, WATER PUMP, Timing Chain, Sprocket*

2. What precautions must be observed when removing the crankshaft? [INTRODUCTION] *Be careful not To Dent or nicked IT*

3. Why do most engines use a torsional vibration damper? [9-1] *To Eliminate VIBRA*

4. What is an advantage and a disadvantage of a cast iron crankshaft as compared to a forged steel crankshaft? [9-1] *Steal is lighter and Stronger*

5. What are identifying features of cast and forged crankshafts? [9-1] *Forge have a wide Separation line*

6. What is the difference in balancing cast and forged crankshafts? [9-1]

7. How many degrees of crankshaft rotation between cylinder firings are there on an inline four-, six-, and eight-cylinder engines? [9-2] *720°*

8. Why did some manufacturers decide to build three- and five-cylinder engines? [9-2] *For more Power*

9. How many degrees of crankshaft rotation between cylinder firings are there on a 90° V-6 engine with a three-throw crankshaft? [9-2] *120°*

10. What is a crank pin with a splay angle? [9-2]

11. What changes were made to make an even firing 90° V-6 engine? [9-2] *Diff order of Fire*

12. How does a 60° V-6 engine differ from a 90° V-6? [9-2]

13. What crankshaft design features give the crankshaft the strength it needs? [9-3]

14. How are crankshafts balanced? [9-3] *Holes ARE Drilled a The Counter*

15. What puts thrust loads on a camshaft? [9-4] *Push Rods*

16. What is the function of the camshaft? [9-4] *Valves*

17. What are the advantages and disadvantages of an overhead camshaft engine compared to a pushrod engine? [9-4]

18. Why don't passenger cars have dual overhead camshafts? [9-4] *Heavy Expensive Noisy*

19. What engine part has the most control over the engine performance characteristics? [9-4] *CAM SHAFT*

20. What are the advantages and disadvantages of timing belts compared to timing chains? [9-4] *Easy to Replace*

21. Where are the highest loads produced in an engine? [9-4] *Low Rings*

22. Where is the most critical lubrication point in an engine? Why is it the most critical? [9-4]

23. When is steel used as the camshaft material? [9-4] *DURABILITY on CAM SHAFT THAT uses Roller Lifter and Heavy Duty Engine*

[handwritten: CAST STEEL]

24. What materials are used for the crankshaft and camshaft timing gears? [9-4]

25. Which parts of a camshaft are machined? [9-4] *[handwritten: LOBES and Journals]*

26. When is the camshaft bearing oil clearance critical? [9-4]

27. What two methods are used to control camshaft end thrust? [9-4]

28. What is the advantage of having a fuel pump placed well forward on the engine? [9-4] *[handwritten: IT CAN Be REPLACED WITHOUT TAKing out CAMSHAFT]*

29. When are auxiliary shafts used in engines? [9-5] *[handwritten: on over Head cam SHAFTS]*

30. How does the required oil supply differ between plain bearings and anti-friction bearings? [9-6] *[handwritten: Plain needs lot of oil, AntiFriction a minimun]*

31. On what basis does an engine designer select engine bearings? [9-6] *[handwritten: DURABiLiTy]*

32. What causes bearing fatigue? [9-6] *[handwritten: BENDing ad FLEXING]*

33. How does fatigue show up in bearings? [9-6] *[handwritten: CRACKS]*

34. How does hardness affect the fatigue life of the bearing? [9-6] *[handwritten: IT LAST LONGER]*

35. Under what conditions is bearing conformability important? [9-6] *[handwritten: DURING ENGINE BREAKIN]*

36. What has been done to engines to minimize crankcase deposits? [9-6]

37. When is the bearing embedability characteristic important? [9-6] *[handwritten: To FULLY emBed Particale]*

38. Under what conditions must the bearing be score resistant? [9-6] *[handwritten: in to THE Bearing]*

39. What are two types of bearing corrosion? [9-6] *[handwritten: ACID - an ADDITIVE Break Down BECause of Heath]*

40. What is a bearing shell? [9-6] *[handwritten: DIFFERENt LAYERS of DIF metals]*

41. When is babbitt used as a bearing material? [9-6] *[handwritten: Under]*

42. When is copper-lead used as a bearing material? [9-6] *[handwritten: Hi Speed ADD LOADS]*

43. When is aluminum used as a bearing material? [9-6]

44. What type of bearing material is most easily damaged by corrosion? [9-6]

45. What is meant by a precision insert-type bearing shell? [9-6] *[handwritten: A very close Tollerance]*

46. What are the methods used to put bearing material on a steel backing? [9-6] *[handwritten: CASTin or SinTERing]*

47. Why is an overlay put on some bearing shells? [9-6] *[handwritten: ELECTRoPLaTing]*

48. How does lubricating oil get between the bearing and journal? [9-6] *[handwritten: OIL Holes are on the lightly loade side of the Bearing]*

49. When does the oil flow in the bearing oil film become turbulent? [9-6]

50. What is the purpose of a bearing bleed hole? [9-6] *[handwritten: used to Lubricate cylinder walls]*

51. What happens to the amount of oil flow from a bearing if the oil clearance is doubled? [9-6] *[handwritten: IT will STARVE OTHER Bearing DOwn THE Line]*

52. What causes oil starved bearings when there is oil pressure? [9-6]

53. What is the importance of bearing spread and crush? [9-6]

[handwritten: SPread The Bearing SHell in Place while THE Engine is Put To Getter when THE Bearing cap is Tightened THE Two SHell ARE Force To Breathe THIS is Bearing CRUSH]

*[left margin handwritten notes:
2nd Palcite
Forming Effort
OiL FiLter
AIR FiLTER
Prevent Bearing damel from 8harp particle zirig To The
Ring OiL Break Down
sner SHells
main Bearing
during High Feed and High P.M.]*

CHAPTER
10

Engine Block and Seals

The engine block, which is the supporting structure for the entire engine, is made of cast iron or from cast or die-cast aluminum alloy. All other engine parts are mounted on it or in it. This large casting supports the crankshaft and camshaft and holds all the parts in alignment. Large-diameter holes in the block casting form the cylinders to guide the pistons. The cylinder holes are called *bores* because they are made by a machining process called boring. Combustion pressure loads are carried from the head to the crankshaft bearings through the block structure. The block has webs, walls, and drilled passages to contain the coolant and lubricating oil, and to keep them separated from each other. Mounting pads or lugs on the block transfer the engine torque reaction to the vehicle frame through attached engine mounts. A large mounting surface at the rear of the engine block is used to fasten a bell housing or transmission case. The modern engine block meets all these requirments, and it has a longer service life than any other part of the engine.

The head, pan, and timing cover are attached to the block. The attaching joints are sealed so that they do not leak. Gaskets are used in the joints to take up machining irregularities and the changes that result from different pressures and temperatures.

10-1 Block Design

Most domestic production automobile engines between 250 and 160 in.³ displacement (4 and 2.6 liters) are overhead-valve six-cylinder engines. They may be either inline or V-type. Larger-displacement engines are V-8's. Four-cylinder engines have displacements of less than 160 in.³ (about 2.6 liters). There was a time when the inline eight-cylinder engine was popular. However, as casting technology developed, production of a one-piece V-block was possible. This design gradually replaced the inline eight for large-displacement engines. The same thing is happening to the six-cylinder engine. It is being replaced by the V-6 engine. A typical V-type engine block is pictured in Figure 10-1.

Cylinder Arrangement. Inline engine cylinders are numbered from the front to the rear, number 1 being at the front. The V-engines present a different problem. In a sense, they are two four-cylinder engines with their bases together and sharing the same crankshaft. Two approaches have been used to number the cylinders of V-engine blocks. One manufacturer (Ford) numbers the right block from 1 to 4 and the left block from 5 to 8. In general, the other manufacturers number their cylinders in the order in which the connecting rods are attached to the crankshaft. They start with the number 1 at the front of the crankshaft and go back to the last number at the rear. A few of these engines have the first cylinder at the right front. Most of the V-8 engines, however, use a numbering system having the number 1 cylinder at the left front (Figure 10-2). One V-8 (Pontiac) has the right front cylinder ahead of the left front cylinder, but the left front cylinder is identified as number 1 and the right front cylinder as number 2. As with all parts of the automobile, *right and left are viewed from the driver's position.*

The four-stroke cycle and crankshaft angles must be considered in the V-block design. For an even-firing engine, the V-8 must have its block at 90° (720° ÷ 8 cylinders). A V-6 uses either 90° or 60° blocks. The crankshaft must be designed, as described in Section 9-2, to have even-firing impulses on the V-6 engine.

The Lower Block. The engine block consists primarily of the cylinders with a web or bulkhead to support the crankshaft and head attachments. The rest of the block consists of a water jacket, a lifter chamber, and mounting flanges. In most engine designs, each main bearing bulkhead supports both a cam bearing and a main bearing. The bulkhead is well ribbed to support and distribute loads applied to it. This gives the block structural rigidity and beam stiffness throughout its useful life.

Two types of lower block designs are in use. One will be called a shallow skirt block. When used on a V-type engine, it is often called a V-block. The base of this block is close to the crankshaft center line. This block base is called the oil pan *rail*. The second type of block will be called a deep skirt block. In this type, the deep skirt extends the oil pan rail well below the crankshaft center line. When used on a V-type engine, it is often called a Y-block.

The *shallow skirt block* is the smallest and lightest of the two engine block types (Figure 10-3). It has the least amount of cast iron and this makes it a small, compact lightweight block. Covers, such as the oil pan and timing cover, are largely lightweight aluminum die castings or sheet steel stampings.

The *deep skirt block* improves the stiffness of the entire engine (Figure 10-4). It provides a wider surface on which to attach the bell housing. This greater rigidity assures smooth, quiet engine operation and durability. The deep skirt must be wide enough to clear the connecting rods as they swing through the block and, therefore, large oil capacity is provided with its use. Because weight is critical, only Buick was building deep-skirt-block V-type

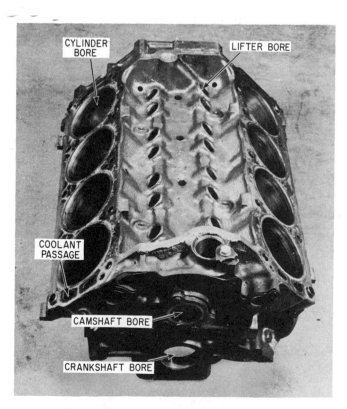

Figure 10-1. A typical V-type engine block.

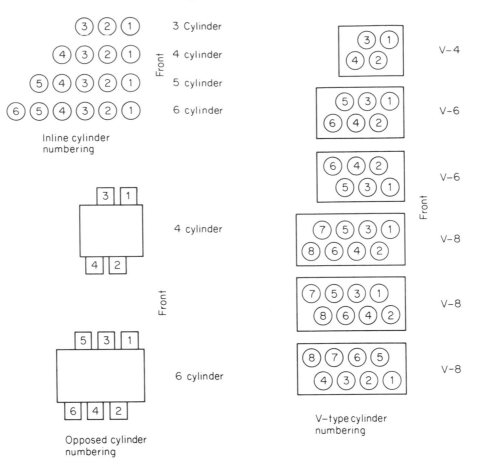

Figure 10-2. Common cylinder numbering arrangements.

Figure 10-3. A typical shallow skirt block with the oil pan rail surface close to the crankshaft centerline. The block pictured here is upside down on a workbench.

Figure 10-4. A typical deep skirt block with the oil pan rail surface that extends well below the crankshaft centerline. The block pictured here is upside down on a work bench.

passenger car engines in 1980. The deep skirt block is used on a number of inline engines.

Block Deck. The cylinder head is fastened to the top surface of the block. This surface is called the *block deck*. The deck has a smooth surface to seal *against* the head gasket. Bolt holes with National Course (NC) or metric threads are positioned around the cylinders to form an even holding pattern. Four, five, or six head bolts are used around each cylinder in automobile engines. These bolt holes go into reinforced areas within the block that carry the combustion pressure load to the main bearing bulkheads. Additional holes in the block are used to transfer coolant and oil.

Cylinder Skirts. The cylinders may be of a *skirtless* design, flush with the interior top of the crankcase (Figure 10-5), or they may have a skirt that extends into the crankcase (Figure 10-6). *Extended skirt* cylinders are used on engines with short connecting

Figure 10-6. A cylinder skirt that extends below the interior top of the crankcase.

rods. In these engines, the pistons move very close to the crankshaft. The cylinder skirt must go as low as possible to support the piston when it is at the lowest point in its stroke. This can be seen in Figure 10-7. The extended cylinder skirt allows the engine to be designed with a low overall engine height, since it has a small block size for its displacement.

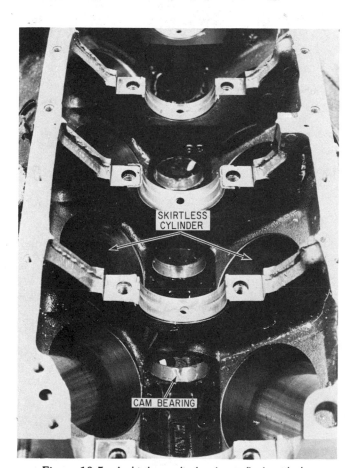

Figure 10-5. A skirtless cylinder that is flush with the interior of the top of the crankcase.

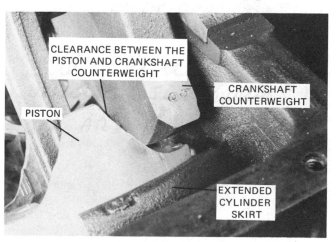

Figure 10-7. The piston comes very close to the crankshaft counterweight when it is at the bottom of the stroke on an engine that has a short connecting rod.

155

Figure 10-8. The block deck has been cut off to show how cooling passages surround the cylinders.

Figure 10-9. An installed convex-type soft plug.

Cooling Passages. Cylinders are surrounded by cooling passages. The coolant passage around the cylinders is often called the cooling jacket (Figure 10-8). In most skirtless cylinder designs, the cooling passages extend nearly to the bottom of the cylinder. In extended skirt cylinder designs, the cooling passages are limited to the upper portion of the cylinder. This can be seen in the cut away engine pictured in Figure 8-3.

During casting, the cores are supported from outside the block. The core supports and casting vents leave holes in the casting. Core holes that remain in the block deck are closed with the gasket and head. Core holes left in the external block wall are machined and sealed with *soft plugs*.

Soft plugs are of two designs. One is a *convex type*. For its use, the core hole is counterbored with a shoulder. The counterbored hole is more expensive to machine than is a straight hole. The convex soft plug is placed in counterbore, convex side out. It is driven in and upset with a fitted seating tool. This causes the edge of the soft plug to enlarge to hold it in place. Figure 10-9 shows an installed convex soft plug. The second type of hole plug is a *cup type*. This type of soft plug is fit into a smooth, straight hole. The outer edge of the cup is slightly bell-mouthed. The bell-mouth causes it to tighten when it is driven in the hole to the correct depth with a seating tool. An installed cup-type soft plug is shown in Figure 10-10. The cup-type soft plug is the most common in use.

Figure 10-10. An installed cup-type soft plug.

Lubricating Passages. An engine block has many oil holes that carry lubricating oil to the required locations. During manufacture, all the *oil holes are drilled* from outside the block. Oil passages are rarely cast in engine blocks. When a curved passage is needed, intersecting drilled holes are used. In some engines, plugs are placed in the oil holes to direct oil to another point before coming back to the original hole, on the opposite side of the plug. Typical oil hole drilling is illustrated in Figure 10-11. After oil holes are drilled, the unneeded open ends may be capped by pipe plugs, steel balls, or cup-type soft plugs. End plugs in the oil passages

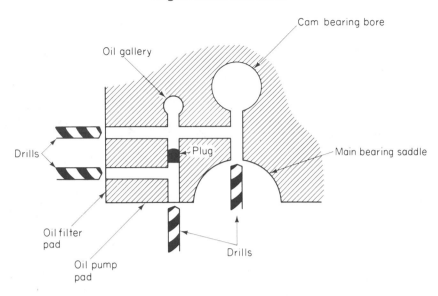

Figure 10-11. Typical oil hole drilling in the main bearing web.

are a source of possible oil leakage in operating engines.

10-2 Block Manufacturing

Cast-iron cylinder block casting technology continues to be improved. The trend is to make blocks with larger cores, using fewer individual pieces. Oil-sand cores, shown in Figure 10-12, are forms that shape the internal openings and passages in the engine block. Before casting, the cores are supported within a core box. The core box also has a liner to shape the outside of the block. Special alloy cast iron is poured into the box. It flows between the cores and the core box liner. As the cast iron cools, the core breaks up. When the cast iron has hardened, it is removed from the core box, and the pieces of sand core are removed through the openings in the block by vigorously shaking the casting.

One way to keep the engine weight as low as possible is to make the block with minimum wall thickness. Engine designers and foundry techniques have made lightweight engines by making the cast-iron block walls and bulkheads only as heavy as necessary to support their required loads. They have omitted as much material from the lifter gallery area as possible. They have even designed small oil filters so that the attachment point size could be reduced. Much of the ability to lighten the block is the result of cores made of a minimum

number of pieces that are secured firmly in place during casting. If they should shift or float in the molten cast iron, the block wall would be too thick in some places and too thin in other places.

Aluminum is used for some cylinder blocks. Early aluminum blocks were cast in a manner similar to cast-iron blocks. These blocks were equipped with a mechanically bonded cast-iron liner in each cylinder. A more recent aluminum block design has the block die-cast from silicon–aluminum alloy with no cylinder liners (Figure 10-13). Pistons with zinc–copper–hard iron coatings are used in the these aluminum bores. Some European engines have die-cast aluminum blocks with replaceable cast-iron cylinder sleeves. The sleeves are sealed at the block deck and at their base. Coolant flows around the cylinder sleeve, so this type of sleeve is called a wet sleeve. Cast-iron main bearing caps are used with aluminum blocks. This is necessary to give the required strength.

Block Machining. After thorough cooling and cleaning, the block casting goes to the machining line. The top, bottom, and end surfaces are cleaned and semifinished with a broach. A *broach* is a large slab with a number of cutting teeth. Each tooth cuts a little more than the preceding tooth. It is somewhat like a large, coarse, contoured file. One pass of the broach will smooth both cylinder decks and the lifter valley cover rail. A second pass will

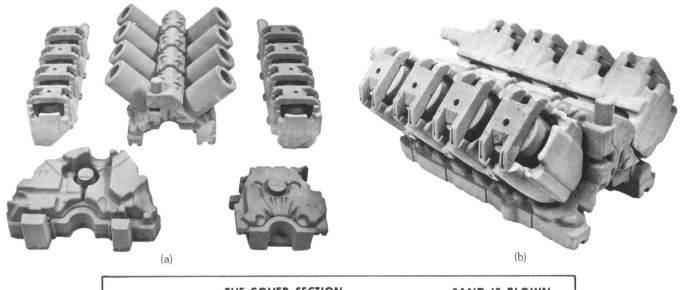

(a)

(b)

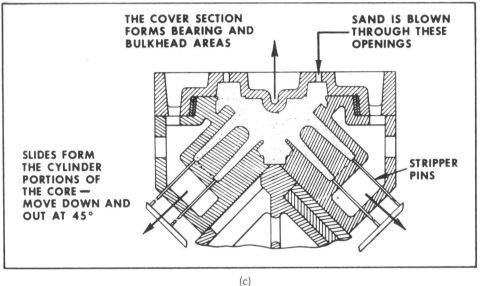

THE COVER SECTION
FORMS BEARING AND
BULKHEAD AREAS

SAND IS BLOWN
THROUGH THESE
OPENINGS

SLIDES FORM
THE CYLINDER
PORTIONS OF
THE CORE —
MOVE DOWN AND
OUT AT 45°

STRIPPER
PINS

(c)

Figure 10-12. Casting cores (Courtesy of Central Foundry Division, General Motors Corporation. (a) Separate cores, (b) assembled cores, and (c) cores in a core box.

smooth the upper main bearing bores and the oil pan rail. The ends of the block may be finished with a third broach. Some of these surfaces are completed with the broach operation; others need to be finished with a mill, a final broach, or a boring operation. Broaching leaves straight lines across the surface while milling leaves curved lines.

The cylinders are bored and honed in a number of operations until they have the required size and finish. Figure 10-14 shows a part of a block production line. A slight notch or *scallop* is cut into the edge of the cylinder on some engines using very large valves (Figure 10-15). All drilling and thread tapping is accomplished on the block line.

Main Bearing Caps. The main bearing caps are cast separately from the block. They are machined and then installed on the block for a final bore finishing operation. With caps installed, the main bearing bores and cam bearing bores are machined to the correct size and alignment. On some engines, these bores are honed to a very fine finish and exact size. *Main bearing caps are not interchangeable* or reversible, because they are individually finished in place. Main bearing caps may have cast numbers indicating their position on the block. If not, they should be marked.

Standard production engines use two bolts to hold the main bearing cap in place (Figure 10-16).

Figure 10-13. A four cylinder block die cast from silicon-aluminum alloy with no cylinder liners (Courtesy of Chevrolet Motor Division, General Motors Corporation).

SCALLOP FOR VALVE CLEARANCE

Figure 10-15. A scallop on the upper edge of the cylinder for valve clearance.

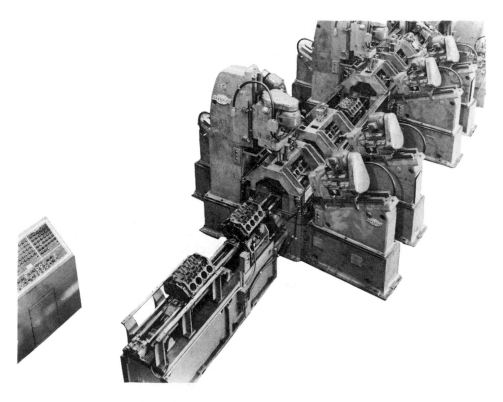

Figure 10-14. One section of an engine production line (Courtesy of Greenlee Brothers & Company).

Figure 10-16. A standard two-bolt main bearing cap.

Heavy-duty and high-performance engines often use additional main bearing support bolts. These could be a cross-bolt design in a deep skirt block or a parallel design four-bolt main cap in the shallow skirt block design (Figure 10-17). Remember that the expansion force of the combustion chamber gases will try to push the head off the top and the crankshaft off the bottom of the block. The engine is held together with the head bolts and main bearing cap bolts screwed into bolt bosses and ribs in the block. The bosses are enlarged places that surround the opening. The extra bolt on the main bearing cap helps to support the crankshaft when there are high combustion pressures and mechanical loads.

10-3 Gaskets and Static Seals

Oil, coolant, and gases flow through passages in the engine block. These are usually kept separate by cast-iron walls, plugs, covers, and caps. They must not leak either internally or externally. Gaskets or static seals are used between engine parts to seal the joint, thus preventing leakage.

Properties of Gaskets. Gasket requirements become greater as engine pressures and temperatures become greater. Four gasket properties must be considered by the engineer when selecting gaskets for each specific application. The gasket must be impermeable, comformable, resilient, and resistant.

Each gasket must be *impermeable* to the fluids it is designed to seal. If the fluid could penetrate the gasket, it would leak and the gasket would be of no value.

Gaskets must *conform* to any existing surface imperfections. This includes surface roughness from machining and slight parting surface warpage.

A *resilient* gasket property will cause the gasket to maintain sealing pressure, even when the joint is slightly loosened as a result of temperature changes or vibration.

The environment of the gasket will change with variations in temperature, pressure, and age. The gasket must be *resistant* to all expected changes in its environment for the engine service life.

Gasket Materials. Many materials are used for gaskets, depending upon the sealing requirements and the cost. One of the oldest gasket materials is *cork,* a natural material from the bark of mediterranean oak and cork pine trees. For gaskets, cork bark chips are held together in sheets with bonding materials, often rubber compounds. The material that bonds gaskets is called a *binder.* This makes a highly impermeable gasket that conforms easily. The use of cork is limited to lightly loaded joints, having uneven surfaces, such as rocker covers and oil pans. Aluminum coatings on cork gaskets helps to reduce heat deterioration. In some cases, the cork gaskets are rubber-coated.

Cork in the gasket is often replaced by *fibers.* These fibers may be cellulose, asbestos, or a mixture of the two. The type of binder material used determines the properties of the gasket. Some gaskets use binders that are impermeable to oil, whereas other gaskets swell on contact with oil. Some gaskets are impermeable to water, whereas other gaskets swell on contact with water. Gaskets that are designed to swell are used where the joint to be sealed cannot be tightened. The swelling of the gasket will seal the joint. An example of this is the side gasket of a rear

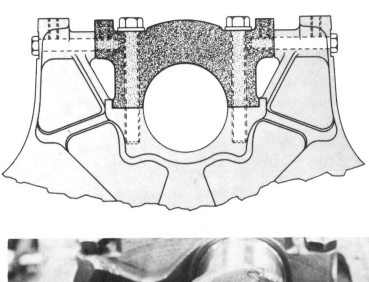

Figure 10-17. Four bolt main bearing caps. (a) Cross bolted (Courtesy of Chrysler Corporation) and (b) parallel bolted design with four bolts.

main oil seal used on a deep skirt block. Gaskets that swell are not used where high pressures are present. Generally, fiber-base gaskets rather than cork are used under high pressures. Fiber gaskets require a better parting surface smoothness and rigidity than is needed for cork gaskets. Typical gasket materials are shown in Figure 10-18.

Molded oil-resistant *synthetic rubber* is often used where the sealing requirements dictate special seal designs. These materials are often used to seal the oil pan ends, corner joints, and intake manifold ends on V-type engines. A new approach to gaskets is a plastic silicone gasket material in a *tube*. It can be used in place of paper and fiber-based gaskets. It completely seals the joint, as it conforms to all surface variations.

Figure 10-18. Common gasket materials. From left to right they are cork, paper, composite and elastomer materials.

Head Gaskets. The most difficult sealing job in the engine is to seal the cylinder head to the block parting surface. The earliest head gaskets were *copper-coated asbestos.* The copper sealed into the machining marks of the head and block and the asbestos provided resilience and conformability. As engine designs were improved, copper on the gaskets was replaced by steel to withstand the higher pressures and temperatures. Steel rings, called *fire rings,* were put on the gaskets around the cylinder openings to seal the combustion chambers. Similar rings were added around some of the other gasket holes as well (Figure 10-19).

Continued development of engine manufacturing techniques provided smoother and flatter parting surfaces, and increased engine power brought on the *embossed steel* head gasket. A raised rib or embossed portion gives steel gaskets the required resiliency. A soft aluminum coating placed on steel gaskets will seal into the parting surface machining marks.

A later head gasket development uses a thin *steel core* with a thin coating of asbestos rolled on the outside to give the gasket the desirable resilient properties. This is needed as the head and block change temperature and as the pressure changes during each cycle.

Most head gaskets must be installed in a specified direction. This is required because the head gasket in most engines helps control engine coolant flow (Figure 10-20). The gasket is marked top or front when the gasket position is critical.

When it is not marked, the gasket should be installed with the stamped identification numbers toward the head. Head gasket types are shown in Figure 10-21.

Cover Gaskets. Timing cover gaskets are usually thin fiber or paper. Cork, fiber, and synthetic rubber are used in different parts of the oil pan. The intake manifold uses embossed steel or reinforced fiber gaskets. Cork or synthetic rubber sections are used on the ends of the lifter valley cover portion of the V-type intake manifold.

The use of formed in-place gasket materials are gradually increasing. The most common is RTV, room-temperature vulcanizing material. This material seals as it air-dries. Another formed-in-place gasket material is called anaerobic. It seals when air is removed from it. RTV materials are

Figure 10-20. Typical head gasket markings.

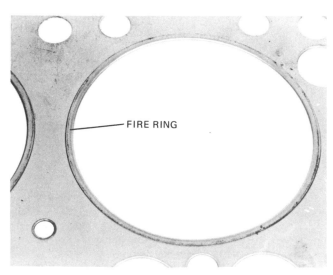

Figure 10-19. A head gasket with a fire ring.

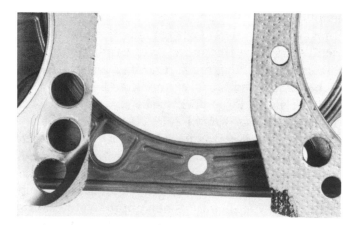

Figure 10-21. Types of head gaskets. The gasket on the left has metal over asbestos. The center gasket is embossed steel. Asbestos is over a steel core on the right gasket. The lower corners are removed to show the core of the gasket.

often used in place of cover gaskets. It works well in the side seals of a main bearing cap used on a deep skirt block. Anaerobic materials seal tight joints such as screw threads.

It is a common practice to use a new gasket each time a part is assembled. The price of a new gasket is small compared to the labor cost of installing it. After use, a gasket will have lost most of its sealing properties. To avoid leaks, *always use new gaskets* during assembly.

10-4 Dynamic Oil Seals

Dynamic oil seals are used between two surfaces that have relative motion between them, such as a shaft and a housing.

Dynamic Seal Operational Requirements. In engines, the seals keep liquids and gases in and keep contaminants out. They must do this with a minimum drag or friction. Oil seals must not press against the moving part so tightly that they cause drag or wear a groove in the moving parts.

Some dynamic seals, such as piston rings, are designed to withstand high pressure. Other seals, such as front and rear crankshaft oil seals, seal against little pressure. Seals in an engine that seal around rotating shafts are classified as radial positive-contact seals.

The selection of the seal is determined by the rubbing speed, fluid pressure, operating temperature, shaft surface requirements, and space available. When these factors are known for a specific application, the oil-seal type can be selected by the engineer. The service technician uses the seal manufactured for the specific application.

Types of Dynamic Seals. Dynamic seals used in automotive engines are most frequently made from a rope-type packing or from synthetic rubber.

Rope-type packing, or braided fabric, is the least expensive type of dynamic seal. It is often used as a rear main bearing seal. In some applications, it is used for the timing cover seal. The rope-type seal provides close contact between the seal and shaft without undue pressure. It therefore has very low friction and low wear characteristics. Figure 10-22 shows a typical rope-type rear main seal in place.

Lip-type dynamic oil seals are used in many engine applications. Some lip-type oil seals are made from leather. Synthetic rubber is generally used for the lip-type seals in automotive engines. They can stand more shaft eccentricity and runout than can the rope-type seals. They can operate at higher shaft speeds, but they require a very smooth shaft finish to provide long-life sealing. Lip-type seals place more load on the shaft then the rope-type seal, and they therefore seal better. This seal load also causes drag from friction. A typical lip-type rear main seal is shown in Figure 10-23.

Dynamic Seal Design Features. Sealing is the result of an interference fit between the shaft and the seal. Rope-type seals must be packed into the seal groove, then trimmed to length. They do not function properly if they are stretched into the

Figure 10-22. Rope-type rear main seal in place.

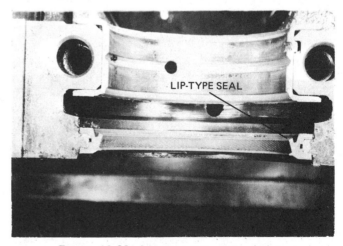

Figure 10-23. Lip-type rear main seal in place.

groove. Stretching will leave a gap behind the seal for the oil to seep through. To help the rope packing to make a good seal, the shaft surface under the packing may have a special finish grind or knurl. This finish will tend to pump the oil back into the engine. A knurled shaft is shown in Figure 10-24.

The lip-type seal is designed so that any increase in the pressure to be sealed will increase the lip pressure against the shaft. This causes better sealing. A spring tension element called a *garter spring* is used to increase the normal lip pressure of the seal. This is required in high-speed applications, where excessive runout occurs, or where the fluid viscosity is low (Figure 10-25).

Lip-type seals must run with a very thin layer of lubrication. Ideal seal operation allows an oil meniscus to form on the outside of the seal with no leakage. If the seal had no lubrication, it would wear the shaft very quickly.

Lip seals are usually held in a steel case or are supported by bonding on to a steel support member. This makes the seal become a one-piece seal. The most common examples of this are the seal installed on the front of the timing cover and the split seal installed on a rear main. The lip seal often has angled ribs to help force the oil back into the engine. The ribs of a lip seal can be seen in Figure 10-26.

The oil seal is aided by an oil *slinger* on both the front and rear ends of the crankshaft. The rear slinger may be a flange on the crankshaft. The front slinger is usually a stamped steel ring or cup located between the damper hub and the crankshaft timing gear. A typical slinger ring is visible in Figure 10-27. The majority of oil that comes along the shaft is thrown clear of the shaft by the slinger. The seal is then able to handle the oil that remains on the shaft.

Figure 10-24. The knurled part of the crankshaft that operates under a rope-type seal.

Figure 10-26. Angled ribs on a lip-type seal.

Figure 10-25. A timing cover lip-type oil seal with garter spring tension.

Figure 10-27. A typical slinger ring behind the front crankshaft oil seal.

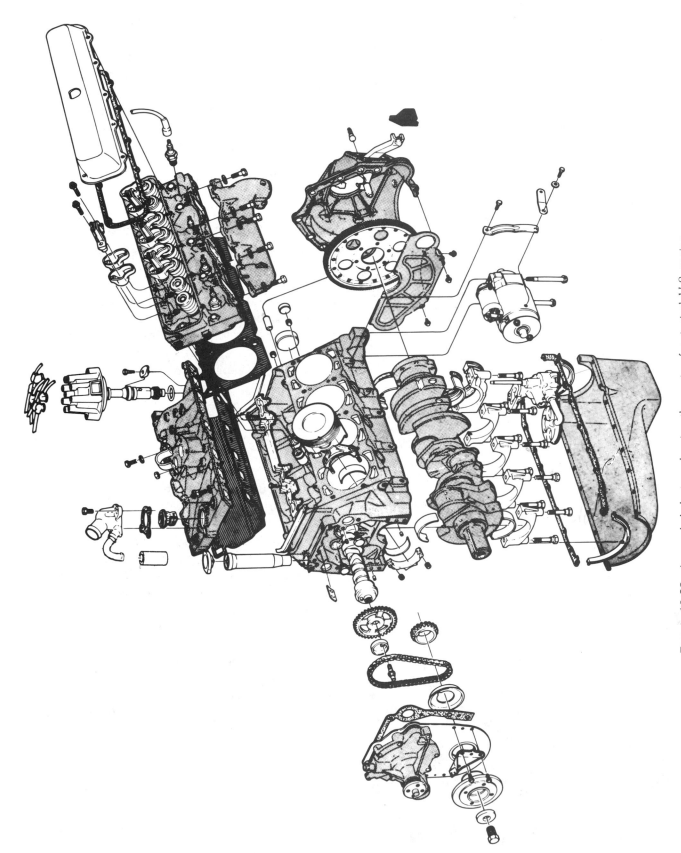

Figure 10-28. An exploded view showing the parts of a typical V-8 engine (Courtesy of Oldsmobile Divison, General Motors Corporation).

165

10-5 Block Attachments

A number of parts are attached to the engine to enclose it and to adapt it to the vehicle. These include covers, housings, and mounts. Most of these can be seen in the exploded view of a typical V-8 engine in Figure 10-28.

Bell Housings. A bell housing enclosing the flywheel and clutch or torque converter is bolted to the rear of the engine block. It is positioned with dowel pins for alignment. These can be seen in Figure 10-29. Offset dowels and shims between the block and bell housing are used to align the bell housing in standard transmission applications. Alignment is necessary so that the clutch shaft matches the pilot bearing in the back of the crankshaft. In original engine production, standard transmission bell housings are attached to the block before final machining of the main bearing bores. In this way the transmission hole in the bell housing is machined to match the alignment of the main bearing bore. This minimizes the chance of drive line misalignment. Alignment of the automatic transmission is simplified by using a flex plate transmission drive. In most automatic transmissions, the case forms the bell housing. Standard transmissions have a separate bell housing with clutch lever attachments. Aluminum bell housings are usually used in passenger car applications to keep the weight as low as possible.

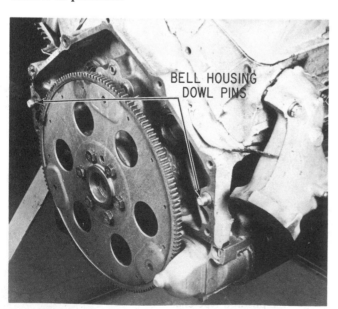

Figure 10-29. Dowel pins for aligning the bell housing or automatic transmission case.

Timing Covers. Manufacturers use a variety of timing covers. The simplest is a stamped sheet metal cover attached with cap screws. Its only purpose is to cover the gears, to keep foreign objects out, and to keep the engine oil in. Some engines use a cast cover in the same way. A cast aluminum cover also tends to muffle some of the timing drive noise.

Some timing covers are die-cast aluminum. These covers work in the same way as do the cast aluminum covers. The die-cast process produces a finished cover. Little or no additional machining is required. Die-cast tooling is much more expensive then tooling for casting; however, a saving is made in machining costs when using die-cast parts.

Some manufacturers have made the timing cover more complicated by including the oil pump and distributor drive together with the fuel pump and water pump. The die-cast process is used so that there is a minimum of machining operations. With these covers, the engine block contains no accessory drives. Timing belts on overhead cam engines are usually covered with a loosely fitting stamped cover to keep foreign materials from the belt. The types of timing covers for pushrod engines are shown in Figure 10-30.

Engine Mounts. Engines are mounted to the vehicle through the rubber insulators. The vibrational characteristics of the engine are checked in the engineering laboratories. The engine mounts are then positioned close as possible to vibration *nodes*. **Nodes are points of minimum vibration.** The rubber used in engine mounts is especially compounded to absorb vibrations. Each specific engine model uses the rubber compound designed to isolate its particular characteristic vibration. The mounts are usually located about halfway back on each side of the block. The rear engine mount is located at the rear of the transmission so that the engine-transmission assembly is supported at three points.

The rubber could separate on old-style engine mounts. When it did, the engine would move in the engine compartment. Figure 10-31 shows one of these mounts that did break. New-type engine mounts are designed with metal surrounding the rubber of the mount. If the rubber breaks, the metal retains the engine so that it can move only slightly. The metal retains the engine and prevent accidents. A mount of this type is pictured in Figure 10-32.

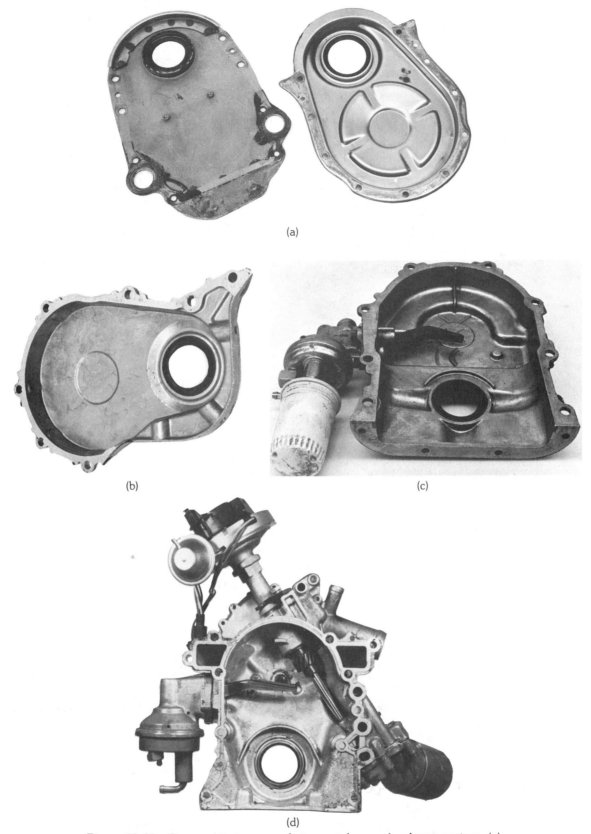

Figure 10-30. Common timing cover designs used on pushrod-type engines. (a) Stamped steel, (b) die cast, (c) die cast with the fuel pump pad, and (d) die cast with fuel pump, water pump, oil pump, and distributor.

Figure 10-31. A broken old style engine mount.   **Figure 10-32.** One design of new type engine mounts.

REVIEW QUESTIONS

1. What is the displacement range of the modern inline and V-type engines? [10-1]
2. How is the right side of a V-type engine identified? [10-1]
3. What purpose is served by the block bulkhead in an engine? [10-1]
4. What are the advantages and disadvantages of a shallow skirt block compared to a deep skirt block? [10-1]
5. What are the advantages and disadvantages of an extended skirt cylinder compared to a skirtless design? [10-1]
6. How are the block casting core holes closed? [10-1]
7. What is the advantage of cup-type soft plugs over the convex-type soft plugs? [10-1]
8. How are oil passages put into an engine block? [10-1]
9. How are the unneeded open ends of oil passages closed on the engine block? [10-1]
10. How does a block get its outside shape in the casting process? [10-2]
11. How are the cores removed from a new block casting? [10-2]
12. What is done during casting to make the block as light as possible? [10-2]
13. Describe three different methods used to make aluminum blocks. [10-2]
14. What is done to give the main bearing caps the strength that they require? [10-2]
15. How are the surfaces of the block casting smoothed or machined? [10-2]
16. Why do some cylinders have a scallop on the top edge of the cylinder? [10-2]
17. How are the main bearing caps made to fit the block? [10-2]

18. Why can't the main bearing caps be reversed on the block? [10-2] *it will not fit*
19. Why do some engines use four bolt main bearing caps? [10-2] *High Performance*
20. Name four gasket properties. [10-3]
21. Name four gasket materials and list two places each may be used in an engine. [10-3] *Cork (oil Pan) Asbestos (Head Gasket) Synthetic (Cover)*
22. What materials are used for bonding and binders in gaskets? [10-3] *Rubber Compound*
23. When is it necessary for a gasket to swell after it is installed? [10-3]
24. What gasket surface is the most difficult to seal in the engine? [10-3] *Head Gasket*
25. Why are some head gaskets marked top or front? [10-3] *Because*
26. Where are synthetic rubber gaskets used on an engine? [10-3] *Intake - oil Pans ends*
27. Where are formed-in-place gaskets used? [10-3] *Cover Gasket*
28. When should gaskets be replaced? [10-3] *All the time*
29. What are the advantages and disadvantages of a rope-type packing oil seal compared to a lip-type dynamic oil seal? [10-4] *Require Smooth Surface*
30. What is the difference in the shaft surface requirements for a rope- and lip-type dynamic seal? [10-4] *Smooth Surface   Groove Surface*
31. How does a slinger operate? [10-4] *it throw oil off the shaft*
32. How is the bell housing bore matched to the engine main bearing bores during the original manufacturing? [10-5] *Ofset Dowel*
33. What are the different types of timing cover designs? [10-5] *Stamped - Die-Cast with oil Pump Distributor*
34. What must be considered by automotive engineers when they select and position the engine mounts? [10-5]

Nodes = Minimum Vibration

Impermeable
Conformable
Resilient
Resistant (changes of Temp.)

CHAPTER
11

Lubrication and Cooling

Lubricating oil is often called the life blood of an engine. It circulates through passages in the engine that carry it to all the engine's rubbing surfaces. Its main job is to form an oil film between these surfaces to keep them from touching. This action minimizes friction and wear within the engine. The lubricant has useful secondary jobs. Cool lubricant picks up heat from hot engine parts and takes it to the oil pan, where it is cooled as air moves past the pan. The oil flow also carries wear particles from the rubbing surfaces to the pan so that the particles will cause no further damage within the engine. Oil between the engine parts cushions the parts from the shock as the combustion pressure forces the piston down.

The lubricant used for motor oil must have properties that will allow it to meet the engine requirements. The motor oil's most important property is its thickness at its normal operating temperature. The thickness of the oil is called *viscosity*. If the oil is too thin it will rapidly leak from the operating clearances. When the oil leaks out too fast, the parts will come in contact, resulting in scoring and wear. When the oil thickness is too great, it will use too much power to overcome drag between the rubbing surfaces. This characteristic is noticeable when comparing the cranking speed of a cold engine with thick oil to a warm engine with thin oil.

Secondary properties in the form of additives, are put in the oil by the motor oil producers. The additives provide the oil with the ability to clean the engine, minimize scuffing, reduce rusting, resist oxidation, and maintain the oil's viscosity characteristics. The oil should be replaced when its properties can no longer protect the engine.

The primary job of the automotive engine cooling system is to keep the normal operating temperature of the block and head. Coolant flow is held at a minimum during warm-up until normal engine temperature is reached; then the coolant flow is gradually increased, as required, to maintain the normal temperature. When the engine operating temperature is too low, the piston and cylinder wall will scuff and wear rapidly. When the temperature is too high, hard carbon deposits will form in the engine. The carbon can cause piston ring and valve sticking and it will cause passage clogging. Operating the engine at normal temperatures will minimize these problems and help to give maximum engine service life.

Two types of cooling systems are used in passenger cars, air and liquid. Some imported passenger cars have used air cooling. Current passenger cars use liquid cooling systems. The coolant removes the excess heat from the hot engine parts and carries it to a radiator, where the heat is given up to the outside air.

11-1 Engine Lubrication System

Automobile engines use a *wet sump* in their lubrication system. **The sump is the lowest part of the lubrication system.** In automobile engines, the sump is in the oil pan. It is called a wet sump because it holds the oil supply. Some racing and industrial engines use a *dry sump*. A dry sump is required when the engine is tilted during operation. A tilt of 37° moves the oil sideways in the oil pan as much as a 75% gravitational pull when rounding a corner. A panic stop can do the same thing, as illustrated in Figure 11-1. When the oil tilts too far in the oil pan, part of the oil pickup is uncovered. This allows air to mix with oil, which will cause oil foaming. A scavenger pump in dry sump engines draws the oil out of a relatively small engine sump and returns it to a separate oil supply tank. The engine oil pump draws oil from the supply tank to feed the engine lubrication system.

All production automobile engines have a full-

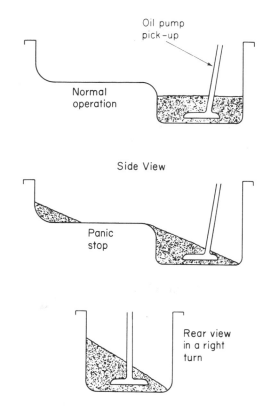

Figure 11-1. Oil movement in the pan during vehicle maneuvers.

pressure oil system. The pressure is maintained by an oil pump. It picks up the motor oil through a passage from an inlet screen in the oil pan. The oil is forced into the lubrication system under pressure. A typical engine lubricating system is shown in Figure 11-2. The oil pump inlet screen may be supplied with a sheet metal plate or *baffle* that keeps the inlet screen covered with oil during sudden stops.

Oil Pump. In most engines, the distributor drive gear meshes with a gear on the camshaft, shown in Figure 11-3. The oil pump is driven from the end of the distributor shaft, often with a hexagon-shaped shaft. Some engines have a short shaft-gear that meshes with the cam gear to drive both the distributor and oil pump. Occasionally, an engine is built that uses separate gears on the distributor and on the oil pump. Both of these gears mate with the same cam gear. This can be seen in Figure 11-4. With these drive methods, the pump turns at one-half engine speed. In other engines, the oil pump, similar to an automatic transmission pump, is driven by the front of the crankshaft so that it turns at the same speed as the crankshaft. An example of

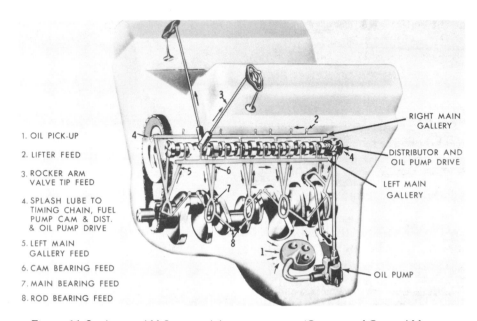

1. OIL PICK-UP

2. LIFTER FEED

3. ROCKER ARM
   VALVE TIP FEED

4. SPLASH LUBE TO
   TIMING CHAIN, FUEL
   PUMP CAM & DIST.
   & OIL PUMP DRIVE

5. LEFT MAIN
   GALLERY FEED

6. CAM BEARING FEED

7. MAIN BEARING FEED

8. ROD BEARING FEED

RIGHT MAIN
GALLERY

DISTRIBUTOR AND
OIL PUMP DRIVE

LEFT MAIN
GALLERY

OIL PUMP

Figure 11-2. A typical V-8 engine lubricating system (Courtesy of General Motors Corporation).

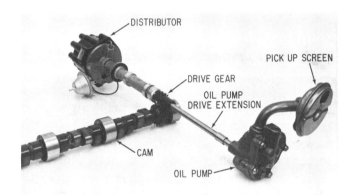

DISTRIBUTOR

PICK UP SCREEN

DRIVE GEAR

OIL PUMP
DRIVE EXTENSION

CAM

OIL PUMP

Figure 11-3. The oil pump is driven by an extension from the distributor drive gear on most engines.

a crankshaft-driven oil pump is shown in Figure 11-5.

Most automotive engines use one of two types of oil pumps: gear and rotor (Figure 11-6). The *gear-type* oil pump consists of two spur gears, in a close-fitting housing. One gear is driven and the other idles. As the gear teeth come out of mesh, they tend to leave a space which is filled by oil drawn through the pump inlet. When pumping, oil is carried around the *outside* of each gear in the space between the gear teeth and the housing. As the teeth mesh in the center, oil is forced from the teeth into an oil passage, thus producing oil pressure. The *rotor-type* oil pump is essentially a special lobe-

OIL PUMP DRIVE GEAR   DISTRIBUTOR DRIVE GEAR

Figure 11-4. Separate gears drive the oil pump and distributor on this engine.

shaped gear meshing with the inside of a lobed rotor. The center lobed section is driven and the outer section idles. As the lobes separate, oil is drawn in just like it is drawn into gear-type pumps. As the pump rotates, it carries oil around *between* the lobes. As the lobes mesh, they force the oil out from between them under pressure in the same man-

Figure 11-5. An oil pump mounted in the front cover of the engine. The oil pump is driven by the crankshaft.

Figure 11-6. A rotor-type oil pump is on the left and a gear-type oil pump is on the right.

ner as does the gear-type pump. The pump is sized so that it will maintain at least 15 lb/in.² (100 kPa) in the oil gallery when the engine is hot and idling. Pressure will increase as the engine speed increases, because the engine-driven pump also rotates faster.

Oil Pressure. In engines with full-pressure lubricating system, maximum pressure is limited with a pressure relief valve. If a pressure relief valve was not used, the engine oil pressure would continue to increase as the engine speed increased. Maximum pressure is usually limited to the lowest pressure that will deliver enough lubricating oil to all engine parts that need to be lubricated. Three to six gallons per minute are required to lubricate the engine. The oil pump is made so that it is large enough to provide pressure at low engine speeds and small enough so that it will not *cavitate* at high speeds. Cavitation occurs when the pump tries to

pull oil faster than it can flow from the pan to the pickup. When it cannot get enough oil, it will pull air. This puts air pockets or caves in the oil stream. **A pump is cavitating when it is pulling air or vapors.**

After the oil leaves the pump, it is delivered to the moving parts through drilled oil passages. It does not have to have any pressures after it reaches the parts that are to be lubricated. The oil film between the parts is developed and maintained by hydrodynamic forces. Hydrodynamic forces occur when a wedge-shaped film develops in the oil that is between two surfaces that have movement between them. This oil film is called a *hydrodynamic oil film.* Excessive oil pressure requires more horsepower and provides no better lubrication than the minimum pressure. High oil pressure and the resulting high rates of oil flow may, in some cases, tend to erode engine bearings in addition to oil pump cavitation.

Oil pressure can only be produced when the oil pump has a larger capacity than all the "leaks" in the engine. The "leaks" are the clearances at end points of the lubrication system. The end points are at the edges of bearings, the rocker arms, the connecting rod spit holes, and so on. These clearances are designed into the engine and are necessary for its proper operation. As the engine parts wear and clearance becomes greater, more oil will leak out. The oil pump *capacity* must be great enough to supply extra oil for these "leaks." The capacity of the oil pump results from its size, rotating speed, and physical condition. If the pump is rotating slowly as the engine is idling, oil pump capacity is low. If the "leaks" are greater than the pump capacity, engine oil pressure is low. As the engine speeds up, the pump capacity increases and tries to force more oil out of the "leaks." This causes the pressure to rise until the pressure reaches the regulated pressure.

A third consideration, the viscosity of the engine oil, is involved in both the pump capacity and the oil leakage. Thin oil or oil of very low viscosity slips past the edges of the pump and flows freely from the "leaks." Hot oil has a low viscosity and, therefore, a hot engine often has low oil pressure. Cold oil is more viscous (thicker) than hot oil. This results in high pressures, even with the cold engine idling. High oil pressure on a cold engine occurs because the oil relief valve must open further to release excess oil than is necessary on a hot engine.

This larger opening increases the spring compression force and this, in turn, increases the oil pressure. Putting higher-viscosity oil in an engine will raise the engine oil pressure to the regulated setting of the relief valve at a lower engine speed.

Pressure Relief Valve. The pressure relief valve is located downstream from the pressure side of the oil pump. It generally consists of a spring-loaded piston and, in a few cases, a spring-loaded ball. Two that have been removed from engines are shown in Figure 11-7. When oil pressure reaches the regulated pressure, it will force the relief valve back against the calibrated spring. The spring is compressed as the valve is forced back. This allows a controlled "leak" from the pressure system at a rate required to maintain the correct regulated pressure. Any change in the relief valve spring tension will change the regulated oil pressure. Higher spring pressures will cause higher maximum oil pressures. In most engines, the oil that is released by the relief valve flows through a passage to the inlet side of the oil pump to be recirculated through the pump, as shown in Figure 11-8. The relief valve is, therefore, usually located in the oil pump housing or pump cover. This method of oil flow from the relief valve prevents oil foaming and excessive oil turbulence in the oil pan. By this method, a solid stream of lubricating oil will be delivered from the pump.

Figure 11-7. A spring loaded piston-and ball-type oil pressure regulator valve.

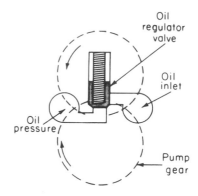

Figure 11-8. Oil pressure regulator valve releases excess oil to the pump inlet.

Oil Filter. Oil leaving the pump flows to the oil filter, where large particles are trapped (Figure 11-9). Clean oil will flow into the engine. Oil filters are designed to trap large particles that could damage engine bearings. Very fine particles flow through the filter. These particles are so fine they can go between engine clearances without doing damage. As the filter traps particles, the holes in the filter become partly plugged. As they plug, the filter traps even smaller particles, thus doing a better filtering job. This better filtering, however, restricts oil flow through the filter. Less oil flow could result in not having enough oil get to the bearings. All filters or filter adapters have a *bypass valve.* Oil can bypass the filter and go directly into the engine when the oil is cold and thick or if the filter plugs. The bypass valve is set to open at 5 to 15 lb/in.² (35 to 100 kPa). The actual pressure depends on the engine model and the normal pressure drop across the filter element. Two-stage or double oil filters usually have greater pressure drops across the filter element than do single-stage filters. This means that there is less oil pressure to carry the oil through the oil passages.

Filters or filter adapters are supplied with a *check valve.* It is designed to keep the filter full when the engine is stopped. The check valve keeps the oil from leaking from the filter back through the oil pump into the pan. This keeps the oil pump primed so that oil pressure will build up rapidly as soon as the engine starts.

Most engine oil filters were made smaller starting in the mid-1970s. Smaller oil filters require smaller adapters to the engine, so the engine weight was reduced. It was possible to use smaller filters because no-lead gasoline produced less deposits in

Figure 11-10. Drilled passages through the block bulkheads allow the oil to go from the main oil gallery to the main and cam bearings. In some engines, oil goes to the cam bearings first, then to the main bearings.

It is important that the oil holes in the bearings match up with the drilled passages in the bearing saddles so that the bearing can be properly lubricated. Over a long period of use, bearings will wear too much. This wear causes excess clearances. The excess clearance will allow too much oil to leak from the side of the bearing. When this happens there will be little or no oil left for bearings located further downstream in the lubricating system. This is a major cause of bearing failure. If a new bearing were installed in place of the oil-starved bearing, it, too, would fail unless the bearing having excess clearance was also replaced. For proper operation, the lubrication system must be balanced so that each bearing uses only the designed amount of oil. It must leave enough oil in the system for the remaining bearings.

Oil Passages in the Crankshaft. The crankshaft is *drilled,* as shown in Figure 11-11, to allow oil from the main bearing oil groove to be directed to the connecting rod bearings. The oil on the bearings forms a hydrodynamic oil film to support bearing loads. Some of the oil may be sprayed through a spit or bleed hole in the connecting rod. This was discussed in Section 8-7. The rest of the oil leaks from the edges of the bearing. It is thrown from the bearing against the inside surfaces of the engine. Some of the oil that is thrown from the crankshaft bearings will land on the camshaft to lubricate the lobes. A part of the throw-off oil splashes on the cylinder wall to lubricate the piston and rings. Oil from the spit hole splashes onto the piston pin, illustrated in Figure 11-12. Oil that lands on the interior walls of the engine drains back into the oil pan for recirculation through the lubricating system. Some large, heavy-duty engines have an oil passage drilled through the connecting rod to carry oil for lubricating the piston pin.

Valve Train Lubrication. The oil gallery may intersect (Figure 11-10) or have drilled passages to the valve lifter bores (Figure 11-2) to lubricate the lifters. When hydraulic lifters are used, the oil

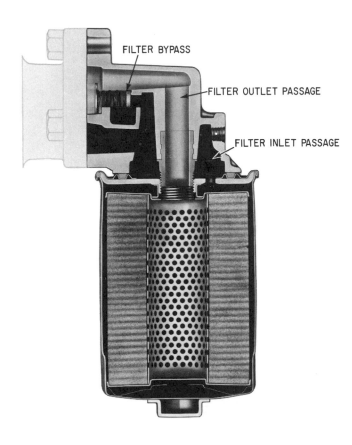

Figure 11-9. A cross section of a typical oil filter (Courtesy of AC Division, General Motors Corporation).

FILTER BYPASS

FILTER OUTLET PASSAGE

FILTER INLET PASSAGE

the oil. Emission-controlled engines also operated at higher temperatures that minimized moisture condensation and acids in the motor oil. By the late 1970s, oil-change periods had been extended to 12,000 miles (19,300 km) by some automobile manufacturers.

Oil Passages in the Block. From the filter, oil goes through a drilled hole that intersects with a drilled main oil *gallery* or longitudinal header. This is a long hole drilled from the front to the back of the block. Inline engines use one oil gallery. V-type engines may use two galleries as pictured in Figure 11-2. One main gallery and two hydraulic valve lifter galleries used on a V-type engine can be seen in

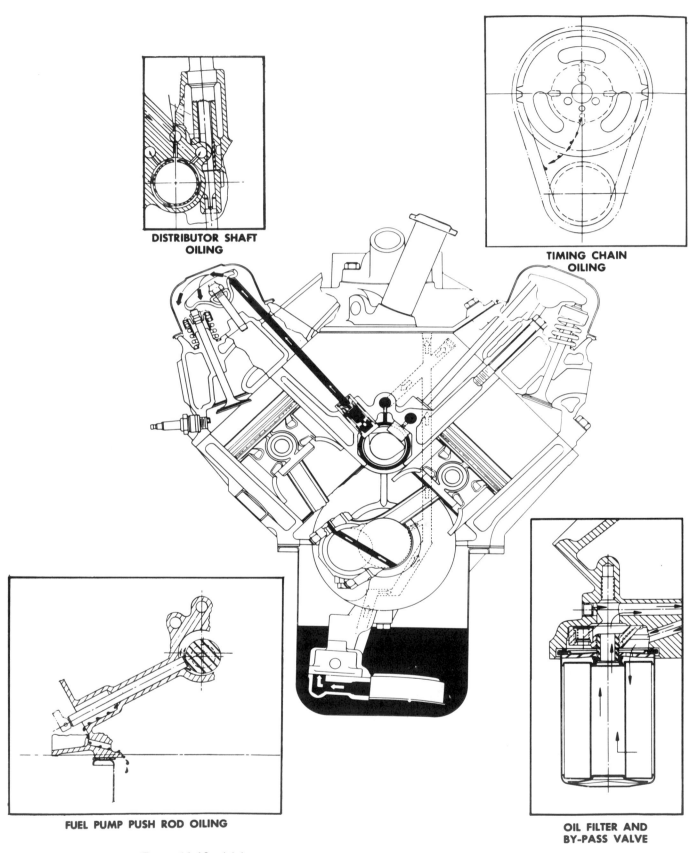

DISTRIBUTOR SHAFT OILING

TIMING CHAIN OILING

FUEL PUMP PUSH ROD OILING

OIL FILTER AND BY-PASS VALVE

Figure 11-10. A lubricating system in a typical V-type engine. This engine has one main oil gallery directly above the camshaft and two lifter gallerys (Courtesy of Chevrolet Motors Division, General Motors Corporation).

Figure 11-11. A drilled oil passage between the crankshaft main and rod bearing journals.

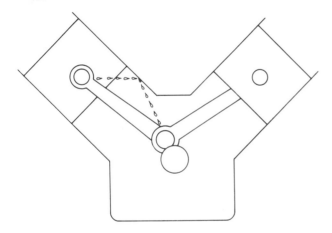

Figure 11-12. The way the piston pin is lubricated from the connecting rod spit hole.

pressure in the gallery keeps refilling them. On some engines, oil from the lifters goes up the center of a hollow pushrod to lubricate the pushrod ends, the rocker arm pivot, and the valve stem tip, as shown in Figure 11-10. In other engines, an oil passage is drilled from the gallery or from a cam bearing to the block deck, where it matches with a gasket hole and a hole drilled in the head to carry the oil to a rocker arm shaft. Some engines use an enlarged head bolt hole to carry lubricating oil around the rocker shaft capscrew to the rocker arm shaft. This design is shown by a line drawing in Figure 11-13. Holes in the bottom of the rocker arm shaft lubricate the rocker arm pivot. Mechanical loads on the valve train hold the rocker arm against the passage in the

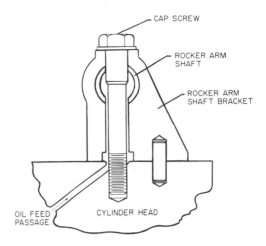

Figure 11-13. Clearance around the rocker shaft bracket cap screw makes a passage for oil to get into the rocker shaft (Courtesy of Dana Corporation).

rocker arm shaft, as shown in Figure 11-14. This prevents excessive oil leakage from the rocker arm shaft. Often, holes are drilled in cast rocker arms to carry oil to the pushrod end and to the valve tip. Rocker arm assemblies need only a surface coating of oil, so the oil flow to rocker assembly is minimized using restrictions or metered openings. The restriction or metering disc is in the lifter when the rocker assembly is lubricated through the pushrod. This can be seen in Figure 7-45. Cam journal holes that index with oil passages are often used to meter oil to the rocker shafts. This was discussed in Section 9-4.

Oil that seeps from the rocker assemblies is

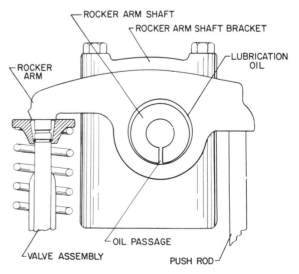

Figure 11-14. The rocker arm pivot is lubricated through the oil passage in the bottom of the rocker shaft.

WINDAGE TRAY

Figure 11-15. A windage tray attached between the crankshaft and oil pan.

returned to the oil pan through drain holes. These oil drain holes are often placed so that the oil drains on the camshaft or on cam drive gears to lubricate them.

Some engines have a positive oil flow directed to the cam drive gears or chain. This may be a nozzle or a chamfer on a bearing parting surface that allows oil to spray on the loaded portion of the cam drive mechanism.

Air in the Oil. Oil in the oil pan is affected by a number of forces. As the car accelerates, brakes, or is turned rapidly, the oil tends to move around in the pan. Pan baffles and oil pan shapes are often used to keep the oil inlet under the oil at all times. As the crankshaft rotates, it acts like a fan and causes air within the crankcase to rotate with it. This can cause a strong draft on the oil, churning it so that air bubbles enter the oil. This causes oil foaming. Oil with air will not lubricate like liquid oil, so oil foaming can cause bearings to fail. A baffle or *windage tray* is sometimes installed in engines to eliminate the oil churning problem. This may be an added part, as shown in Figure 11-15, or it may be a part of the oil pan. Windage trays have a good side effect by reducing the amount of air disturbed by the crankshaft, so that less power is drained from the engine at high crankshaft speeds.

11-2 Crankcase Ventilation

All engines have crankcase ventilation systems to remove blow-by gases from within the engine. Some older engines used a screened inlet on the oil fill cap and a draft tube outlet under the engine. This system would draw vapors from the crankcase while the car was in motion. Vehicle emission studies have shown that these blow-by gases contribute to air pollution. Vehicle emission laws made it necessary to equip engines with *positive crankcase ventilation* (PCV) systems. The draft tube was replaced with a PCV valve and connecting hoses. This system will pull the crankcase vapors into the intake manifold. The vapors are sent to the cylinders with the intake charge to be burned in the combustion chamber. Under some operating conditions, the blow-by gases would be forced back through the inlet filter. A line connecting the crankcase inlet to the carburetor air filter solved this problem. It allows back-up vapors to be drawn through the carburetor with the incoming air. This line also lets only filtered air enter the engine to help keep the oil clean. The modern PCV system, shown in the line drawing of Figure 11-16, eliminates all crankcase emissions. At the same time, it provides enough crankcase ventilation to reduce oil contamination and deposit buildup.

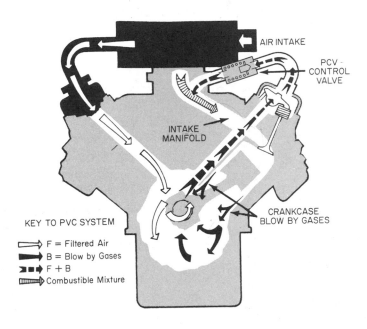

Figure 11-16. A line drawing showing the principles of positive crankcase ventilation (Courtesy of General Motors Research).

11-3 Engine Cooling System

Coolant recirculates from the radiator to the engine and back to the radiator. Low-temperature coolant leaves the bottom outlet of the radiator. It is pumped into the warm engine block, where it picks up some heat. From the block, the warm coolant flows to the hot cylinder head, where it picks up more heat. Figure 11-17 shows the coolant flow inside a typical engine. The hot coolant is returned to the top inlet of the radiator. Air flowing across the radiator cools the high-temperature coolant. As it cools, the coolant settles to the bottom of the radiator to be recycled through the engine.

Coolant Pump. The coolant pump is a *centrifugal pump;* it pulls coolant in at the center of the *impeller.* Centrifugal force throws the coolant outward so that it is discharged at the impeller tips. This can be seen in Figure 11-18. The pump is sized and the impeller is designed to absorb no more

Figure 11-17. Coolant flow through a typical V-type engine (Courtesy of Oldsmobile Division, General Motors Corporation).

Figure 11-18. Coolant flow through the impeller and scroll of a coolant pump.

Figure 11-19. An engine installation showing a belt driven coolant pump.

power than is required to provide enough coolant flow. The coolant pump is driven by a belt from the crankshaft. This can be seen in the typical engine installation shown in Figure 11-19. The belt is tightened with an idler. On most engines, the alternator serves

as the belt-tightening idler. As engine speeds increase, more heat is produced by the engine and more cooling capacity is required. The belt-driven pump increases the pump impeller speed as the engine speed increases to provide extra coolant flow at the very time it is needed.

Coolant leaving the pump impeller is fed through a *scroll*. **The scroll is a smoothly curved passage that changes the fluid flow direction with minimum loss in velocity.** The scroll is connected to the front of the engine so as to direct the coolant into the engine block. On V-type engines, two outlets are used, one for each cylinder bank. Occasionally, diverters are necessary in the coolant pump scroll to equalize coolant flow between the cylinder banks of V-type engines for even cooling of the engine.

Coolant Flow in the Engine. Coolant flows through the engine in two ways, parallel or series. In the *parallel flow system,* coolant flows into the block under pressure, then crosses the gasket to the head through main coolant passages beside *each* cylinder. The gasket openings of a parallel system are shown in Figure 11-20. In the *series flow system,* the coolant flows around all the cylinders on each bank. All of the coolant flows to the *rear* of the block, where large main coolant passages allow the coolant to flow across the gasket. Figure 11-21 shows the main coolant passages. The coolant then enters the rear of the heads. In the heads, the coolant flows forward to an outlet at the *highest point* in the engine cooling passage. This is usually

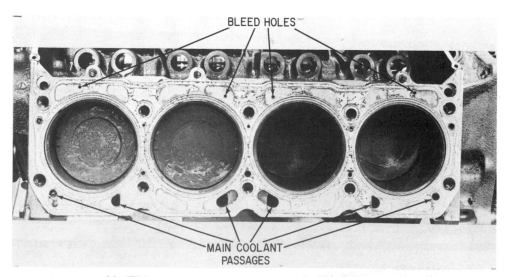

Figure 11-20. The gasket openings are shown for a parallel-type flow cooling system.

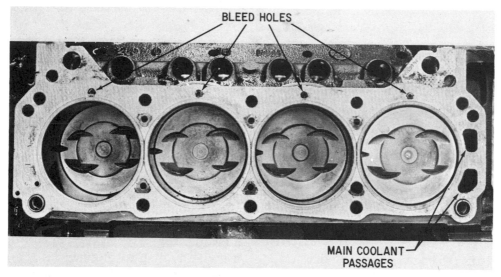

Figure 11-21. The gasket openings are shown for a series-type flow cooling system.

located at the front of the engine. The outlet is either on the heads or in the intake manifold. Some engines use a combination of these two coolant flow systems and call it a series–parallel coolant flow.

The cooling passages inside the engine are designed so that the whole system can be drained. It is also designed so that there are no pockets in which steam can form. Any steam that develops will go directly to the top of the radiator. In series flow systems, *bleed holes* or *steam slits* in the gasket, block, and head provide this function. They short circuit a very small amount of coolant at the same time. Often, this short-circuited coolant is directed against hot areas in the head, such as exhaust valves or spark plugs.

Bypass. A thermostat is located at the engine outlet to restrict coolant flow until the engine reaches the operating temperature of the thermostat. A *bypass* around the closed thermostat allows a small part of the coolant to circulate within the engine during warm-up. It is a small passage that leads from the engine side of the thermostat to the inlet side of the coolant pump. Coolant will flow through the bypass, short-circuiting the radiator whenever the coolant pressure is different on the ends of the bypass. This happens even when the thermostat is open. The bypass may be cast or drilled into the engine and pump parts. One example of this is shown in Figure 11-22. This is called an internal bypass. An external bypass is visible as a hose on the front of many engines. Figure 11-23 shows an exter-

Figure 11-22. One type of internal cooling system bypass.

Figure 11-23. One type of external cooling system bypass.

Figure 11-24. Typical automotive cooling system thermostats.

nal bypass. It connects the engine coolant outlet to the coolant pump. In some inline engines, coolant is bypassed through the intake manifold and heater core. The bypass aids in uniform engine warm-up. This eliminates hot spots and prevents excessive coolant pressure in the engine when the thermostat is closed.

Thermostat. The thermostat is a temperature-controlled valve placed at the engine coolant outlet. An encapsulated wax-based plastic pellet or aneroid bellows heat sensor is located on the engine side of the thermostatic valve. Typical automotive cooling system thermostats are pictured in Figure 11-24. The thermostat is linked to the valve to open or close it. As the engine warms, heat swells the heat sensor. A mechanical link, connected to the heat sensor, opens the thermostat valve. As the ther-mostat begins to open, it allows some coolant to flow to the radiator. This is cooled while the re-maining part of the coolant continues to flow through the bypass. In normal operation, the thermostat will be partially open. It adjusts to the engine cooling requirements that are needed to hold normal engine operating temperatures.

The coolant pump forces the coolant through the cooling system. The flow of coolant is restricted by a closed thermostat. The thermostat restriction causes the cooling pressure to rise in the engine. As the thermostat gradually opens, the coolant flow rate increases and the pressure lowers. The thermostat is wide open with maximum coolant flow only under extreme heat conditions. Idling in traf-fic, or pulling a load up a long, steep grade in warm weather are examples of operating conditions in which the thermostat may be wide open.

REVIEW QUESTIONS

1. What are the jobs that must be done by the lubricating oil within the engine? [INTRODUCTION]
2. What is motor oil's most important property? [INTRODUCTION]
3. Why are additives put in motor oil? [INTRODUCTION]
4. What is the job of the engine cooling system? [INTRODUCTION]
5. What are the advantages and disadvantages of a wet sump compared to a dry sump? [11-1]
6. Make a drawing to show how oil is pumped through a gear-type oil pump. Show the direction of oil flow and gear rotation. [11-1]
7. What problems will be caused if the oil pump is either too large or too small? [11-1]
8. What purpose does the oil pressure serve when the oil reaches the parts that are to be lubricated? [11-1]
9. What is necessary to have oil pressure in an engine? [11-1]
10. What things can be done to bring the engine oil pressure up to the regulated pressure? [11-1]
11. Where does the oil go when it flows through the relief valve? What is the advantage of using this flow method? [11-1]

12. What happens to the oil as the filter becomes partly plugged? [11-1]
13. When does oil go through the filter bypass valve? [11-1]
14. When does the filter check valve keep the oil from flowing? [11-1]
15. Why have filters been made smaller? [11-1]
16. What does an oil gallery do in an engine? [11-1]
17. What is the major cause of bearing failure? [11-1]
18. What is meant by the term "balanced oil system"? [11-1]
19. How are the cam lobes lubricated? [11-1]
20. Name two ways oil gets to the rocker arm pivots. [11-1]
21. What causes oil to foam? [11-1]
22. Why have vehicle emission laws made it necessary to equip engines with PCV systems? [11-2]
23. Why is the coolant pump called a centrifugal pump? [11-3]
24. Why does a coolant pump need a scroll? [11-3]
25. How does a parallel coolant flow differ from a series flow system? [11-3]
26. Where is the thermostat and coolant outlet located in an engine cooling passage? [11-3]
27. What is the purpose of bleed holes and steam slits in a head gasket? [11-3]
28. Why do cooling systems need a thermostat bypass? [11-3]
29. What is the job of the thermostat? [11-3]
30. Why does the thermostat cause the coolant pressure to rise in the engine? [11-3]

CHAPTER
12

Engine Assembly

The way an engine is assembled is very important to its operation and service life. First, soil and contamination should be removed from the parts before they are assembled. Second, parts should be thoroughly coated with fresh lubricant. Third, assembly bolts, cap screws, and nuts must be tightened to the correct torque.

Engine parts are cleaned by rinsing the parts in a petroleum-based cleaning fluid and drying with compressed air. The parts should then be wiped with a rag dampened with clean motor oil. This will coat them with enough surface oil to prevent rust.

During assembly, the bearings and journals should be coated with a very thin layer of an assembly lubricant that will mix with oil. White grease, which is commonly used in the auto shop, works well as an *assembly lubricant*. Heavy motor oil can be used as an assembly lubricant if there is no delay in assembling and starting the engine. Motor oil is used to coat the piston assemblies and cylinder walls before the pistons are installed in the cylinders.

This chapter on engine assembly has been placed here as a guide for students who are reassembling a study engine. The engine parts have been in the running engine together in the study engine so it is not critical to check all clearances. An assembled study engine can be given a short run without coolant to make sure that the engine will operate normally. Running the study engine also gives a reason to assemble the engine correctly. The general reassembly procedures given in this chapter also apply to an engine being assembled after overhaul. One exception is the measurement of all operating clearances. Another exception is gaskets. Study engines do not require new gaskets or seals, even when they are given a short run after assembly. A little oil leakage from the engine can be cleaned up easily. Reusing the gaskets keeps the cost low each time the study engine is run after it has been reassembled.

There are a number of differences in engines. The procedures given in this chapter can be followed when assembling most engines. Other assembly procedures will have to be used when assembling engines that differ from these.

It is assumed that the study engine has not had extensive service. The parts were operating in the engine, so they will fit and operate when the engine is reassembled. Make sure that bolts of the correct length are used. Bolts that are too long can jam into other parts. Bolts that are too short may strip threads. It is also assumed that the engine will be disassembled again, so sealing directions are not discussed. The procedures given in Chapter 18 should be followed, including sealing directions, when an engine is being assembled to be put into service.

12-1 Shaft Assembly

The engine block and shafts should be thoroughly cleaned to remove any dust or other surface contamination. The cylinders can then be coated with a thin film of motor oil to prevent surface rust. All of the bearing journals are coated with a thin film of assembly lubricant just before the shafts are installed. This will provide an initial lubrication for engine startup.

Installing the Camshaft. In this assembly procedure, it is assumed that the cam bearings have not been removed. In pushrod engines, the cam is carefully slid into place, taking extreme care to avoid scratching the cam bearings. This is done most easily with the block placed on end so that the camshaft can be lowered into the block. On overhead cam engines, the valves are usually put in place before the cam is installed (see Section 12-3). Do not allow the camshaft to hit the rear expansion or soft plug. If it does, the soft plug could loosen and cause an oil leak.

Installing the Crankshaft. It is assumed that the rear oil seal was not removed and is still in place. The main bearing saddles, caps, and the back of all the main bearing shells should be wiped clean; then the bearing shells can be put in place. It is important that each bearing tang lines up with the slot in the bearing support. The bearing shells must have some spread to hold them in the bearing saddles and caps during assembly. The surface of the bearings is then given a thin coating of assembly lubricant to provide initial lubrication for engine startup. If new bearings are to be installed, they should be checked with gauging plastic.

The crankshaft with lubricant on the journals is carefully placed in the bearings to avoid damage to the thrust bearing surfaces. Figure 12-1 shows thrust surface damage resulting from careless assembly. The bearing caps are installed with their identification numbers correctly positioned. The caps were originally machined in place, so they can only fit correctly in their original position. The main bearing cap bolts are tightened finger-tight and the crankshaft is rotated. It should rotate freely. Pry the crankshaft forward and rearward as shown in Figure 12-2. This will align the cap half with the block saddle half of the thrust bearing. The dial gauge will indicate the amount of thrust bearing clearance.

Tightening Procedure—Main Bearing. The following assembly procedure will identify any tight bearings. Tighten the front main bearing cap to the specified assembly torque; then rotate the

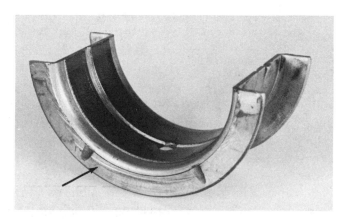

Figure 12-1. Damage to the thrust surfaces of the thrust bearing from careless assembly.

Figure 12-2. Prying the crankshaft forward and backward before tightening the main bearing caps will align the thrust bearings. A properly positioned dial gauge will indicate the crankshaft end clearance.

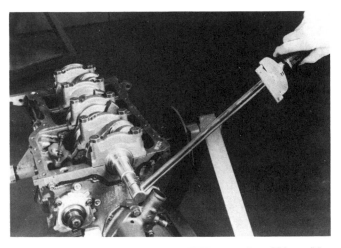

Figure 12-3. Measuring the crankshaft turning torque after each main bearing cap is properly tightened. An abnormal increase in torque indicates a problem that should be corrected before additional assembly.

crankshaft with a torque wrench (Figure 12-3). Use either the front damper bolt or one of the flywheel bolts. The exact amount of torque is not important. The important thing to watch for is any large increase in torque required to rotate the crankshaft. Observe and record the torque required to rotate the crankshaft with only the front main cap torqued. Proceed to the next bearing and tighten it in the same way. Again, measure the torque required to rotate the crankshaft. Follow this same pro-

Figure 12-4. A particle not cleaned from behind a bearing shell during a previous assembly.

cedure on all the main bearings. If tightening any one of the main bearing caps causes a large increase in the torque required to rotate the crankshaft, immediately stop the assembly process. Determine the cause of the increase in torque before any further assembly. It may be necessary to remove the crankshaft and bearings to find the cause of the torque increase. An increase in the torque needed to rotate the crankshaft is often caused by a foreign particle that was not removed during cleanup. It may be on the bearing surface, the crankshaft journal, or between the bearing and saddle. An example of a particle left in a study engine in a school auto shop is pictured in Figure 12-4. The damage it did to the bearing is shown in Figure 12-5. The shaft should turn *freely* after fully torquing all the main bearing cap bolts. It should never require over 5 ft-lb (6.75 N•m) of torque to rotate the crankshaft.

Timing Drives—Pushrod Engines. On pushrod engines, the timing gears or chain and sprocket should be installed next. The timing marks on the gears must line up when installed, as shown in Figure 12-6. The same is true of the sprockets, as shown in Figure 12-7. When used, the replaceable fuel pump eccentric is installed as the cam sprocket is fastened to the cam. The crankshaft should be rotated several times to see that the camshaft and timing gears or chain rotates freely. The timing mark alignment should be rechecked at this time. If the engine is equipped with a slinger ring, it should

Figure 12-5. Damage done to the bearing as a result of the particle shown in Figure 12-4.

Figure 12-6. Timing marks on the timing gears are aligned to time the camshaft to the crankshaft.

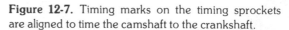

Figure 12-7. Timing marks on the timing sprockets are aligned to time the camshaft to the crankshaft.

also be installed on the crankshaft in front of the crankshaft gear. The slinger ring is positioned so that the outer edge faces away from the gear or sprocket, as shown in Figure 12-8.

It is assumed that the front oil seal is still in place. The timing cover and gasket are placed over the timing gears and/or chain and sprockets. The attaching bolts are loosely installed to allow the damper hub to align with the cover as it fits in the seal. The damper is installed on the crankshaft. On some engines, it is a press fit and on others it is held with a large center bolt. After the damper is secured, the timing cover with attaching bolts can be tightened to the specified torque.

12-2 Piston and Rod Assembly

The pistons are reinstalled on the rod if they have been removed. Care must be taken to make sure that the pistons and rods are in the correct cylinder. They must face the correct direction. There is usually a notch on the piston head indicating the front. This will correctly position the piston pin offset toward the right side of the engine. The connecting rod identification marks on pushrod inline engines are normally placed on the camshaft side. This is also the oil filter side of most engines. The notch and numbers on a piston and rod assembly can be seen in Figure 12-9. On V-type engines, the connecting rod cylinder identification marks are on the side of the rods that can be seen

Figure 12-8. The slinger ring is positioned with the outer edge facing away from the timing sprocket.

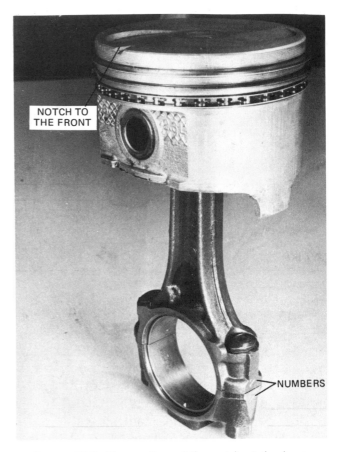

Figure 12-9. The position of the notch at the front of the piston and the connecting rod numbers are shown in this illustration.

from the bottom of the engine when the piston and rod assemblies are installed in the engine. For assembly, the piston and rod are supported on an assembly fixture in a press. The piston pin is inserted in the piston and aligned with the rod hole. The pin is pressed into the rod hole until it is properly centered in the piston and rod using fixtures similar to the one pictured in Figure 12-10. The service manual should be checked for any special piston and rod assembly instructions.

Piston Rings. Each piston ring, one at a time, should be placed backward in the groove in which it is to be run. Its *side clearance* in the groove is checked with a feeler gauge, as shown in Figure 12-11. If a ring is tight at any spot, check for deposits or burrs in the ring groove. Each piston ring, one at a time, is then placed in the cylinder in which it is to operate.

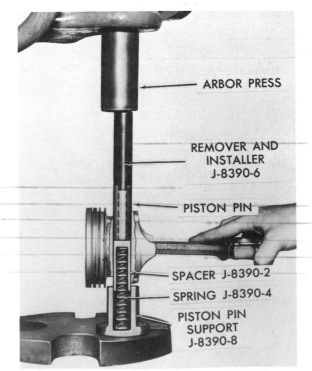

Figure 12-10. Typical fixture used to press a piston pin into a rod (Courtesy of Cadillac Division, General Motors Corporation).

Figure 12-11. The side clearance of the piston ring is checked with a feeler gauge.

Inverting the piston, push each ring into the lower quarter of the cylinder (Figure 12-12), then measure the *ring gap* (Figure 12-13). It should be approximately 0.004 in. for each inch of bore diameter

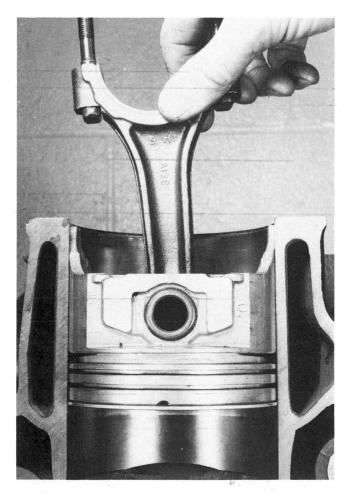

Figure 12-12. A piston is used to push the ring squarely into the cylinder.

(0.004 mm for each centimeter of bore diameter). Leave the rings in the cylinder until they are to be installed on the piston. This will keep them from being mixed up or damaged.

The oil rings are installed first. The expander-spacer of the oil ring is placed in the lower ring groove. One oil ring rail is carefully placed above the expander-spacer by winding into the groove. The other rail is placed below the expander-spacer. The ring should be rotated in the groove to make sure the expander-spacer ends have not overlapped. If they have, the ring must be removed and reassembled correctly.

Figure 12-13. The ring gap is measured with a feeler gauge with the ring positioned squarely in the cylinder.

The compression rings require the use of a *piston ring expander* that will only open the ring gap enough to slip the ring on the piston. Figure 12-14 shows one of the best types of piston ring expanders in use for engine repair. Be careful to install the ring with the correct side up. The top of the compression ring is marked with a dot, the letter T, or the word TOP (Figure 12-15). After the rings are installed, they should be rotated in the groove to make sure that they move freely. The rings should be checked to make sure they will go fully into the groove so that the ring face is flush with the surface of the piston ring lands. Usually, the rings are placed on all of the pistons before any of the pistons are installed in the cylinders.

Installing Piston and Rod Assemblies. The cylinder is wiped with a cleaning cloth. It is then given a liberal coating of clean motor oil. This oil is spread over the entire cylinder wall surface by hand.

Figure 12-14. A good type of ring expander being used to install a piston ring.

Figure 12-15. Identification marks used to indicate the side of the piston ring to be placed toward the head.

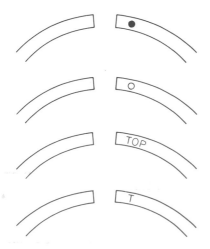

The connecting rod bearings are prepared for assembly in the same way as are the main bearings. The piston is then inverted and slushed around in clean petroleum-based cleaning fluid to remove all particles of dirt that may have gotten on them while installing the rings. The piston is air-dried, then dipped in a bath of clean motor oil.

When the piston is lifted from the oil, it is held to drip a few seconds. This will allow the largest part of the oil to run out of the piston and ring grooves. The *piston ring compressor* is then put on the piston to hold the rings in their grooves.

The bearing cap is removed from the rod and protectors are placed over the rod bolts. The crankshaft is rotated so that the crankpin is at the *bottom* center. The upper rod bearing should be in the rod and the piston should be turned so the notch on the piston head is facing the front of the engine. The piston and rod assembly is placed in the cylinder through the block deck. The ring compressor must be kept tightly against the block deck as the piston is pushed into the cylinder. This procedure is illustrated on a cutaway V-type engine pictured in Figure 12-16. The ring compressor holds the rings into their grooves so that they will enter the cylinder. Care should be taken to make sure that the rod or the bolts do not hit the cylinder wall or the crank pin journal. If they hit, they will cause a small nick. The nick will cause scoring. Figure 12-17 shows rod bearing damage caused by a crankshaft that was nicked during assembly. The piston is pushed into the cylinder until the rod bearing is fully

Figure 12-16. Installing a piston and rod assembly using a cast iron ring compressor and short pieces of hose placed over the connecting rod bolts.

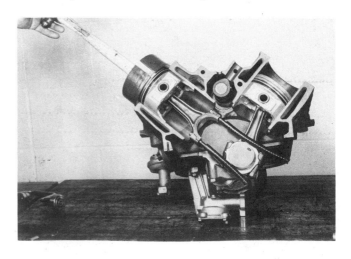

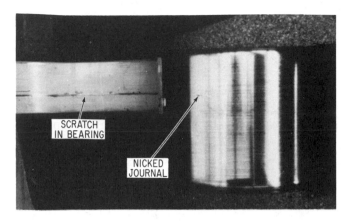

Figure 12-17. A crankshaft journal that was slightly nicked during assembly and the damage it did to the bearing as the engine ran.

seated on the journal. The rod cap, with the bearing in place, is put on the rod. The retaining nuts or cap screws are installed only finger-tight at this time. All of the piston assemblies should be installed before tightening any of the connecting rod retaining nuts or cap screws.

Tightening Procedure—Rod Bearing. Check and record the torque required to rotate the crankshaft with all of the piston rings dragging on the cylinder walls. Next, the retaining nuts on one bearing should be torqued, then the torque required to rotate the crankshaft should be rechecked and recorded. Follow the same procedure on all the rod bearings. If tightening any one of the rod bearing caps causes a large increase in the torque required to rotate the crankshaft, immediately stop the tightening process. Determine the cause of the increased rotating torque the same way as was done on the main bearings. When all the rod bearing retaining bolts are properly torqued, the total increase in the crankshaft turning torque should not be more than 5 ft/lb above the torque first recorded with the rod bearing retaining bolts finger-tight.

Rotate the crankshaft several revolutions to make sure that the assembly is turning freely and there are no tight spots. The bottom of the engine is then ready for final assembly.

Oil Pan. It is recommended that the oil pump cover be removed and the pump gears heavily coated with assembly lubricant. This will provide initial lubrication and oil pump priming. The cover is installed

and the retaining screws properly torqued. pump is installed, making sure that the pump drive is placed in position. If this is left out, the pan will have to be removed when the faulty assembly is discovered. A number of engines have a short oil pump drive shaft that fits between the oil pump and the bearing boss in the block. A ring pressed on the top end of the shaft keeps the shaft from coming out of the pump when the distributor is taken out of the engine after it has run for some time.

On some engines, the oil pump pickup screen and pickup tube is attached to a passage on the block. Care must be taken to position the screen so that it fits flat in the bottom of the pan. It is also important that there is no leakage at the joint where the pickup tube attaches to the block. This is one of the places where there *must* be a good gasket if the engine is to be run.

With the oil pump in place, the oil pan gaskets are properly positioned. The oil pan is carefully placed over the gaskets. All of the oil pan bolts should be started into their holes before any are tightened. The bolts should be alternately snugged up; then they should be properly torqued.

The oil filter should be installed with the proper gasket. The surface of the gasket is coated with oil. The filter is turned on the adapter until the gasket touches the seat. The filter is tightened an additional one-fourth to one-half turn by hand. Do not use wrenches to tighten the filter.

12-3 Valve Assembly

The head, valves, and valve spring parts should be rinsed with petroleum-based cleaning fluid, then dried with compressed air. The valve guides should be given special cleaning attention because it is easy to overlook dirt inside a hole.

Installing Valves. Valves are assembled in the head, one at a time. The valve guide and stem are given a liberal coating of motor oil and the valve is installed in its guide. The valve is held against the seat as the valve spring seat or spacer, valve spring, valve seals, and retainer are placed over the valve stem. One end of the *valve spring compressor* pushes on the retainer to compress the spring. The valve locks are installed while the valve spring is compressed, as shown on a cutaway head in Figure 12-18. The valve spring compressor is slowly and carefully released while making sure that the valve locks seat properly between the valve stem grooves and the retainer. The assembled valve should *not* be hit with a hammer. Hitting might start a very small crack that could lead to premature valve failure. Each valve is assembled in the same manner.

Installing the Camshaft—Overhead-Cam Engines.

The camshaft is usually installed on overhead-cam engines before the head is fastened to the block deck. Some engines have the camshaft located directly over the valves. The cam bearings on these engines can be either one-piece or split. A fixture is required on one engine that has a one piece cam bearing to hold the valves open as the cam is slid endways into the cam bearings. This is shown in Figure 12-19. The cam bearings and journals are lubricated before assembly. In others, the camshaft bearings are split to allow the camshaft to be installed without depressing the valves. This type is shown in Figure 12-20. The caps are tightened evenly to avoid bending the camshaft. The valve clearance or lash is

Figure 12-18. Installing valve locks while the valve spring is compressed.

checked with the overhead camshaft in place. Some bucket-type cam followers can be adjusted by turning a wedge-shaped screw with an internal hex wrench (Figure 12-21). Others have shims under a follower disc. One example of this type can be seen in Figure 12-20. On these, the camshaft is turned so that the follower is on the base circle of its cam. The clearance of each bucket follower can then be checked with a feeler gauge. The amount of clearance is recorded and compared to the specified clearance. The cam is then removed and shims of the required thickness are put in the top of the bucket followers. If the clearances are within specifications, the cylinder head is ready to be installed. If not, the clearances must be corrected.

Figure 12-19.—A fixture being used to hold bucket-type cam followers down as the camshaft is being slid endways into the cam bearings on an overhead cam engine.

The follower pivot should be put in place after the cam is installed on the finger follower type of overhead cam mechanism. Each valve spring is slightly depressed to install the finger follower. This can be done with a special lever tool designed for this job or it can be done using a common valve spring compressor with a flat plate across the combustion chamber, as shown in Figure 12-22. Mechanical pivots must be adjusted to provide the specific clearance or lash while the follower is on the base circle of the cam. Hydraulic pivots will automatically adjust to zero clearance.

12-4 Head and Valve Train

The block deck and head surfaces should be rechecked for any handling nicks that could cause a gasket leak. There are usually alignment pins or

Figure 12-20. Split type cam bearings on an overhead cam engine using bucket type cam followers. One bucket follower is removed to show the valve tip, retainer, and lock.

Figure 12-22. Installing a finger follower while the valve assembly is held open with a valve spring compressor.

Figure 12-21. Cam follower adjustment by turning a wedge-shaped screw.

rear and side to side. This is illustrated in Figure 12-23. The bolts are usually tightened to approximately half the specified torque following the tightening sequence. They are then retorqued to the specified torque following the same tightening sequence.

Timing Drives—Overhead-Cam Engines. With the head bolts properly torqued, the cam drive should be installed on overhead cam engines. This is done by aligning the timing marks of the crankshaft and camshaft drive sprockets with their respective timing marks. The location of these marks differs between engines, but the marks can be identified by looking carefully at the sprockets. The timing belt or chain must be tight on the left side of the engine.

dowels at the front and rear of the block deck to position the gasket and head. Care should be taken to properly position any marked head gasket (up, top, front, etc.). The gasket and head are placed on the block deck. All of the head bolts are loosely installed. Very often, the head bolts have different lengths. Make sure that the bolt of correct length is put into each location.

Head bolts are tightened following a sequence specified in the service manual. In general, the tightening sequence starts at the center of the head and moves outward, alternating between front to

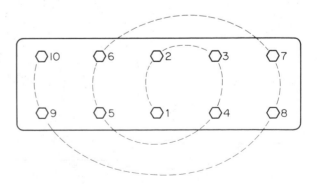

Figure 12-23. A typical cylinder head tightening sequence.

The crankshaft sprocket pulls the cam sprocket around from this side. It is on *your* right as you look at the front of the engine. This can be seen in Figure 12-24. The tightening idler may be on either or both sides of the timing belt or chain. After the camshaft drive is engaged, rotate the crankshaft through two full revolutions. On the first full revolution, you should see the exhaust valve almost close and the intake valve just starting to open when the *crankshaft* timing mark aligns. At the end of the second revolution, both valves should be closed and all of the timing marks should align. This is the position the crankshaft should have when cylinder number 1 is to fire.

12-5 Valve Lash

The following discussion assumes that no valve servicing has been done on a study engine. If valve and seat servicing have been done, check Chapter 18 for additional assembly procedures that must be

Figure 12-24. A timing belt that drives the cam and auxiliary shaft.

followed. The outside of the lifters and the lifter bores in the block should be cleaned and coated with assembly lubricant. The lifters are installed in the lifter bores and the pushrods put in place. There are different-length pushrods on some engines. Make sure that the pushrods are installed in the proper location. The rocker arms are then put in place, aligning with the valves and pushrods. Rocker arm shafts should have their retaining bolts tightened a little at a time, alternating between the retaining bolts. This keeps the shaft from bending as the rocker arm pushes some of the valves open.

Hydraulic Lifters. The retaining nut on some rocker arms mounted on studs can be tightened to a specified torque. The rocker arm will be adjusted correctly at this torque when the valve tip has the correct height. Other types require tightening the nut to a position that will center the hydraulic lifter. The general procedure for adjusting the hydraulic lifter types is to tighten the retaining nut to the point that all the free lash is gone. With the manifold or lifter cover plate off the engine, it is easy to see this point, as shown in Figure 12-25. The lifter plunger starts to move down after the lash is gone. From this point, the retaining nut is tightened a specified amount, such as three-fourths of a turn, or one-and-one-half turns.

Solid Lifters. The valve clearance or *lash* must be set on a solid lifter engine. This is required so that the valves can positively seat. Some service manuals give an adjustment sequence that can be followed to set the lash. If this is not available than the following procedure can be used on all engines requiring valve lash adjustment. The valve lash is adjusted

Figure 12-25. Hydraulic lifter can be seen with the intake manifold off from the engine.

with the valves completely closed. The crankshaft is rotated in the normal direction until both valves of cylinder number 1 are closed and the timing marks on the damper align with the top dead center (TDC) indicator. This places the engine at the start of the power stroke when ignition should take place. In this crankshaft position, the combustion pressure would be high. Both valves are on the base circle of their cam lobes, so the valves are positively closed. If the valve lash is specific for an operating engine, the initial valve lash on a cold engine should be adjusted to 0.002 in. (0.05 mm) greater than the operating valve lash specification. This cold valve lash adjustment will result in an operating valve lash very near the specified valve lash after the engine has been warmed. Two methods used to adjust the valve lash are shown in Figure 12-26.

After setting the valve lash on cylinder number 1, the crankshaft is rotated in its normal direction of rotation to the next cylinder in the firing order. This is done by turning the crankshaft 90° on eight-cylinder engines, 120° on even-firing six-cylinder engines, and 180° on four-cylinder engines. The valves on this next cylinder are adjusted in the same manner as those on cylinder number 1. This procedure is repeated on each cylinder *following the engine firing order* until all the valves have been adjusted. The service technician will usually go through the valve lash check a second time to make sure that the lash has been adjusted correctly.

The same valve lash adjustment sequence is used on overhead cam engines. These engines with rocker arms or with adjustable finger follower pivots are adjusted in the same way the rocker arms

are adjusted on pushrod engines. Engines with bucket cam followers have a different adjustment procedure. Some are adjusted by placing a shim with correct thickness in the shim pocket. Others are adjusted with a tapered adjusting screw. These adjustments can be seen in Figure 12-27.

12-6 Manifolds and Covers

When used, lifter covers with gaskets are installed on the engine. The intake manifold gasket for a V-type engine may be a one-piece gasket or it may have several pieces. The gasket types were discussed in Section 10-3. V-type engines with open-type manifolds have a cover over the lifter valley. The cover may be a separate part or it may be part of a one-piece intake manifold gasket. Closed-type intake manifolds on V-type engines require gasket pieces at the front and rear of the intake manifold. Inline engines usually have a one-piece intake manifold gasket.

The intake manifold is put in place over the gaskets, making sure the gaskets do not slip out of place. The correct-length manifold bolts are installed. The bolts are tightened to the specified torque following the tightening sequence specified in the applicable service manual.

Some exhaust manifolds do not use gaskets; others do. The gaskets and exhaust manifolds are installed. The exhaust operates at very high temperatures, so there is usually some expansion-and-contraction movement in the manifold to head joint. It is very important to use attachment bolts, cap screws, and clamps of the correct type and

Figure 12-26. Two methods used to adjust valve locks. (a) Adjusting the rocker arm pivot and (b) making the adjustment at the push rod end of the rocker arm.

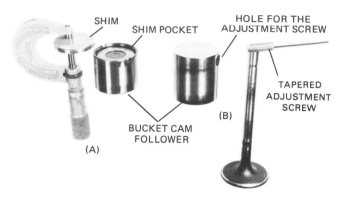

Figure 12-27. Two methods used to make adjustments on bucket-type cam followers. (a) Using different shims on the left and (b) using a tapered adjustment screw on the right.

length. They must be properly torqued to avoid both leakage and cracks.

All that remains to enclose the engine is to install rocker covers on pushrod engines and valve assembly covers on overhead cam engines. With gaskets properly attached, these covers are installed and properly fastened.

The water pump with its gaskets is attached to the front of the engine. The thermostat, housing, and gaskets are installed. Where used, a bypass hose is fastened in place.

If the engine is to be stored all openings should be plugged or taped closed. If the engine is to be run, the required accessories should be installed and connected, with the exception of the distributor.

12-7 Preparation For Operation

It is desirable to run the study engine. It can be run without coolant if a serviceable starter, flywheel or flex plate, carburetor, fuel pump, distributor, coil, spark plugs, and oil pressure indicator are installed. The engine does not have to be mounted in a running stand, although this is desirable. A temporary setup can be used to run the engine. The engine can be run if it is resting on blocks on the floor and supported above by a sling and chain fall or hoist. Jumper cables can be used to make connections from the battery to the engine frame and starter. Small jumper wires can be used to connect the ignition system. A small fuel tank can be attached to the fuel pump. The carburetor can be operated by hand.

Oil Pressure. With oil in the engine and the distributor out of the engine, oil pressure should be established before the engine is started. This can be done on most engines by rotating the oil pump by hand. A socket speed handle makes an ideal crank to turn the oil pump. A flat blade adapter that fits the speed handle will operate on General Motors engines. The V-type Chrysler engine requires the use of the same flat blade adapter, but it also requires an oil pump drive. One can be made by removing the gear from an old oil pump hex drive shaft. A one-quarter-inch drive socket can be used on Ford engines. Examples of these are pictured in Figure 12-28. Engines that do not drive the oil pump with the distributor will have to be cranked with the spark plugs removed to establish oil pressure. The load on the starter and battery is reduced with the spark plugs out so that the engine will have a higher cranking speed.

Initial Ignition Timing. After oil pressure is established, the distributor can be installed. Rotate the crankshaft in its normal direction of rotation until there is compression on cylinder number 1. This can be done with the starter or by using a wrench on the damper bolt. The compression stroke can be determined by covering the number 1 spark plug opening with a finger as the crankshaft is rotated. Continue to rotate the crankshaft slowly as compression is felt until the timing marks on the damper line up with the timing indicator on the tim-

Figure 12-28. Drivers used to rotate oil pumps to build up oil pressure before installing the distributor.

ing cover. You will recall that at this crankshaft position, both valves are closed. With the timing marks aligned, the crankshaft is set at the point ignition *should* occur. Do not turn the crankshaft from this position until the distributor has been installed.

The direction the distributor rotates can be easily determined by looking at the vacuum advance unit. The arm on the vacuum advance unit points in the direction of distributor rotation. This can be seen in Figure 12-29. The position of the vacuum advance unit must be located properly on the engine. The distributor cap tower in which the number 1 spark plug cable is to be placed must also be determined. This is called the number 1 tower. With the distributor cap lifted straight up from the distributor, rotate the distributor in its normal direction until the rotor points toward the number 1 cap tower. In a breaker point distributor, the points should be just ready to open. In the pulse inductor distributors, a tooth must line up with the timer core. If the distributor is turned any farther in the normal direction of rotation, the points would open or trigger the inductor. At this point, the distributor *will* fire the number 1 spark plug. The angle of distributor gear drive will cause the distributor rotor to turn a few degrees when installed. Before installing the distributor, the shaft must be positioned to compensate for the gear angle. If, after installation, the rotor is in the wrong position, the distributor will have to be lifted just enough to clear the gear

teeth. The rotor is turned just enough to allow the distributor gear to mesh with the next drive gear tooth. A little practice will enable you to estimate the amount of turn to be expected between the gear teeth.

Sometimes, the distributor does not go all the way down into the engine because the end of the distributor shaft does not mesh with the oil pump drive. A simple method to get the distributor shaft to line up with the oil pump drive is to crank the engine with the distributor installed as far as possible in the engine. As the engine is cranked, the rotating distributor shaft will line up with the oil pump drive and then the distributor can be fully seated. The engine will have to be cranked through *two full revolutions* to correctly realign the timing marks on the compression stroke. The breaker points or inductor should be set visually. With the distributor hold-down clamp loose, rotate the distributor housing slightly in a direction opposite to the direction of rotor rotation. The breaker points must be set so that they are just ready to be opened by the distributor cam (Figure 12-30). On inductive pickup-type distributors, the tooth of the rotor and the timer core must line up (Figure 12-31). This distributor position is close enough to the basic timing to start the engine. If the distributor hold down clamp is slightly loose, the distributor housing can be adjusted to make the engine run smoothly after the engine has been started.

Figure 12-29. The location of the vacuum advance unit indicates the direction of the distributor rotor rotation.

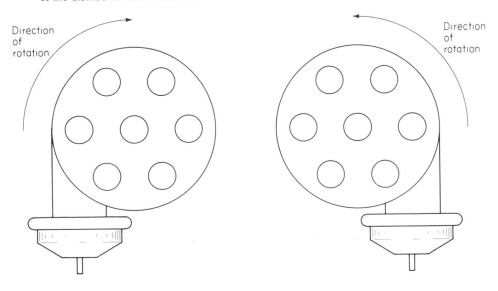

Figure 12-30. The breaker points are just ready to open to fire the spark plug as the cam turns counterclockwise.

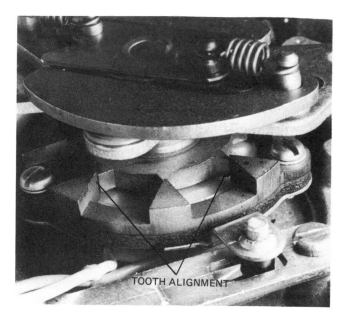

TOOTH ALIGNMENT

Figure 12-31. The rotor and timer core are lined up, ready to fire the spark plug.

REVIEW QUESTIONS

1. What precautions should be observed when installing a camshaft in a push-rod engine? [12-1]
2. What precautions should be taken when installing the main bearing shells? [12-1]
3. What precautions should be taken when installing the crankshaft? [12-1]
4. Why should a crankshaft be pried forward and backward before tightening the main bearing caps? [12-1]
5. Describe the procedure recommended for tightening the main bearing caps to prevent tight bearings. [12-1]
6. What indicates tight bearings during the main bearing tightening procedure? [12-1]
7. How should the crankshaft rotate after fully tightening all of the main bearing cap bolts? [12-1]
8. Where does the front slinger ring fit and how is it properly positioned? [12-1]
9. What precautions should be taken to make sure the pistons and connecting rods are correctly positioned in the cylinder? [12-2]
10. How is the piston ring side clearance checked? [12-2]
11. How is the ring gap properly measured? [12-2]
12. What precautions should be taken when installing the oil rings? [12-2]
13. What precautions should be taken when installing the compression rings? [12-2]
14. How is the piston and connecting rod assembly prepared for installing in the cylinder? [12-2]

15. How is the piston and connecting rod assembly installed in the engine? [12-2]
16. What procedures should be followed to make sure the connecting rod cap nuts are tightened with no binding of the bearings? [12-2]
17. What should be done to make sure the oil pump is properly primed? [12-2]
18. Why is the oil pump drive on many engines installed before the oil pump is fastened in place? [12-2]
19. What precautions should be observed when installing the oil pump pick-up screen? [12-2]
20. How are the valves, guides, and valve assemblies prepared for assembly? [12-3]
21. How are the valve assemblies put in the engine? [12-3]
22. Why should you avoid hitting the valve tip after assembly? [12-3]
23. How is the camshaft installed in an overhead cam engine? [12-3]
24. How is the valve lash adjusted? [12-3]
25. How are finger followers installed? [12-3]
26. What precautions should be observed when installing the head gasket? [12-4]
27. What procedures should be followed for installing and torquing the head? [12-4]
28. Which side must be tight when installing the timing belt? [12-4]
29. How can you tell when the crankshaft is in position for cylinder number 1 to fire? [12-4]
30. What precautions should be observed when assembling the valve train? [12-5]
31. What is the general procedure used to tighten rocker arm retaining nuts so that the hydraulic lifter travel will be centered? [12-5]
32. How can you be sure the valves are on the base circle of the cam when making the valve lash adjustment? [12-5]
33. What precautions should be observed when installing the manifolds? [12-6]
34. How is the oil pressure established before the engine is started? [12-7]
35. Describe the procedure used to set the basic ignition timing before the engine is started. [12-7]

PART
III

ENGINE REPAIR

This part of this book describes the reconditioning of an engine for use in an automobile. The material is presented in the same order as the engine work is normally performed. The discussion starts with the way to determine the engine problem. This is followed by a discussion of how to recognize the cause of the problem. The discussion then goes into methods used to recondition the parts. The next section discusses the techniques and procedures of engine assembly. Engine repair problems that can be done without removing the engine from the chassis are discussed in Chapter 19.

This text is not designed to be used in place of vehicle service manuals. Rather, it is designed to supplement them by discussing the reasons for the reconditioning procedures. The engine section of the service manual is needed for special disassembly and reassembly procedures, for size and clearance specifications, and for tightening torques.

Every effort is made to recommend correct reconditioning practices. Repair to ''get by for awhile'' is discouraged. There are, however, some jobs that can be done correctly without reconditioning the entire engine. Replacing connecting rod bearings or replacing the timing chain are examples of these jobs. It should be remembered that disassembly, cleaning, resealing, and reassembly costs the same, whether the job is done correctly or if it is patched up.

Two reconditioning standards are discussed: commercial standards and precision standards. *Commercial standards* are those followed in repairs made for the general motoring public. They provide a satisfactory repair at the lowest cost. *Precision standards* are followed by the hobbyist and the racing mechanic. These standards are as close to perfection as they can be. Precision standards require the use of expensive shop equipment and a lot of labor time by highly paid skilled technicians. It is, therefore, very expensive to recondition an engine to precision standards. When it is done, it is called engine *blueprinting*. Technicians who are used to working with precision standards have difficulty accepting commercial standards. This section of the book discusses both commercial and precision standards.

The reason for the repair and the standards to which parts should be reconditioned are stressed, regardless of the type of engine rebuilding equipment available for use. No effort is made to teach equipment operation. When describing a process, the reason is discussed first. This is followed by what is done to make the repair and what it should be like when the job is completed. The reader or the automotive instructor will decide on the specific repair procedure to be used.

It is important to keep records of everything you find as you work on an engine. First, keep track of all abnormal conditions and where they are in the engine. Keep a record of all measurements taken. In general, maximum service limits are twice the limits of new parts. The records you keep will be very useful when deciding on the repairs to be made.

It is important to stress safety again. Safety glasses should be worn at *all* times when working in the shop. Keep the work area and equipment clean. The disassembled engine parts should be kept in order. They should be marked or tagged where their placement in the engine is critical.

CHAPTER
13

Determining the Engine Condition

There are three reasons engines require service. First, periodic service should be done to maintain the vehicle to reduce the chance of failure. An oil change and a valve lash adjustment on a solid lifter valve system are examples of periodic service. Second, engine service is required when the customer wants a change in the appearance or performance of the engine. Examples of this type of service include changing manifolds or a carburetor. Third, an engine is serviced when a problem exists and needs to be corrected. Examples of this type of service include replacing valves, piston rings, and bearings. When service is required only three things are done: adjust, repair, or replace the part. The job of the automotive service technician is to (1) verify the complaint, (2) decide what is to be done to correct the problem, (3) make the repairs, and (4) recheck the engine to make sure that the repairs have corrected the problem. This final step is often called verifying the repair.

Repairs are routine once a decision has been made to recondition an engine. Disassembly procedures are followed for the engine being repaired. Serviceable parts of the engine are reconditioned. This is usually done at an automotive machine shop. The engine is then carefully reassembled and put back into service. Determining if the engine should be reconditioned is *not* routine. The following discussion will be a guide to you in helping your customer decide on which alternative way the engine problem should be corrected.

An engine gradually wears out as it runs. Most engines with

normal maintenance will run well over 100,000 miles (160,000 km). As the engine reaches high mileage, some part failure will begin to occur. This failure is considered to be *mature failure*. At high mileage, the performance of the engine has gradually been reduced and the cost to keep it operating has increased.

When an engine problem is caused by an unusual condition, rapid wear, or unexpected failure, it is called a *premature failure*. While the engine is being repaired, the cause of premature failure should be identified and corrected before the engine is reassembled and returned to service.

Most engine repair work is done to correct a part that has prematurely failed, rather than to recondition the entire engine. Examples of this include the repair of failed valves, timing gear, and oil leaks.

Going back to basics, several questions must be answered by the service technician. How does the customer know there is a problem in the engine? How do you determine if the engine should be reconditioned? Can the problem be corrected without disassembling all or part of the engine? Should the engine be replaced or be repaired?

13-1 Recognizing the Problem

Performance gradually decreases as the engine increases in mileage. The change occurs so slowly that the operator may not even notice it. In most cases, however, the operator is the first person to be aware of an engine problem. The operator will observe the vehicle instruments, especially indicator lights. The change in engine sound, unusual smells, vapors and smoke, and a reduction in performance all indicate to the operator that the engine is not functioning normally.

Engine Instruments. The operator depends on the oil pressure indicator to suggest that the engine has a problem. The engine should be turned off *immediately* if the engine loses oil pressure. If the engine is run without oil pressure for any length of time, the bearings will score. This will generally cause heavy crankshaft and bearing wear. An example of this is shown in Figure 13-1. If operation is continued, the worn bearing clearance becomes so large that it will produce a knocking sound. The

(a)

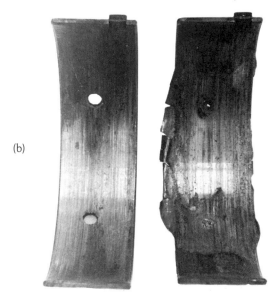

(b)

Figure 13-1. One result of oil pressure loss. (a) extreme wear of the connecting rod journal and (b) overheating and failure of the bearings.

knock may cause one of the connecting rods to break. When no oil is thrown on the cylinder wall by the connecting rods, the pistons will scuff and seize. If the engine is damaged to this extent, there will be very few usable parts remaining in the block.

The engine heat indicator is another important instrument. Problems can occur if the coolant temperature is either too hot or too cold. If the temperature is too hot, the oil on the cylinder wall will become too thin. Excessive temperature will also cause the pistons to expand more than they

Figure 13-2. A badly scuffed piston caused by an overheated engine.

parts wear and the performance changes slowly so the operator does not notice these changes. The performance has to become very bad or change suddenly before the operator is aware that an engine problem exists. Once the problem is noticed, the operator will have the automobile checked by the service technician.

Operating conditions that indicate a problem include hard starting, lack of power, and rough running. Also included are some of the sounds an engine produces, such as a ticking sound, a knock, or a squeak. The valve train and combustion will make sounds at one-half engine speed. Sounds coming from rotating parts are generally continuous. A dead miss on one cylinder is both heard and felt. It could be caused by a mechanical fault in one cylinder. It could also be caused simply by one spark plug cable that has fallen off from the spark plug.

were designed to do. These two conditions will cause the piston metal to contact the cylinder walls and this will produce severe scuffing. Figure 13-2 shows a scuffed piston. In extreme cases, the pistons will seize in the cylinders.

Low-temperature problems do not cause failure as rapidly as do high-temperature problems. Low-temperature engine operation does not produce enough heat to evaporate all of the moisture that accumulates in the oil pan as a result of the combustion process. The moisture stays in the oil pan and mixes with the combustion by-products. This will form sludge and cause corrosion and rust. The sludge will plug passages and restrict the normal flow of the motor oil through the engine. These conditions lead to rapid wear and premature engine failure. Low operating temperature does not usually cause sudden unexpected engine failure.

The other vehicle instruments do not indicate problem inside the engine. They are, however, important to the continued operation of the engine. They indicate problems that can be corrected on the outside of the engine. The charge indicator and fuel-level gauge are examples of this type of instrument.

Engine Operation. The operator becomes accustomed to the operating characteristics of an automobile when it is driven for some time. The

Engine Fluid Leaks. The operator generally takes care of fuel, oil, and coolant requirements of the engine. A problem is indicated if an engine begins to use more of these fluids. The service technician can correct many fluid consumption problems from outside the engine. In other cases, they are an indication of a more serious problem that requires internal engine repairs.

Engines are designed to keep the fluids (coolant, fuel, and oil) in the engine. Leaking seals allow the fluids to seep over the outside surface of the engine (Figure 13-3). If the leakage is slight, dirt will gradually mix with the leaking fluid to form a heavy black deposit over the outside of the engine. This condition is common on high-mileage automobiles that have had little routine maintenance. When leakage becomes severe, the fluid will wash the surface, causing it to be wet and streaked. Fluid leaks are a problem in themselves. They may also indicate an impending serious engine problem.

The location of small leaks can often be found by placing clean paper under the vehicle while it is parked overnight. Fluid will fall straight down from the lowest point as shown in Figure 13-4. With a good visual inspection, the fluid flow can be followed from directly above the drops. Sometimes, it is necessary to clean the engine before the cause of the leak can be determined.

In most cases, oil leaks can be repaired by

Figure 13-3. Typical example of oil seeping out through the rocker cover gasket and coating the outside of the engine.

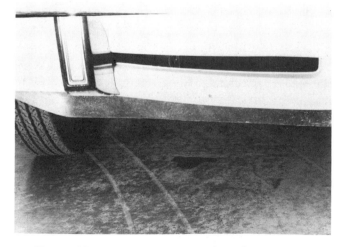

Figure 13-4. An oil spot on a clean floor directly below an oil leak.

replacing the gasket or an oil seal. A cover over the leaking gasket will have to be removed to replace it. In most cases, the gasket or seal can be replaced without removing the engine from the chassis.

Coolant leaks come from hose connections, the thermostat housing, or the coolant pump. Oil leaks usually come from the rocker covers, timing cover, V-type intake manifolds, or oil pan. Fuel leaks come from the fuel pump, fuel lines, or carburetor. Some of the engine assembly bolts go into the coolant system and some go into the oiled part of the engine interior. Sometimes, leaks occur around these bolts. They can be removed and sealant put on the threads, then reinstalled.

Broken Parts. Broken parts that can be seen on an assembled engine are problems that are easy to identify. They generally can be repaired without disassembling the engine. An internal broken part will usually make noise if the engine can be run. By listening carefully, the general location of the noise can be determined. The engine will have to be opened in the area of the noise for more detailed inspection. The rocker cover should be removed if a light tapping noise at one-half engine speed is heard on the top of the engine. If a heavy noise is heard on the bottom of the engine, the oil pan should be removed. The repair procedure to be used to correct the problem will be selected after the broken part is identified.

13-2 Determining the Mechanical Condition

Once the operator recognizes that there is a problem in the engine, a service technician will check it to see what the problem is. The technician should *first* get a *full* description of the problem from the operator. This step is often overlooked, so the service technician spends too much time locating the problem. Once the service technician understands the operator's complaint, the problem should be checked and identified. The steps to be taken include a visual inspection, listening to the sound, and using shop test equipment.

In checking the engine, the first thing the service technician will do is to give the engine a good external visual inspection. The visual inspection should include observing the deposits inside the tailpipe.

The tailpipe should be relatively clean on catalytic-converter-equipped vehicles. If the vehicle used leaded gasoline, the interior of the tailpipe should be coated with a gray to brown powder. Black soot in the tailpipe indicates excessive fuel consumption. A wet, oily tailpipe inner surface indicates excessive oil consumption.

The technician will start and run the engine after giving it a visual inspection. The sounds the engine makes, fluid leaks, and visible emissions coming from the engine and tailpipe are noted.

Compression. The source of engine power is combustion pressure. If the combustion chambers leak, the engine cannot produce the normal power and economy with low emissions. Checking combustion chamber leakage can be done in several ways. One of these is the *compression test*. Compression should only be checked on a warm engine.

Deposits on the spark plugs that are taken out to make a compression test give an indication of how each cylinder has been running. During the compression test, the engine is usually cranked with a remote starter switch so the work can be done from under the hood. A typical compression tester is shown in use in Figure 13-5.

Compression of each cylinder is measured by cranking the engine with a compression gauge connected in the spark plug hole. With the throttle fully open, the engine is cranked through exactly *five* compressions. The gauge reading should rise to nearly full pressure as the piston comes up on the first compression stroke. Cylinder compression pressure shown on the gauge after the fifth compression stroke should be recorded as compression pressure. Gauge pressure is then released, the gauge pickup moved to the next cylinder, and the sequence repeated. This is done on each of the engine cylinders. A check of the engine compression in this way is called a *dry compression test*. The gauge readings obtained can be checked against specifications to determine if they are up to standard. Engines with compression ratios of 8.0:1 to 8.5:1 have minimum compression pressures as low as 100 lb/in.^{2} (6895 Pa). Engines with 8.5:1 to 9.5:1 compression ratios have minimum pressures of 125 lb/in.^{2} (8618 Pa). Engines with compression ratios of 9.5:1 to 10.5:1 have minimum compression pressures of 140 lb/in.^{2} (9653 Pa). Normal compres-

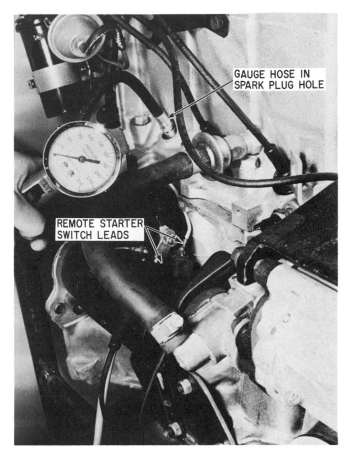

GAUGE HOSE IN SPARK PLUG HOLE

REMOTE STARTER SWITCH LEADS

Figure 13-5. A typical compression tester in use.

sion pressures are 30 to 40 lb/in.^{2} (2068 to 2758 Pa) higher than the minimum compression pressures. A smooth-running engine depends upon *equal* compression pressures between cylinders. If the lowest cylinder pressure is at least 70% of the highest, the engine will be able to run smoothly. Some manufacturers allow 20 lb/in.^{2} (1379 Pa); others allow as high as 40 lb/in.^{2} (2758 Pa) variation between cylinders and these pressures are still considered normal compression.

Low-compression pressure does not indicate the specific cause of the low pressure. One way to help pinpoint the problem is to follow the compression test with a *wet compression test*. About 1 tablespoon (15 ml) of motor oil is put into each cylinder through the spark plug hole. The engine is cranked a few revolutions helping the oil to flow around the piston rings sealing them. The compression test sequence is rerun with the cylinders wet with oil. If the pressures on the wet compression test are much higher than those taken on the dry compression test, the low-compression pressure is due

Figure 13-6. A head gasket burned out between two cylinders causing low compression on both cylinders.

to piston ring leakage. If the wet compression test shows only a slight increase in pressure, the low-compression-pressure problem is due to valve leakage. If two cylinders side by side have low compression on both wet and dry compression tests, it may be due to head gasket failure, allowing compression to leak between them. The head gasket pictured in Figure 13-6 caused this problem. Low compression may be the result of a temporary deposit buildup on the valve stems caused by low-speed city driving. Repeated hard accelerations in freeway driving will often remove this type of deposit. If a wet compression test shows the same low readings, the engine will have to be repaired to give satisfactory performance.

Sometimes it is difficult to pinpoint the cause of low compression. A *cylinder leakage test* that puts air into a cylinder at a measured rate is useful in pinpointing the cause of low compression. Excessive cylinder leakage is indicated by a low reading on the gauge of the tester. This is really a low-pressure reading because the air leaks out as fast as it is put in. A normal cylinder will have a gauge reading above 85%. If air can be heard leaking from the tailpipe, an exhaust valve is leaking. An intake valve leak can be heard at the carburetor. Air coming from the oil filter cap is the result of the rings leaking. Some ring leakage is normal during the cylinder leakage test. This happens because the rings are not moving as they normally do in a running engine. Leakage into the cooling system shows up as bubbles in the coolant. The bubbles can be seen when the radiator filler cap is removed and the coolant filled up into the filler neck.

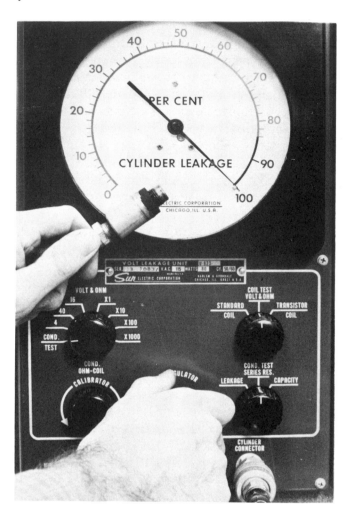

Figure 13-7. A cylinder leakage tester with one spark plug hole adapter.

The cylinder leakage test is run on a warmed-up engine by cranking the engine to bring one piston to top center. A hose is attached between the cylinder leakage instrument and an adapter that is screwed into the spark plug hole. One adapter is shown in Figure 13-7. The instrument is calibrated by adjusting the air pressure to zero leakage (100% on the gauge) with the hose blocked. It is connected to the adapter to allow air to flow into the combustion chamber. If the cylinder does not leak, the pressure in the cylinder will also go to a 100% leakage reading. With normal leakage past the rings, the cylinder will show 85% or more. A problem is indicated if a cylinder has a gauge reading less than 85%.

After the top center of the first cylinder is found, a timing wheel can be attached to the distributor rotor to be used to help bring each piston to top center. Cylinder leakage tests are only run with the piston on top center of the compression stroke. The test sequence is run on each cylinder following the cylinder firing order.

An engine in good mechanical condition will run with high manifold vacuum. **Manifold vacuum is developed by the pistons as they move down on the intake stroke to draw the charge from the carburetor and intake manifold.** Air to refill the manifold comes past the throttle plate into the manifold. Vacuum will increase any time the engine turns faster or has better cylinder sealing while the throttle plate remains in a fixed position. Manifold vacuum will decrease when the engine turns more slowly or when the cylinders no longer do an efficient job of pumping.

Another test of engine performance can be done without special equipment. While watching the engine speed change on a tachometer connected to a running engine, remove a spark plug cable from one spark plug at a time. Reconnect each spark plug cable before removing the next one. Insulated pliers are useful to remove the ignition cables, to avoid getting an electric shock. One type of insulated pliers used to remove a spark plug cable is shown in Figure 13-8. The engine speed should drop an equal amount as each spark plug cable is removed. When the speed drop is not the same the cylinder with the

least drop is not producing its share of the engine power. This cylinder has some malfunctioning component. Many brands of test equipment have a built-in electronic circuit that can short out or kill one or more cylinders at a time. This is done by a switch without removing the ignition cables. An expanded-scale tachometer is used to magnify the reading so that the technician can easily and quickly see small changes in the engine speed. Figure 13-9 shows one of these testers. When this test is performed on an engine with a catalytic converter, each cylinder is only shorted out long enough to take a reading. Fuel from the nonfiring cylinder will ignite in the converter. This will produce excessive temperatures that will ruin the catalyst if the engine is allowed to misfire for an extended period of time.

Lubrication. The condition of the motor oil and the oil pressure are useful in determining the condition of the engine. Oil pressure is maintained as the oil pump pushes oil through small bearing clearances. If the bearings have become worn, the clearances will increase. This larger clearance lets the oil flow through the bearings easier, and this, in turn, lowers the oil pressure. Low pressure could also be caused by a worn-out oil pump. When the pump wears, it allows some oil to flow backward around the pump gears. The result of this is that the pump is unable to move as much oil as it was designed to move, so the

Figure 13-8. Using an insulated pliers to remove a spark plug cable terminal from one spark plug.

Figure 13-9. A "kill" switch being used on an engine analyzer to measure the drop in engine speed when the cylinder is not firing.

Figure 13-10. Badly worn rotor-type oil pump gears.

oil pressure is lower. Badly worn rotor-type oil pump gears are pictured in Figure 13-10.

The engine oil consumption is another indicator of the engine condition. Oil consumption is measured by the amount of oil the operator has to add between oil changes, no matter how it gets out of the engine. If the oil leaks out through faulty gaskets or seals, these should be replaced. If the oil goes out through the combustion chamber and tailpipe, at least part of the engine will have to be disassembled to make the repair. It may not be possible to determine if the oil is getting into the combustion chamber past the rings, between the valve stem and guide, or through the positive crankcase ventilation system.

The rate at which the oil becomes contaminated is another indication of the engine condition. Often, the operator will neglect doing periodic maintenance on a high-mileage engine. Lack of maintenance, such as oil and filter changes, will allow contaminants to build up in the engine. These contaminants cause rapid engine wear.

Coolant. Engine coolant should be clean and the radiator full. An engine that requires service will usually have dirty coolant. Coolant that has picked up rust from the cooling system will be brown. Sometimes, combustion gases seep into the coolant and sometimes oil will leak into the coolant through a faulty gasket. Contaminated coolant indicates that there may be a problem requiring engine repair.

Turning Torque. It should take a certain amount of effort to rotate the crankshaft of an engine. If the engine has little compression, the crankshaft will rotate easily. If something is binding in the engine, it will be hard to rotate the crankshaft. Cylinder problems can cause the crankshaft to rotate easily during part of the turn and rotate hard during other parts of the turn. The turning torque can be quickly determined by listening to the starter as it cranks the engine. If it cranks rapidly and smoothly, there is little turning torque. If it cranks slowly, the engine may be at fault. This slow turning is the same sound produced by the starter cranking an engine at subzero temperatures. There are faults in the engine if the engine speeds up and slows down as it cranks.

13-3 Engine Repair Decision

The decision to repair an engine should be based on all of the information about the engine that is available to the service technician. This information must be examined and interrelated to determine the actual condition of the engine. There are a number of different ways an engine problem can be corrected. These different ways to correct the engine are called *repair options*. The whole engine can be replaced, part of the engine can be replaced, the whole engine can be reconditioned, or only the faulty part can be repaired. In some cases the engine might not be worth repairing. It is the responsibility of the technician to discuss advantages and disadvantages of the different repair options with the customer. The customer who is paying for the repair, must make the final decision on the reconditioning procedure to be used. The decision will be based on the recommendation of the service technician.

Component Repair. Most customers want to spend the least money, so they only have the faulty component repaired. This decision may be pennywise and dollar foolish. If a part fails, it may only be the result of the faulty part. An example of this would be a failed timing sprocket and chain. The problem will be corrected when a new timing sprocket and chain are installed. On the other hand, a part may have failed because some other part is not functioning correctly. An example of this could be a rocker arm pivot that was badly scored because it did not

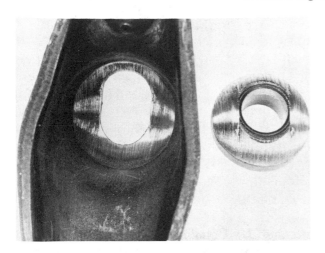

Figure 13-11. Scored rocker arm and pivot caused by loss of lubrication.

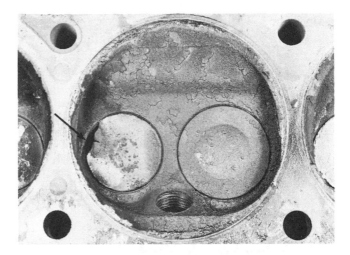

Figure 13-12. A badly burned exhaust valve can be easily seen when the head is first removed.

get proper lubrication. The lubrication could have been restricted by a faulty metering valve inside the hydraulic lifter. Only the rocker arm and pivot could be replaced. The new rocker arm and pivot soon failed because the lifter problem was also not repaired. Figure 13-11 shows a rocker arm and pivot from an engine where this was done. It is the responsibility of the service technician to explain to the customer why the part failed and to suggest different repairs that could be made to keep the part from failing again.

Valve Job. Failure of a valve to seal on the seat is one of the most common engine problems caused by premature failure. When there is a leak between the valve and seat, the combustion pressure is lowered. This will reduce both the engine economy and power. The high-pressure combustion gases escaping between the leaking valve and seat will burn the valve face, and this, in turn, increases the leakage. Figure 13-12 shows a badly burned exhaust valve in a head that was removed because of low cylinder compression. The top of the pistons are also visible once the head is removed for a valve job. The pistons may indicate that the pistons should also be serviced.

Failure of a valve to seal on the seat can result from abnormal combustion, incorrect valve lash, valve stem deposits, abusive engine operation, and so on. Valve leakage is corrected by doing a *valve job.* This does not necessarily correct the malfunc-

tion that caused the valve to leak. Stopping the valve leakage improves manifold vacuum. The greater manifold vacuum may draw the oil past worn piston rings and into the combustion chamber during the intake stroke causing the oil consumption to increase.

Overhaul. In this discussion, engine overhaul includes both a ring and a valve job. New connecting rod bearings are usually installed during an overhaul. Sometimes, this type of reconditioning is called a *minor overhaul.* A minor overhaul can usually be done without removing the engine from the chassis. It does require removal of both the head and oil pan. The overhaul is usually done when the engine lacks power, has poor fuel economy, uses an excessive amount of oil, produces visible tailpipe emissions, runs rough, or starts hard. It is still only a repair procedure. Many worn parts remain in the engine. Other engine problems may be noticed after the oil pan is removed and the piston and rod assemblies are taken out. The customer should be informed about any other engine problem so that the service the engine requires can be authorized.

Rebuild. A complete engine reconditioning job is called rebuilding. Sometimes, this type of reconditioning is called a *major overhaul.* To rebuild the engine, it must be removed from the chassis and be completely disassembled. All serviceable parts are reconditioned to either new or to service standards. All bearings, gaskets, and seals are replaced. When the reconditioning is done properly, a rebuilt engine should operate as long as a new engine.

A special form of precision engine rebuilding is called *blueprinting*. Blueprinting is usually done to give the engine maximum performance. The clearances are all set near the maximum specifications and combustion chambers are adjusted to minimum and equal volumes. All servicing details are done with extreme care. The engine is carefully balanced. Blueprinting requires skilled and detailed labor and so it is very expensive.

The vehicle is out of service while the engine is being rebuilt. This is inconvenient to some customers and very expensive to others. If the vehicle is used to produce income, the loss of income while it is out of service for repair must be added to the rebuilding cost. This will increase the total cost to the customer. Because of the total cost, the customer might decide to replace the engine instead of rebuilding it.

Replacement. The quickest way to get a vehicle back in service is to exchange the faulty engine for a different one. In an older vehicle, the engine may be replaced with a used engine from a salvage yard. In some cases, only a reconditioned block, including the crank, rods and pistons, is used. This replacement assembly is called a *short block*. The original heads and valve train are reconditioned and used on the short block. The replacement assembly is called a *long block* when the reconditioned assembly includes the heads and valve train. Many automotive machine shops maintain a stock of short and long blocks of popular engines. Usually, the original engine parts, called a *core,* are exchanged for the reconditioned assembly. The core parts are reconditioned by the automotive machine shop and put back in stock for the next customer.

Some engines are *remanufactured*. The engine cores are completely disassembled and each serviceable part is reconditioned with specialized machinery. Engines are then assembled on an engine assembly line similar to the original manufacturer's assembly line. The parts that are assembled together as an engine have not come out of the same engine. The remanufactured engine usually has new pistons, valves, and lifters, together with other parts that are normally replaced in a rebuilt engine. All clearances and fits in the remanufactured engine are the same as in a new engine. A remanufactured engine should give as good service as a new engine and it will cost about half as much. Remanufactured engines usually carry a warranty. This means they will be replaced if they fail during the period of the warranty. They may even cost less than a rebuilt engine because much of the reconditioning is done by specialized machines rather than by expensive skilled labor.

Options to Be Considered. The average customer wants the engine repaired for the lowest possible cost and the vehicle left in the shop for the least amount of time. The service technician wants to perform the repair correctly so that it will give the service the customer expects. It is rarely possible to do both at the same time. There are a number of questions that must be answered before selecting the repair procedure to be used.

The market value of the vehicle with the faulty engine is one of the first things to consider. Would the market value of the repaired vehicle be as much as the repair cost? Would it make better sense to replace the vehicle rather than to recondition the engine? Does the customer think the repair is worth doing, regardless of the market value of the vehicle? What is the condition of the rest of the vehicle? Would a reconditioned engine put greater loads on the transmission and axle and cause them to fail? The engine accessories are worn after high mileage. How long will they be expected to function on the reconditioned engine or should the accessories also be reconditioned? This will add to the repair cost.

After the problem is known and the repair options considered, the service technician and the customer agree on the repair procedure to be followed. The customer will then authorize the technician to make the repairs. If the service technician runs into any unexpected conditions that require additional service, the customer should be contacted to authorize the additional service work.

13-4 Engine Removal

The normal procedure when doing upper engine service is to leave the engine in the chassis, removing only the parts that are required to do the service that is needed. The engine exterior and the engine compartment should be cleaned before work is begun. A clean engine is easier to work on, it helps keep dirt out of the engine, and it minimizes accidental damage from slipping tools. The battery ground cable is disconnected to avoid the chance of sparks. A better procedure is to remove the battery from the vehicle.

Working on the top of the engine is made easier if the hood is removed. With fender covers in place, the hood is loosened from the hinges. With a person on each side of the hood to support it, the hood is lifted off as the bolts that hold the hood are removed. The hood is usually stored on fender covers placed on the top of the automobile, where it is least likely to be damaged.

The coolant is drained from the radiator and engine block to minimize the chance of getting coolant into the cylinders when the head is removed. The exhaust manifold is disconnected. On some engines, it is easier to remove the exhaust pipe from the manifold. On others, it is easier to separate the exhaust manifold from the head and leave the manifold attached to the exhaust pipe. On V-type engines, the intake manifold must be removed before the heads can be taken off. In most cases, a number of wires, accessories, hoses, and tubing must be removed before the manifold and head can be removed. If the engine is not familiar, it is a good practice to put tape on each of the items removed, as shown in Figure 13-13. The tape can be marked with the location of each item so they can be easily

replaced during engine assembly. The spark plugs should be removed so that they are not broken.

The oil pan can be removed from most engines without removing the engine from the chassis. The exhaust pipes and steering linkages often have to be loosened and lowered for oil pan clearance. On some engines, the engine mounts will have to be loosened so the engine can be jacked up an inch or two to allow the oil pan to be slipped out. It may be necessary to rotate the crankshaft to a specific position so that the oil pan will clear the bottom of the connecting rods.

The engine heads and manifolds can be removed with the engine staying in the chassis. Figure 13-14 shows a V-8 engine block after the heads have been removed for a valve job. The engine can usually remain in the chassis for a minor overhaul. The engine has to be removed from the chassis when the engine is to be rebuilt. When this is required, the manifolds and heads are not removed until the engine is out of the chassis. This makes engine disassembly much easier.

There are two ways to remove the engine. The engine can be lifted out of the chassis with the transmission attached, or the transmission can be separated from the engine and left in the chassis. The method to be used must be determined before the engine is disconnected and removed. In the following discussion, it is assumed that the engine is to be removed for overhaul with the transmission attached.

First, the battery is removed to avoid sparks and

Figure 13-13. Each hose, wire, and control is tagged before removal for easy reassembly.

Figure 13-14. A V-8 engine block after the heads have been removed for a valve job.

spilled acid. If the engine compartment is steam-cleaned, it will be much easier to work on the engine. The engine oil and coolants are drained. They should be examined to determine their condition before they are discarded. Their condition may be useful in identifying engine problems. With fender covers in place, the hood is removed. All coolant hoses are removed and the transmission oil cooler lines are disconnected from the radiator. The radiator mounting bolts are removed and the radiator is lifted from the engine compartment. This gets the radiator out of the way so that it will not be damaged while working on the engine. This is a good time to have the radiator cleaned while it is out of the chassis. If it does not require cleaning, it is placed where it will not be damaged as the overhaul progresses. The automobile trunk makes a good storage place for the radiator.

If the engine is air conditioned, the compressor can usually be separated from the engine, leaving all air conditioning hoses securely connected to the compressor and lines. The compressor can be fastened to the side of the engine compartment, where it will not interfere with engine removal. It may also be necessary to loosen the air conditioning condenser and fasten it aside. If it is necessary to disconnect the air conditioning lines, the system will have to be bled down before opening the system. Follow standard air conditioning bleed-down procedures. All air conditioning line openings should be securely plugged immediately after they are disconnected to keep dirt and moisture from the system. They should remain plugged until immediately before reassembly. The air conditioning system will have to be evacuated and recharged after the engine is reinstalled and the system assembled.

All wires, hoses, tubing, and controls that connect from the automobile to the engine are tagged for location and then disconnected. This will aid in reassembly. It is a good practice to remove the spark plugs, distributor, carburetor, alternator, and fan to avoid accidental damage during engine removal. It is also a good practice to recondition these units while the engine is being rebuilt so that the finished engine will operate trouble-free for a long time.

Under the car, the propeller shaft is removed and the exhaust pipes disconnected. In some installations, it may be necessary to loosen the steering linkage idler arm to give clearance. The transmission controls, speedometer cable, and clutch linkages are disconnected and tagged. They should be moved out of the way so that they will not be bent as the engine is removed. The automatic transmission cooling lines and the starter wires are removed.

A sling, either a chain or lift cable, is attached to the manifold or head cap screws on top of the engine. A hoist is attached to the sling and snugged to take up most of the weight. The engine has three mounts, one on each side and one at the back of the transmission. The mount bolts are removed. This leaves the engine resting on the mounts. The rear cross member is removed and the transmission is lowered. The hoist is tightened to lift the engine. The engine will have to nose up as it is removed. The front of the engine must come almost straight up as the transmission slides from under the floor pan, as illustrated in Figure 13-15. The engine and transmission are hoisted free of the automobile, swung clear, and lowered on an open floor area. Blocks are placed under the engine to steady it. The hoist is loosened but kept attached to the engine as a safety measure until the transmission is removed.

The exterior of the engine should be given a thorough visual examination before any other work is done. After noting any fluid leaks and obvious

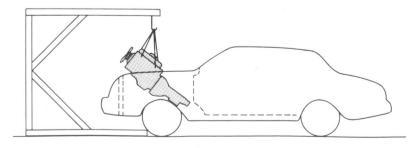

Figure 13-15. An engine must be tipped as it is pulled from the chassis.

faults, the exterior of the engine should be thoroughly cleaned. Steam cleaning is a good way to clean the exterior of the engine.

13-5 Engine Disassembly

The bolts attaching the standard transmission to the bell housing are removed. the transmission is pulled straight back until it is free. The bell housing and clutch can then be removed.

If it is done carefully, the automatic transmission can be removed without draining the fluid. The torque converter cover is removed. The torque converter is disconnected from the drive plate (Figure 13-16). With both the engine and transmission properly supported, the transmission is disconnected from the engine block. Carefully pry between the

torque converter and drive plate. This will separate the torque converter from the engine (Figure 13-17). The torque converter will stay on the transmission, so no fluid is lost (Figure 13-18).

All remaining accessores are removed from the engine and stored where they will not be damaged. Engine disassembly is best done on an engine stand (Figure 13-19). It can also be done on a clean floor with the engine on blocks. It is easier when the engine is placed with the flywheel end on blocks (Figure 13-20).

The following disassembly procedure applies primarily to pushrod engines. The procedure will have to be modified somewhat when working on overhead cam engines. These disassembly pro-

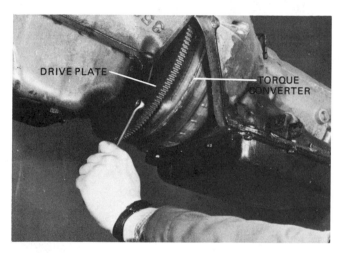

Figure 13-16. The torque converter is disconnected from the drive plate.

Figure 13-17. The transmission and engine are supported and the transmission case is unbolted from the engine. The torque converter is then separated from the drive plate.

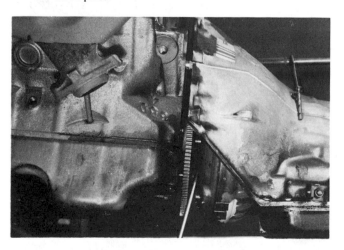

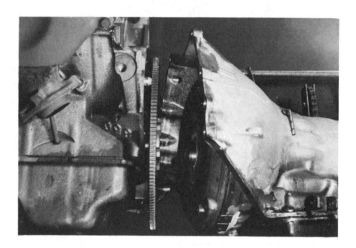

Figure 13-18. The torque converter stays with the transmission as it is separated from the engine.

Figure 13-19. A V-8 engine on an engine stand for disassembly.

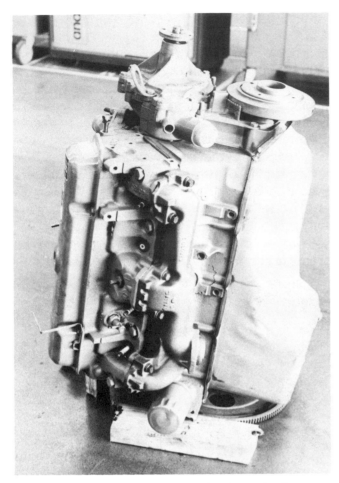

Figure 13-20. A V-8 engine on blocks on the floor for disassembly.

Figure 13-21. Typical deposits inside a rocker cover.

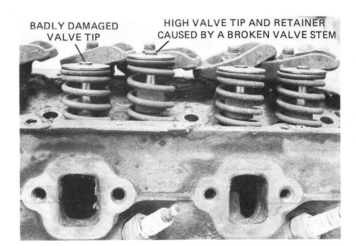

Figure 13-22. Valve faults of the type that can be seen when the rocker cover is removed.

cedures have been described at the beginning of Chapters 6 through 9.

Removal of the rocker covers gives the technician the first opportunity to see inside a part of the engine. A good visual examination of this area should be made to identify and determine the cause of any abnormal condition. Pay special attention to the type and quantity of deposits (Figure 13-21). Examine the rocker arms, valve springs, and valve tips for obvious defects. **Obvious defects are those that are easily seen.** Figure 13-22 shows one obvious defect. The valve tips are worn and one valve stem is high. In this case, the valve stem was broken.

Remove the manifold hold-down cap screws and nuts and lift off the manifolds. If the gaskets are stuck, a flat blade, such as a putty knife, can be worked beside the gasket to loosen it. Care must be taken to avoid damaging the parting surface as the gasket is loosened. When the manifold and lifter valley cover are off V-type engines, the technician

has another opportunity to examine the interior of the engine. Here again, check the deposits. The pushrods and lifters are exposed for a visual inspection. Typical deposits in a lifter valley of a high-mileage engine is shown in Figure 13-23. On some V-type engines, it is possible to see the condition of the cam at the bottom of the lifter valley.

With the manifold off the V-type engine, any obvious abnormal conditions are noted, the rocker arms are loosened, and the pushrods removed. The usual practice is to leave the lifters in place when doing only a valve job. The lifters can be removed at this time if they are causing the problem or if the engine valve train is to be serviced.

The head cap screws are removed and the head is lifted from the block deck. If the head gasket is stuck, carefully pry the head to loosen the gasket.

Figure 13-23. Typical deposits in a lifter valley of a high mileage V-8 engine.

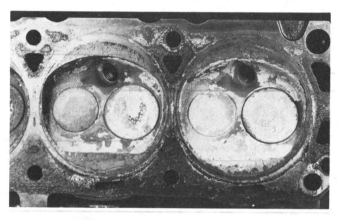

Figure 13-24. Typical combustion chamber deposits.

Special precautions should be taken to pry only on edges of the head that will not break. Parting surfaces should *not* be scratched. Scratched or burred surfaces will lead to leaks on the repaired engine.

The combustion chamber is exposed when the head is removed. The combustion chamber pocket in the head and the top of the piston should be given a thorough visual examination. The most obvious condition involves deposits such as those shown in Figure 13-24. A normal combustion chamber is coated with a layer of hard, light-colored deposits. If the combustion chamber has been running too hot, the deposits will be very thin and light-colored. If the combustion chamber has been running too cold, the deposits will be thick, dry, and black. Heavy, wet deposits result from oil that is entering the combustion chamber. If oil consumption is excessive, the large quantity of oil going into the combustion chamber will wash the carbon deposit from the place where the oil comes in. Because of this, the cause of excessive oil consumption can usually be seen after the head is off.

At this point, the cylinder taper and out-of-round should be checked just below the ridge and just above the piston when it is at the bottom of the stroke, as shown on the cutaway cylinder in Figure 13-25. These measurements will indicate how much cylinder wall work is required. If the cylinders are worn beyond the specified limits, they will have to

(a)

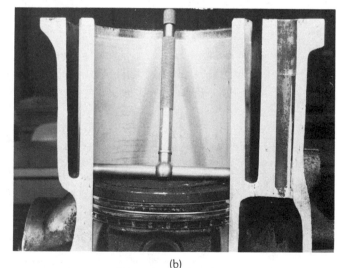

(b)

Figure 13-25. When the head is first removed the cylinder taper and out-of-round should be checked below the ridge and above the piston when it is at the bottom of the stroke.

Figure 13-26. Typical deposits in the bottom of a high mileage V-type engine.

Figure 13-27. Typical deposits on the oil pump pick up screen of a clean engine.

be rebored to return them to a satisfactory condition.

The engine is turned upside down and the oil pan is removed. This is the first opportunity to see the working parts in the bottom end of the engine. Deposits are again a good indication of the engine condition and the care it has had (Figure 13-26). Heavy sludge indicates infrequent oil changes. Hard carbon indicates overheating. The oil pump pick-up screen should be checked to see how much plugging exists (Figure 13-27). The connecting rods, caps, and main bearing caps should be checked to make sure they are *numbered*. If not, they should be numbered with number stamps or a prick punch. The parts are marked so they can be reassembled in exactly the same position.

Piston and Rod Removal. The ridge above the top ring travel must be removed before the piston and connecting rod assembly is removed. Cylinder wear that leaves an upper ridge is shown in Figure 13-28. This is necessary to avoid catching a ring on the ridge and breaking the piston. Failure to remove the ridge is likely to cause the second piston land to break when the engine is run after reassembly with new rings as pictured in Figure 13-29. The ridge is removed with a cutting tool that is fed into the metal

Figure 13-28. Most of the cylinder wear is on the top inch just below the cylinder ridge (Courtesy of Dana Corporation).

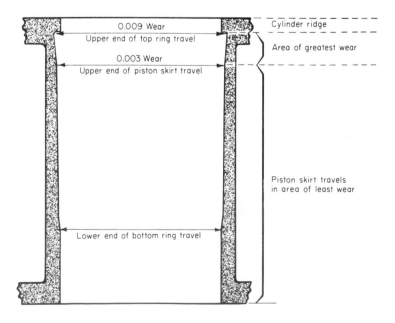

0.009 Wear
Upper end of top ring travel

0.003 Wear
Upper end of piston skirt travel

Cylinder ridge

Area of greatest wear

Piston skirt travels in area of least wear

Lower end of bottom ring travel

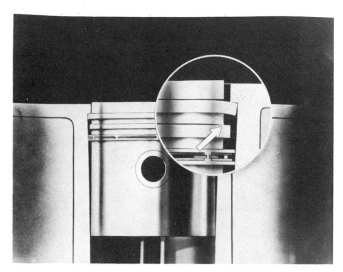

Figure 13-29. Failure to remove the ridge is likely to cause the second piston ring land to break when the engine is run after reassembly with new rings. (Courtesy of Sealed Power Corporation).

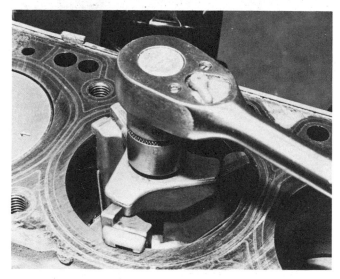

Figure 13-30. The ridge being removed with one type ridge reamer before the piston assemblies are removed from the engine.

ridge. A guide on the tool prevents accidental cutting below the ridge. The ridge reaming job should be done carefully with frequent checks of the work so that no more material than necessary is removed. One type of ridge reamer is shown in Figure 13-30.

Connecting rod nuts are taken off the rod so the rod cap with its bearing half can be removed. The rod bolts are fitted with protectors and the piston and rod assemblies are removed as described in section 8-1.

Shaft Removal. The camshaft can be removed with the engine in the chassis on most automobiles. It is necessary to remove the radiator and grill to do this. The engine must be out of the chassis if the crankshaft is to be removed.

The next step in disassembly is to remove the water pump and crankshaft vibration damper. The bolt and washer that hold the damper are removed. The damper should be removed only with a threaded puller similar to the one in Figure 13-31. If a hook-type puller is used around the edge of the damper, it may pull the damper ring from the hub. If this happens, the damper assembly will have to be replaced with a new assembly. With the damper assembly off, the timing cover can be removed, exposing the timing gear or timing chain. Examine these parts for excessive wear and looseness. A worn timing chain on a high-mileage engine is shown in Figure 13-32. Bolted cam sprockets can be removed to free the timing chain. On some engines this will require removal of the crankshaft gear at the same time. Pressed-on gears and sprockets are removed from the shaft only if they are faulty. They are removed after the camshaft is removed from the block. It is necessary to remove the camshaft thrust-plate retaining screws when they are used. You should refer to camshaft drive details in Section 9-4.

The camshaft can be removed at this time or it can be removed after the crankshaft is out. It must

Figure 13-31. A puller being used to pull the vibration damper from the crankshaft.

Figure 13-32. A worn timing chain on a high mileage engine.

Figure 13-33. A valve spring compressor being used to remove the valve locks.

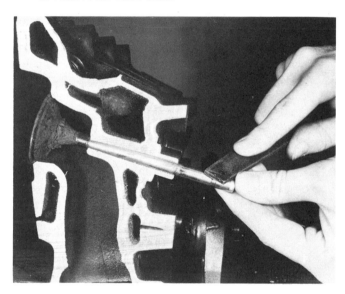

Figure 13-34. Removing burrs from around the valve lock grooves and tip before removing the valves from the head.

be carefully eased from the engine to avoid damaging the cam bearings or cam lobes. This is done most easily with the front of the engine pointing up. Bearing surfaces are soft and scratch easily, and the cam lobes are hard and chip easily.

The main bearing caps should be checked for position markings before they are removed. They have been machined in place and will not perfectly fit in any other location. After marking, they can be removed to free the crankshaft. When the crankshaft is removed, the main bearing caps and bearings are reinstalled on the block to reduce the chance of damage.

Valve Removal. After the heads are removed and placed on the bench, the valves are removed. A C-type valve spring compressor, similar to the one being used in Figure 13-33, is used to free the valve locks or keepers. The valve spring compressor is air-powered in production shops where valve jobs are being done on a regular basis. Mechanical valve spring compressors are used where valve work is done only occasionally. After the valve lock is removed, the compressor is released to free the valve retainer and spring. These are lifted from the head together with any spacers being used under the valve spring. Here again, the parts should be kept in order to aid in diagnosing the exact cause of any malfunction that shows up. The valve tip edge and lock area should be lightly filed or stoned, as shown in Figure 13-34, to remove any burrs *before* sliding the valve from the head. Burrs will scratch the valve guide.

When all valves are removed following the same procedure, the valve springs, retainers, locks, guides, and seats should be given another visual examination. Any obvious faults should be noted.

Parts that are obviously not repairable should be marked and set aside for later reference and fault diagnosis. No further labor time should be spent on them.

REVIEW QUESTIONS

1. How does premature failure differ from mature failure? [INTRODUCTION]
2. Who is usually the first person to be aware of an engine problem? [13-1]
3. What engine damage might be caused if the engine is run with no oil pressure? [13-1]
4. What engine damage might be caused if the engine is run with high coolant temperature? With low coolant temperature? [13-1]
5. What does the operator notice that may suggest that the engine needs to be repaired? [13-1]
6. How is the cause of an external fluid leak found? [13-1]
7. How is a broken part located? [13-1]
8. What visual checks are made on an engine with a problem? [13-2]
9. How is a compression test done? [13-2]
10. What information is indicated to the technician by the dry and wet compression tests? [13-2]
11. What is the normal difference in compression pressures between cylinders? [13-2]
12. What information can a cylinder leakage test indicate to the technician that a compression test does not indicate? [13-2]
13. Where is the piston in the four-stroke cycle when the cylinder leakage test is run? [13-2]
14. What makes the manifold vacuum increase? [13-2]
15. How is a manifold cranking vacuum test done? [13-2]
16. What information can you get from a manifold vacuum test? [13-2]
17. What is indicated when the engine rpm drops slightly as a spark plug cable is removed? What is indicated when it drops a large amount? [13-2]
18. List possible causes of low oil pressure. [13-2]
19. How does the operator know that an engine has excessive oil consumption? [13-2]
20. What does the condition of the coolant indicate about the condition of the engine? [13-2]
21. What does the sound of the engine cranking indicate about the condition of the engine? [13-2]
22. What is meant by the term "repair options"? [13-3]
23. Who must make the final decision about which repair option is to be taken? [13-3]
24. When is a valve job done? [13-3]
25. What is included in the engine overhaul, as discussed in this book? [13-3]
26. Why is an overhaul only a repair procedure? [13-3]
27. How does rebuilding differ from an overhaul? [13-3]
28. What is engine blueprinting? [13-3]

29. When might an engine be replaced rather than be rebuilt? [13-3]
30. What is the difference between a short block and a long block? [13-3]
31. How does remanufacturing differ from rebuilding? [13-3]
32. What things must be considered before selecting the option to be used to repair the engine? [13-3]
33. Why should the engine and engine compartment be cleaned before the engine is worked on? [13-3]
34. Describe the procedure used to remove an engine. [13-4]
35. What are two ways that the engine can be removed from the chassis? [13-4]
36. List the steps, in order, that are taken to remove the engine and transmission assembly from a vehicle. [13-4]
37. What should be done to the radiator and engine accessories while the engine is being reconditioned? [13-4]
38. How is a part removed when the gasket is stuck to both parts? [13-4]
39. Why is it a good practice to keep all of the engine parts in order when the engine is disassembled? [13-4]
40. What are the two theories that apply to cleaning an engine while servicing? [13-4]
41. What do combustion chamber deposits indicate about an engine? [13-4]
42. Why should the cylinders be checked right after the head is removed? [13-4]
43. Why is the position of each part marked before disassembly? [13-4]
44. Why is it necessary to remove the ridge from a cylinder? [13-4]
45. What precautions should be observed so that no damage will occur as the piston and rod assemblies are removed? [13-4]
46. Why shouldn't a hook-type puller be used on a vibration damper? [13-4]
47. What precautions should be taken when removing the camshaft? [13-4]
48. What precautions should be taken when removing valves? [13-4]

CHAPTER
14

Part Failure Analysis

There are two basic reasons for disassembling engines for repairs: either a part has *worn out* or *premature failure* of a part has occurred. Most engines are run until a part fails, even when they have high mileage. It is a common practice to do a valve job every time the heads are removed, even if premature piston failure was the cause of head removal. It is also common practice to do an overhaul when any premature part failure occurs on a high-mileage engine.

During engine disassembly, the defective parts that cannot be reconditioned are noted. It is a waste of time to clean or attempt to recondition them. After the engine is completely disassembled, the parts related to the defective part must be given a very careful visual inspection. This is done to identify the cause of the premature failure. The cause should be found before any cleaning or reconditioning removes any indication of the problems leading to the failure. These indications of failure have been called *fatigue fingerprints*.

14-1 Valve Condition

A careful examination of the valves that have been removed from the engine will indicate causes of valve failure. Valves fail because their operating limits have been exceeded. This means that the valves have been expected to do more than they were designed to do. The cause of the failure must be corrected so that the new or reconditioned valves will perform satisfactorily. Valve failure results from misaligned valve seating, excessive temperature, high-velocity seating, and high mileage. These conditions are usually interrelated.

Valve Seating. Valve face *burning* (Figure 14-1) and valve face *guttering* (Figure 14-2) are common types of valve failure. They result from poor seating that allows the high-temperature and high-pressure combustion gases to leak between the valve and seat. Poor seating results from too small a valve lash, hard carbon deposits, valve stem deposits, excessive valve stem-to-guide clearances, or out-of-square valve guide and seat.

A valve lash that is too small can result from improper valve lash adjustment on solid lifter systems. It can also result from misadjustments in a valve train using hydraulic lifters. The clearance will also be reduced from valve head cupping shown in Figure 14-3, or valve face and seating wear. Figure 14-4 shows typical intake valve and seat wear.

Hard carbon deposits such as those shown in Figure 14-5 are loosened from the combustion chamber. Sometimes, these flaking deposits stick between the valve face and seat to hold the valve slightly off its seat. This reduces valve cooling through the seat

Figure 14-1. Valve face burning.

Figure 14-2. Valve face guttering.

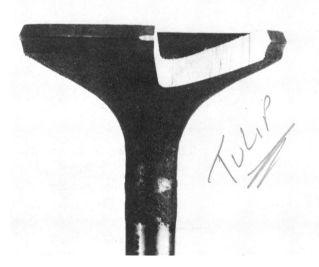

Figure 14-3. A cupped valve head.

Figure 14-4. Typical intake valve and seat wear.

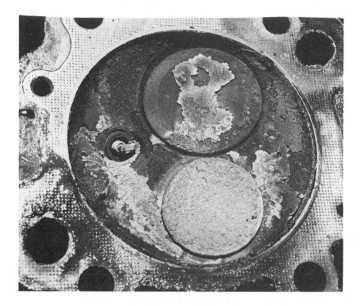

Figure 14-5. Normal carbon flaking in a combustion chamber.

Figure 14-6. Valve face peening.

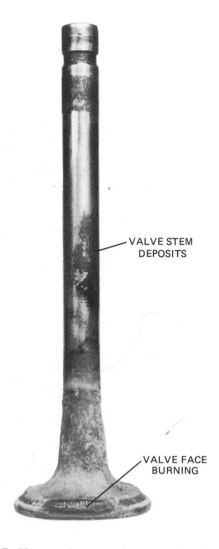

Figure 14-7. Heavy valve stem deposits and valve face burning.

and allows some of the combustion gases to escape. Continued pounding on hard carbon particles gives the valve face a *peened* appearance, pictured in Figure 14-6. Positive valve rotors help to combat the effects of deposits to prevent valve burning.

Fuel and oil on the hot valve will break down to become hard carbon and varnish deposits that build up on the valve stem. Heavy valve stem deposits are shown in Figure 14-7. These deposits cause the valve to stick in the guide so that the valve does not completely close on the seat and the valve face to burn. This is one of the most common causes of valve face burning.

If there is a large clearance between the valve stem and guide, too much oil will go down the stem. This will increase deposits, as shown on the intake valve in Figure 14-8. In addition, a large valve guide clearance will allow the valve to cock or lean sideways, especially with the effect of the rocker arm action. Continued cocking keeps the valve from seating properly and it will leak, burning the valve face.

Sometimes, the cylinder head will slightly warp as the head is tightened to the block deck during assembly. Other times, heating and cooling will cause warpage. When head warpage causes valve guide and seat misalignment, the valve cannot seat properly.

Excessive Temperatures. High valve temperature occurs when the valve does not seat properly; however, it can occur even when the valve is seating

Figure 14-8. Heavy intake valve deposits.

Figure 14-10. Badly guttered valve face.

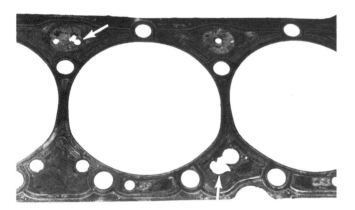

Figure 14-9. A corroded head gasket that has enlarged the coolant metering holes.

Figure 14-11. Hoop stress cracks in a valve head.

properly. Cooling system passages in the head may be partially blocked by faulty casting or by deposits built up from the coolant. A corroded head gasket, as pictured in Figure 14-9, will change the coolant flow. This can cause overheating when the coolant is allowed to flow to the wrong places. Extremely high temperatures are also produced by preignition and by detonation. These are forms of abnormal combustion. Both of these produce a very rapid increase in temperature that can cause uneven heating. The rapid increase in temperature will give a thermal shock to the valve. **A thermal shock is a sudden change in temperature.** The shock will often cause radial cracks in the valve. The cracks will allow the combustion gases to escape and *gutter* the

valve face. A badly guttered valve face is shown in Figure 14-10. If the radial cracks intersect, a pie-shaped piece will break away from the valve head. A thermal shock can also result from rapid cycling the engine from full throttle to closed throttle and back again. Valves with a hard metal facing have special problems. Excess heat causes the base metal to expand more than the hoop of the hard face metal. The hard face metal hoop is stressed until it cracks. The crack allows gases to gutter the base metal, as shown in Figure 14-11.

High engine speeds require high gas velocities. The high-velocity exhaust gases hit on the valve stem and tend to erode or wear away the metal mechanically. The gases are also corrosive so the valve stem will tend to corrode. Corrosion removes

Figure 14-12. Necked valve stem.

Figure 14-14. A piston ruined by a valve head that broke from the stem.

the metal chemically. The corrosion rate doubles for each 25 °F (14 °C) increase in temperature. Erosion and corrosion of the valve stem cause *necking* that weakens the stem and leads to breakage. Necking is shown in Figure 14-12.

Misaligned Valve Seats. When the valve-to-seat alignment is improper, the valve head must twist to seat each time the valve closes. If twisting or bending becomes excessive, it fatigues the stem and the valve head will break from the stem. An example of this can be seen in Figure 14-13. The break appears as lines arcing around a starting point. The head usually damages the piston when it gets trapped between the piston and the head.

High-Velocity Seating. High-velocity seating is indicated by excessive valve face wear, valve seat recessing, and impact failure. It can be caused by

excessive lash in mechanical lifters and by collapsed hydraulic lifters. Lash allows the valve to hit the seat without the effect of the cam ramp to ease the valve onto its seat. Excessive lash may also be caused by wear of parts, such as the cam, lifter base, pushrod ends, rocker arm pivot, and valve tip. Weak or broken valve springs allow the valves to float away from the cam lobes so that the valves are uncontrolled as they hit the seat. The normal tendency of hydraulic lifters is to pump up under valve float conditions, and this reduces valve impact damage.

Impact breakage may occur under the valve head or at the valve lock grooves. The break lines radiate from the starting point. The valve head from impact breakage will fall into the combustion chamber, too. In most cases, it will ruin the piston before the engine can be stopped, as pictured in Figure 14-14.

High Mileage. Excessive wear of the valve stem (Figure 14-15), guide, valve face, and seat are the result of high mileage. They usually have a great deal of deposits. The valves will, however, still be seating and they will show no sign of cracking or burning.

When the valve stems do not have enough lubricant, they scuff. To scuff, they temporarily weld to the guide when the valve is closed. The weld breaks as the valve is forced to open. Welded metal tears

Figure 14-13. A valve head broken from the stem.

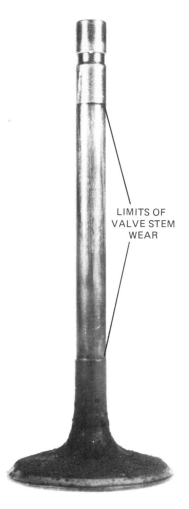

Figure 14-15. High mileage valve stem wear.

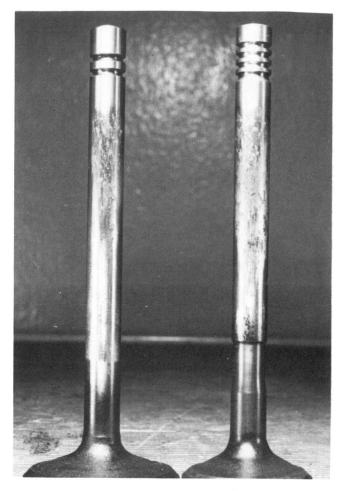

Figure 14-16. Scuffed valve stem as a result of loss of valve train lubrication. One of these valves was taken from the guide shown in Figure 14-17.

Figure 14-17. Scuffed valve guides.

from the guide and sticks to the valve stem. An example of valve stem scuffing is shown in Figure 14-16. The metal knobs on the valve stem scratch the valve guide as it operates. This also scuffs the valve guide, as pictured in Figure 14-17. In a short time, the valve will stick in the guide and not close. This will stop combustion in that cylinder. Both valve and valve guide will have to be replaced.

Often, valve tips become damaged. This damage can be seen before the valves are removed from the head. Figure 14-18 shows a valve tip that has become mushroomed from excessive valve lash. This caused pounding that flattened and widened the valve tip. Other valve-tip problems are caused by rapid rotation as the valve is being opened. This causes circles on the valve tip. Still other valves do not rotate at all. These valve wear in the direction of the rocker arm or finger follower movement. Ex-

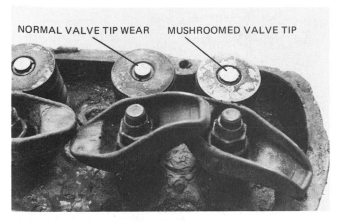

Figure 14-18. A mushroomed valve tip.

Figure 14-19. Excessive valve tip wear.

Figure 14-20. Normal piston thrust surface wear.

Figure 14-21. A piston burned as a result of detonation.

amples of excessive valve tip wear can be seen in Figure 14-19.

14-2 Piston Condition

Normal piston wear shows up as even wear on the thrust surfaces of the piston. The wear is from the top to the bottom in the center of the thrust surfaces, as shown in Figure 14-20. The top ring is slightly loose in the groove. This type of piston can usually be reconditioned for additional useful service.

Heat Damage. Holes in pistons, burned areas, severely damaged ring lands, and scoring are obviously abnormal conditions. The exact nature of the abnormal condition should be determined. This is necessary so that the cause can be corrected. The driver can then be given advice that will minimize the possibility of the damage recurring.

Combustion knock or detonation will burn the edge of the piston from the head down in behind the rings, as pictured in Figure 14-21. This burning usually occurs at a point farthest from the spark plug, where the hot end gases rapidly release their heat energy during detonation. In some cases, the heat breaks through the piston head. Preignition results from multiple flame fronts that build up temperature and pressure early in the combustion cycle. High temperature softens the piston, allowing combustion pressure to burn through, usually

Figure 14-22. A piston head burned as a result of pre-ignition.

Figure 14-23. A scuffed piston skirt from overheating and lubrication breakdown.

Figure 14-24. Piston ring scuffing from overheating.

Figure 14-25. Piston ring stuck in their grooves with hard carbon.

near the middle of the piston head, as shown in Figure 14-22. The piston metal will often show some spattering.

Another form of heat damage is scuffing, similar to that pictured in Figure 14-23. This happens when excess heat causes the piston to expand until it becomes tight in the cylinder bore. The lubricant is thinned from heat and it is squeezed from the cylinder wall. This causes metal-to-metal contact. Excessive heat can come from a malfunctioning

cooling system as well as from abnormal combustion.

Piston rings may get hot spots on their face from a lack of lubrication, from high combustion temperatures, or from ineffective cooling systems. Metal from the ring hot spots will transfer to the cylinder wall, scuffing the ring and piston (Figure 14-24).

Worn piston rings allow hot combustion gases to blow by the piston. The worn rings will also allow oil to come up from the crankcase to the combustion chamber. Hot combustion gases meet the oil in the area of the rings where the heat will partially burn the oil. This produces hard carbon around the rings, causing them to stick in the grooves, as shown in Figure 14-25. If this is the only piston problem, it can be corrected by cleaning.

Corrosion Damage. Low operating temperatures will produce a corrosive mixture in the oil. Coolant leakage into the combustion chamber increases the rate of corrosion. Corrosion produces mottled gray pits on the aluminum piston. Low operating temperatures are caused by short-trip driving or by a faulty or missing cooling system thermostat.

Mechanical Damage. Piston damage can result from mechanical problems. Connecting rod misalignment will show up as a diagonal thrust surface wear pattern across the piston skirt, which indicates that the piston is not operating straight in the cylinder (Figure 14-26). This, in turn, means that the rings are not running squarely on the walls, so that they cannot seal properly.

Piston damage can come from the loss of a piston pin lock ring. The lock will come out if the lock grooves are damaged or if the lock ring is weak. It will also come out if the rod is bent so that a side load is placed on the piston pin, forcing it against the lock ring. The piston and possibly the cylinder will be badly damaged as the lock ring slides between them. A new piston will be required when this type of damage occurs, as shown in Figure 14-27.

Figure 14-27. Piston damage when a pin lock ring came out

Figure 14-28. A cracked piston.

Pistons can crack, usually on the skirt or near the piston pin boss. Cracks generally occur at high mileage, because of overloading or because the pistons were improperly designed. Any crack in a piston is cause for rejection. A typical piston skirt crack is shown in Figure 14-28.

Dirt entering the engine greatly increases the amount of wear. Dirt will scratch and wear the face of the piston rings and wear the side of the ring and groove. The side clearance of the top piston ring

DIAGONAL WEAR

HEAVY BEARING WEAR

Figure 14-26. Angled piston skirt wear caused by a misaligned connecting rod.

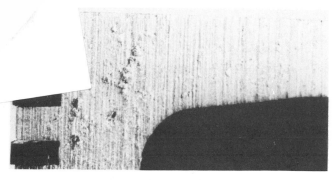

Figure 14-29. A badly worn piston ring and piston ring groove.

Figure 14-30. Piston lands damaged from broken rings.

Figure 14-31. Camshaft lobe failure starting at the edge of the lobe.

Figure 14-32. Badly worn cam lobes.

must not exceed 0.006 in. (0.15 mm). Dirt will come in with the air through a leaking air filter element or through an air leak between the filter and the carburetor air horn. Dirt in the oil will cause abnormal wear on the piston skirts, sides of the ring groove, and the oil ring. A badly worn piston ring and groove are shown in Figure 14-29. Worn piston rings must be replaced. Piston ring groove side wear is usually uneven. This causes the ring to hit only the high spots. The high spots cause high localized side loads on the ring when the ring is forced against the side of the groove. These loads may break the ring, which, in turn, generally cause excessive piston land damage similar to that shown in Figure 14-30. Worn piston skirts cause excess clearance, which allows the piston to slap. Piston slap can lead to cracked pistons.

14-3 Valve-Train Condition

Failure of a valve may be caused by the condition of the valve train. Valve-train wear can usually be seen as wear. It can result from exceeding operating limits and from loss of lubrication.

Camshaft. Camshaft failure shows up as cam lobe or cam bearing journal failure. Lobe failure starts as small pits caused by heavy edge wear of the lobe, illustrated in Figure 14-31. The lobe will gradually break down as the engine is operated until it almost looks like a round journal. A badly worn cam, such as the one pictured in Figure 14-32, will cause badly worn lifters. Occasionally, the lobe will chip as the result of impact damage, usually from careless handling.

Lifters. Camshaft lobe failure will damage the lifter face. Face failure may start with pits (Figure 14-33) or with smooth wear (Figure 14-34). The worn lifter face gradually becomes concave (Figure 14-35). In severe cases, the lifter has worn all the way through the face. This will also wear the cam lobe. Pitted or concave lifter faces are not repairable. The lifter must be replaced. Always install new lifters with a new cam. A worn cam *may* damage new lifter faces and thus give a short lifter service life. Worn lifters always damage the lobes of a new cam shaft.

232

Figure 14-33. Pitted lifter face.

Figure 14-35. Excessive lifter wear causing a concave face. This lifter was operating on the camshaft pictured in Figure 14-32.

Figure 14-34. Smooth wear caused by a spinning lifter.

Rocker Arms. Worn rocker shafts and matching rocker arms should be replaced. Some older cast rocker arms could have their valve tip end faces ground, but this practice has been discontinued. Stamped rocker arms, pivot balls, or shafts should be replaced when any wear can be seen. Two different types of rocker arm failure are pictured in Figure 14-36.

Pushrods. Pushrod failure results from lack of lubrication or from overstressing. Lack of lubrication results in excessive pushrod end wear. Overstressing will cause bent pushrods. Rod bend can be detected by rolling the pushrod across a flat surface. Any wobble indicates a bent rod. Pushrods that are bent or worn must be replaced.

Valve Springs. Valve springs may rotate in use and become shiny on the end, as shown in Figure 14-37. They may develop pits or take a set. Wear and pits are checked visually. Valve spring set results in warping the spring or in loss of tension.

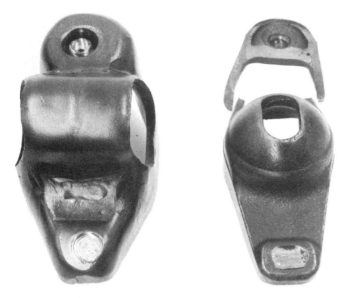

Figure 14-36. Failed rocker arms. A badly worn rocker arm is on the left and a broken rocker arm is on the right.

Figure 14-37. Valve spring's showing shiny ends from rotation.

Valve Retainers and Locks. Checking valve retainers and locks is limited to a visual inspection. Cracks and wear or questionable serviceability are reasons for replacement. If the lock or retainer should fail, the valve could fall into the combustion chamber and do extensive damage. Figure 14-38 shows what happened to a valve when the locks failed.

Figure 14-38. A valve bent when a lock failed allowing the valve to fall into the combustion chamber.

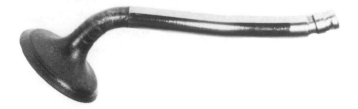

Often, valve locks are replaced on a routine basis during a valve job.

14-4 Shaft and Bearing Condition

Shaft damage includes scored bearing journals, bends or warpage, and cracks. Damaged shafts must be reconditioned or replaced.

Crankshaft. The crankshaft is one of the most highly stressed engine parts. The stress increases four times every time the engine speed doubles. Any sign of a crack is a cause to reject the crankshaft. If a cracked crankshaft is continued in service, it will break, as did the shaft shown in Figure 14-39. Cracks in high-production passenger car engine crankshafts can be seen during a close visual inspection. High-rpm racing crankshafts should be checked with Magnaflux to show up any very small cracks that would lead to failure.

Bearing journal scoring is one of the most common crankshaft defects. Scoring appears as scratches around the bearing journal surface. Generally, there is more scoring near the center of the bearing journal. This can be seen in Figure 14-40. Dirt and grit carried in the oil enter between the journal and bearing. If the particles are too large to get through the oil clearance, they will partially embed in the bearing surface and scratch the journal. Dirt can also be left on the journal during assembly if the parts are not thoroughly cleaned.

Crankshaft journals can have nicks or pits in them. Nicks are caused by carelessness when the

Figure 14-39. A broken cast crankshaft.

Figure 14-40. A scored connecting rod bearing journal.

Figure 14-42. The pitted part of the main bearing journal rides in the main bearing oil groove.

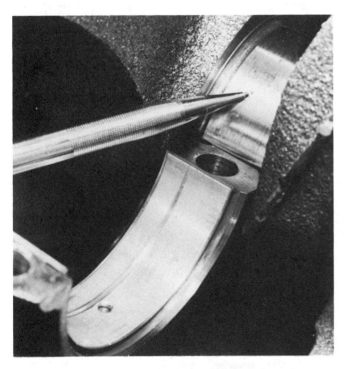

Figure 14-41. A nicked crankshaft journal that scratched the bearing.

Figure 14-43. A damaged connecting rod journal.

journal is bumped with another part while exposed or during assembly (Figure 14-41). Pits such as those pictured in Figure 14-42 are caused by corrosion.

Rough journals and slight bends can be corrected by grinding the journals on true centers. Forged shafts with excess bend should be straightened before grinding.

If the relatively inexpensive standard production crankshaft is damaged beyond grinding limits, it should be replaced. Excessive damage is pictured in Figure 14-43. It was caused when a connecting rod bolt came out of the assembly shown in Figure 14-44. More expensive racing crankshafts, or crankshafts that are to be modified, can have the journals built up by welding or by special metal spray techniques. They are then straightened and reground. This process is expensive and, therefore, is done only when it is less costly than purchasing a

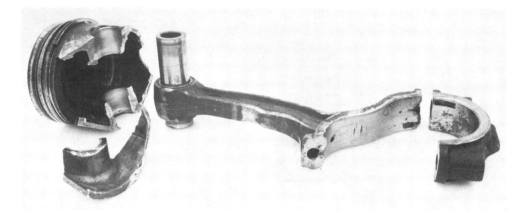

Figure 14-44. A connecting rod bolt came out. This caused the damage to the piston and rod. It also damaged the crankshaft journal pictured in Figure 14-43.

new special crankshaft. The connecting rod journal shown in Figure 14-45 would have to be built up and reground to be used again.

Camshaft. Cam lobe damage, similar to that shown in Figure 14-46, is the most usual type of camshaft failure. The most critical period for cam wear is during the first 15 minutes of operation after the engine is assembled. Cam lubrication is critical during this time. Manufacturers recommend specific cam surface lubricants, to be used during assembly. They run all the way from SE motor oil to especially

prepared lubricants available at their own dealer parts department.

Cam lobes are designed with a slight taper across the surface, and the tappet or lifter has a slight crown. This places the lifter contact point slightly offset from the center of the lifter. The offset causes slight lifter rotation. This produces a wear pattern that is nearly even. A flat based lifter or a used lifter on a new camshaft will contact the high edge of the cam lobe, producing edge wear that will lead to surface breakdown.

Like crankshafts, passenger car camshafts that have abnormal wear are replaced. Camshafts can be reground on special profile cam grinders. This can be done as a repair procedure or as a means of changing the cam lobe shape to change the engine performance. It is generally more expensive to repair a camshaft by grinding than it is to purchase a new standard production camshaft. Repair grinding is, therefore, only done on special camshafts and on some low-production heavy-duty camshafts.

Figure 14-45. A connecting rod journal badly worn from lack of lubricant.

Figure 14-46. A camshaft lobe failure from pitting caused by fatigue.

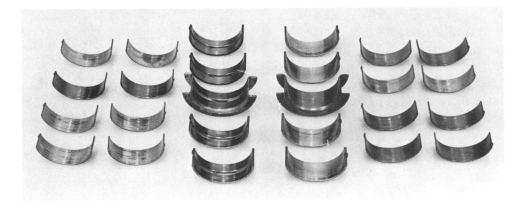

Figure 14-47. Typical lay out for inspecting bearings to determine the cause of failure.

Grinding to repair a camshaft increases the distance between the top of the lifter and the rocker arm. Some rocker arms can be adjusted to correct this distance. Nonadjustable rocker arms with hydraulic lifters require a longer pushrod. The new pushrod adds to the expense of camshaft grinding.

Bearings. Bearing *distress* or failure results from foreign particles, from lubrication breakdown, from overloading, and from corrosion. Any type of bearing failure is cause for replacement. The failed bearing should be given a close visual examination to determine the cause of failure that produced bearing distress. The *cause* must be corrected before the engine is reassembled, to prevent recurring failure.

The best procedure for examining bearings is to remove the bearing shells from the block, rods, and caps. They should be placed on a table in the same order as their location in the engine, bearing face up, as pictured in Figure 14-47. The bearing distress of each bearing can then be related to the other bearings and to the shaft journals.

Foreign particles are dirt, metal chips, or abrasives. Dirt may have been left in the engine when it was assembled, even between the bearing and cap. An example of this is shown in Figure 14-48. It may enter through breather openings or through faulty filters. The dirt is carried in the oil to the bearings. Metal chips may have lodged in the engine during machining operations. They are also produced as internal engine parts wear. Normally, the small metal particles from wear will settle to the bottom of the oil pan, get trapped in the oil filter, or they are drained from the engine during an oil

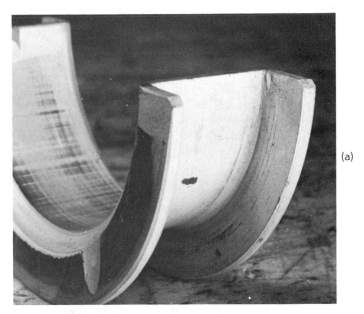

(a)

Figure 14-48. Failure caused by a particle left behind the bearing shell during assembly (a). This caused a worn spot on the bearing surface (b).

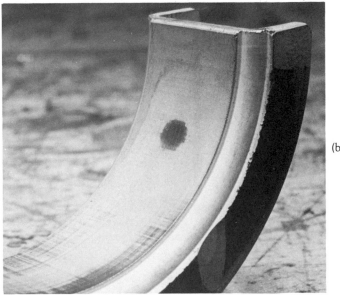

(b)

change. When the engine parts rapidly breakdown, as shown in many of the illustrations in this chapter, the particles may move through the lubrication system to the bearing. The most likely causes of small iron particles are oil pump gear wear, cam and lifter wear, rocker arm wear, and timing gear and chain wear. Aluminum particles come from piston skirt wear. Abrasives are left in the engine after reconditioning any time the parts are not thoroughly cleaned using proper cleaning methods. They also get into the engine through faulty filters. Abrasives tend to embed in soft bearing metals such as aluminum, brass, copper, and babbitt. When this happens, the abrasives rapidly cut into the hard bearing journal that is running against them. Abrasives should never be used to clean the surface of soft metals.

Foreign particles *embedded* in bearings, pictured in Figure 14-49, should be checked to determine the type of material. Bearing materials tend to cover these particles so they will not damage the shaft. Large particles will not embed. They will score the bearing and shaft. The type of embedded particle can be determined by scraping some of the embedded material from the bearing surface near the end of the bearing shell, where most of the foreign material will collect. Steel particles can be picked up with a magnet. Aluminum particles will dissolve and produce bubbles when a drop of 10% solution of sodium hydroxide is placed on them. Other embedded particles, such as dirt, brass, and copper, can best be identified by their color.

Lubrication breakdown may result from the

Figure 14-50. Bearing material starting to leave the steel backing.

motor oil not getting to the bearing, from heat that excessively thins the motor oil, and from overloading the bearing so that oil is squeezed from the bearing surface. These problems are interrelated and may occur at the same time. Oil starvation wipes or extrudes low-melting-temperature bearing material from the steel backing. Figure 14-50 shows the bearing material starting to leave the steel backing. Temperatures may even increase enough to cause the steel bearing backing to turn to a blue color from overheating. This can be seen on the right hand bearing in Figure 14-51.

Figure 14-51. The bearing material is completely gone from the bearing shells on the right from overheating.

Figure 14-49. Foreign particles embedded in the bearing.

The main reason oil does not get to a bearing is excessive oil leakage or throw-off from other bearings that are closer to the oil pump. When leakage from bearing clearances becomes greater than the oil pump's capacity, at the speed the pump is running, oil starvation will occur. Worn cam bearings often cause oil leakage that starves the crankshaft bearings. The valve train holds the camshaft against the bottom of the bearing all the time. Because of this, wear occurs only on the lower side of the bearing. The excess bearing clearance that is formed above the cam does not produce a knocking sound because the load is always in the downward direction. If oil is delivered to the upper side of the worn cam bearing it can freely escape to the pan. This would reduce the amount of oil going to the main and rod bearings, where oil starvation may occur. Fortunately, most cam lubrication passages are on the lower side of the bearing. Blocked oil passages are usually the result of misaligned oil holes in a bearing shell. They are *not* the result of plugging with deposits. The mismatch can be seen on the outer surface of the bearing shell.

Oil starvation can occur at very high engine speeds in some engine designs. High crankshaft rotational speed on these engines cause centrifugal forces on the oil in the crankshaft oil holes that are greater than oil pump pressure. To lubricate, the oil must have a laminar or smooth layer flow to form a film. At excessive speeds, the oil film can no longer flow smoothly and turbulent flow develops in the oil. This causes voids or open spaces in the oil film. Oil starvation occurs in these voids.

As the lubrication film breaks down, friction and heat increase. Heat thins the oil, which, in turn, reduces the lubrication film. This process is unstable and the film breakdown increases rapidly until the bearings fail.

Full-throttle low-speed operation is called lugging. This type of operation puts very high loads on the bearings. The high loads tend to squeeze oil from the bearing. Engine and engine parts manufacturers caution against lugging because it causes rapid damage to engine parts. Lugging can occur only with vehicles equipped with a drive line using standard transmissions. Automatic transmissions allow enough slippage so that the engine speed is high enough at full-throttle, low-vehicle-speed operation that lugging cannot occur.

Each time a load is put on a bearing, the bearing surface flexes or bends slightly. Bearings will flex during all engine operation. The flexing effect increases as the engine loads increase, especially during lugging. Gradually, the flexing will cause bearing *fatigue,* which will cause fine cracks in the bearing surface. In time, the cracks will join at the backing bond to loosen pieces of the bearing material from the backing. The loosened piece of bearing material overheats, then melts. The rate of bearing failure will increase when there is less bearing material on the shell. Fatigue failure results in bearings that appear similar to the one pictured in Figure 14-52.

Bearing overloading can occur on a part of a misaligned bearing or journal. This may be the result of a bent rod or shaft or a warped engine block. Journal taper usually results from journal wear; however, it is always possible to have faulty grinding. It results in bearing wear, as shown in Figure 14-53.

Figure 14-52. Bearing material missing from the bearing as a result of fatigue failure.

Figure 14-53. Bearing wear caused by a misaligned journal.

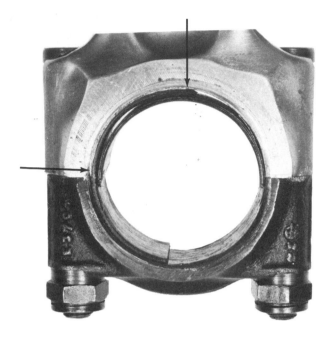

Figure 14-54. A spun bearing. The lower cap bearing has rotated under the upper rod bearing.

Lightly loaded short-trip driving does not produce enough heat to drive off condensed water and gases. These collect in the oil, causing acid and sludge. The acid attacks the bearings and the bearings corrode.

Faulty bearing installation can lead to premature failure. Bearings that fit too tightly will not leave enough oil clearance, so they do not have enough oil. Bearing shells that do not have enough crush may rotate with the shaft. This is called a *spun bearing,* as pictured in Figure 14-54. Dirt that becomes trapped behind the bearing shell causes a tight spot in the bearing, which will fail.

Careful interpretation will pinpoint the cause of the bearing distress. The cause must be corrected so that the reconditioned engine will have a normal service life.

14-5 Block Condition

Block faults occur in the cylinder wall, cooling system, and shaft bore alignment, and as broken parts. All the other engine parts depend upon the block for support, alignment, and operating climate.

Cylinder wall wear, of the type in Figure 14-55, is one of the most noticeable abnormal block condition. Cylinder walls, in normal use, have a smooth

Figure 14-55. A cylinder wall scored from an oil ring installed incorrectly.

glaze from smoothing effects of the piston and rings during operation. This happens even when the cylinders are tapered, out-of-round, and have a wavy wall. Dirt and broken rings will cause scratches on the wall. If the scratches remain in the reconditioned engine, they will let some oil go into the combustion chamber on the intake stroke. The scratches will also allow combustion gases to blow into the crankcase during the power stroke.

Sometimes the connecting rod is allowed to strike the bottom edge of the cylinder as the piston and rod assembly are removed or installed. This will nick the bottom edge of the cylinder and raise sharp points. If these points are not removed, they will scratch the piston skirt of the reconditioned engine, as shown in Figure 14-56.

Figure 14-56. Piston skirt scratched as a result of nicks on the bottom edge of the cylinder.

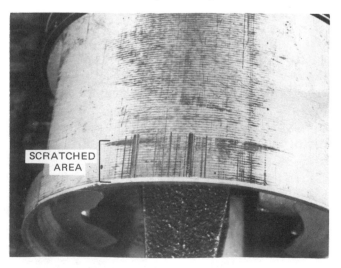

SCRATCHED AREA

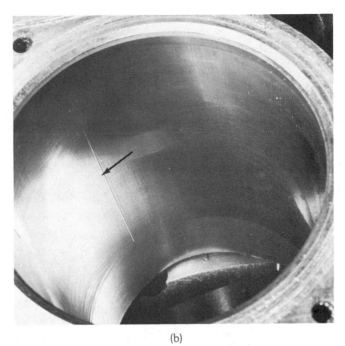

(a)   (b)

Figure 14-57. An overlapped oil ring (a) and scratched cylinder wall (b).

Piston pin locks that come out of the piston will usually score the cylinder wall. If the piston pin moves out of its bore, it, too, will score the wall. If the oil ring spacer is overlapped during assembly it will score the cylinder wall, as shown in Figure 14-57. Occasionally, such deep scores are produced that they break into the coolant jacket. When this happens, a thin-walled cast iron sleeve can be installed to repair the cylinder. To do this the cylinder is bored oversize to a diameter to accept the sleeve. The sleeve is pressed in, then bored to the required cylinder diameter.

Gears, sprocket, and pulley hubs crack. Parts with visible cracks must be replaced. Figure 14-58 shows a cracked crankshaft timing sprocket. A cracked damper hub is pictured in Figure 14-59.

Figure 14-58. A cracked crankshaft timing sprocket.

Figure 14-59. A cracked damper hub.

REVIEW QUESTIONS

1. What are the two basic reasons for engines to be disassembled for repairs? [INTRODUCTION]
2. Why must parts that have failed be given a careful visual examination before cleaning? [INTRODUCTION]
3. Why do valves fail? [14-1]
4. What problems cause poor valve seating? [14-1]
5. What is one of the most common causes of valve face burning? [14-1]
6. What valve problems result from excessive heat? [14-1]
7. What causes high velocity valve seating? [14-1]
8. What can cause valve necking? [14-1]
9. What are the indications of high mileage on valves? [14-1]
10. What causes valve tip wear? [14-1]
11. What causes holes to be melted through pistons? [14-2]
12. What causes piston scuffing? [14-2]
13. What causes the piston rings to stick in the grooves? [14-2]
14. What mechanical damage can occur to a piston? [14-2]
15. How can the piston ring land damage from a broken ring be identified? [14-2]
16. How can cam lobe wear be determined? [14-3]
17. Why should new lifters always be installed with a new camshaft? [14-3]
18. When are the valve train failures bad enough to require the parts to be replaced? [14-3]
19. What should be done with a small crack in a crankshaft? [14-4]
20. What is one of the most common crankshaft defects? [14-4]
21. What can be done when a camshaft is faulty? [14-4]
22. Why is it important to determine the cause of bearing failure? [14-4]
23. What are the most likely causes of small iron particles in the engine oil? [14-4]
24. What are the most likely causes of small aluminum particles in the engine oil? [14-4]
25. Why should foreign particles imbedded in a bearing be checked to determine the type of material? [14-4]
26. How are embedded particle materials identified? [14-4]
27. What can cause lubrication breakdown in a bearing? [14-4]
28. What engine faults keeps the oil from getting to the bearings? [14-4]
29. What is the usual cause of blocked oil passages? [14-4]
30. Why should the oil in a bearing have a laminar flow? [14-4]
31. What damage will engine lugging cause? [14-4]
32. What type of engine operation will cause bearing fatigue failure? [14-4]
33. What bearing problems are caused by faulty installation? [14-4]
34. What faults can be found on a block? [14-4]

CHAPTER
15

Preparing For Service

The engine parts are cleaned after the engine has been disassembled and the cause of obviously abnormal and failed parts has been identified. More abnormal conditions may show up after the parts are cleaned.

It is normal for an engine to build up deposits as it operates. More deposits can be expected with a longer time between oil changes and with greater mileage. All the deposits are removed during engine reconditioning. Any deposits remaining may cover over abnormal conditions that could lead to premature failure of the reconditioned engine. Deposits not cleaned from the parts as the engine is being reconditioned may loosen in service and get into the oil. They can plug the oil pump inlet screen and cause premature wear of engine parts.

Scale deposits build up in the engine cooling system. Cleaning the cooling system is one of the operations that is often overlooked when reconditioning the engine. The cooling system dries out as the engine is being reconditioned. The dry scale will usually flake off and fall to the bottom of the block cooling passage. If this scale is not cleaned out of the block it will keep the coolant from circulating around the lower part of the cylinders. This will cause overheating that can lead to premature scuffing of the rings and pistons. The soft plugs or core plugs will have to be removed to thoroughly clean the cooling passages in the block and head. Deposits left in the exhaust crossover of the intake manifold can cause poor warm-up and slow choke opening on V-type engines. They must also be cleaned from the engine.

One of the main reasons that a thorough cleaning is often neglected is the time that must be taken to do a good cleaning job. It is estimated that it takes as long to properly clean the engine as it does to recondition the engine parts. This suggests that, in the long run, the most efficient cleaning methods are the least costly. This is especially true if the cleaning equipment can be kept in continuous operation. Simple, less costly equipment with a small capital investment is more economical for an occasional cleaning job, even if it is less efficient.

After the parts are thoroughly cleaned, they are inspected. Inspection starts with another visual inspection. Some problems, such as a fine piston crack, may not have shown up on a dirty piston. Critical iron and steel parts, such as a crankshaft or connecting rod, may be Magnafluxed to locate defects that cannot otherwise be seen.

Parts that appear to be serviceable should be given a dimensional inspection. This will indicate the amount of wear on the part. From the inspection, the service technician can determine what parts can be reconditioned for further service.

15-1 Cleaning Requirements

The deposits on the outside of the engine are usually a mixture of oil and dirt. Sometimes, antifreeze will leak out of the engine and mix with the other exterior deposits. Some of the exterior surfaces will be rusty. The surface deposits inside the engine are a mixture of soft and hard carbon. These deposits are caused by the breakdown of hot motor oil and by blow-by gases. Scale and rust will build up in the cooling passages of the block.

Cleaning chemicals applied to the parts will mix with and disolve the deposits. The chemicals loosen the deposits so that they can be brushed or rinsed from the surface. **A deposit is said to be soluble when it can be dissolved with a chemical.**

It is always a good practice to wear safety glasses when working on machinery. It is especially important to wear them when cleaning parts. Soil, water, steam, and chemicals will splash off as the parts are cleaned. *Eye protection must be used.*

Soil. Any material that collects on the surface of the engine metal can be called soil as well as being called a deposit. Some of the soil that is deposited on an engine is *water-soluble*. **Water soluble means that it will mix with or dissolve in water.** Exterior dirt on the engine is an example of water-soluble soil.

Chemical cleaners are required to dissolve organic soils, such as oil, carbon deposits, and protective coatings. Still other chemical cleaners are required to clean rust and scale. Heat will always increase the rate at which the cleaning chemical will act to dissolve the soil.

It is important to know what kind of material is being cleaned when chemicals are used. Iron and steel will rust. Soft metals, such as aluminum and bearing metal, are dissolved by some types of cleaning chemicals. *Inhibitors* must be used in the chemical solution when cleaning these materials. Plastic parts are usually destroyed by chemical cleaning materials that are designed to clean metal.

Engines do not collect much water-soluble soil on the outside and none on the inside. Most of the soil deposit on the engine parts is mixed with oil and grease. These deposits can be cleaned with a steam cleaner or with a chemical cleaner mixed in a distillate, such as diesel fuel. The deposit is coated with the mixture. After a soaking period, the deposit and chemical cleaner are flushed off with the high-pressure water from a hose. The chemical cleaner will loosen the deposits and the water will rinse it from the surface.

Oil and Grease. Oil and grease are often removed with a petroleum-based cleaning solvent. This type of deposit can also be removed using steam cleaning or chemicals. Petroleum solvents will mix with the oil and grease in the deposit to thin them. When enough petroleum solvent is used, the dilute mixture will run from the surface. A thin oil coating will remain on the surface to prevent rusting. **Never use gasoline for cleaning.** The gasoline evaporates so fast that it cannot carry the soil from the parts. Gasoline vapors ignite easily. When they ignite, they will burn rapidly, causing a very hot fire.

Steam cleaning uses a mixture of high-temperature steam and high-pressure water. Usually, a *chemical cleaner* is added to the water-steam mixture to speed the cleaning process. The steam heat thins the oil and grease while the high-pressure water rinses it from the surface. During steam cleaning, the parts become hot. This heat helps to

Figure 15-1. Typical high mileage combustion chamber deposits.

dry the parts rapidly after they are cleaned. In some cases, it may be necessary to coat the clean parts with a thin film of clean oil to keep them from rusting. Chemicals *emulsify* oil and grease soil so that it can be rinsed from the surface. **Emulsification is indicated when the oily material turns milky white as it mixes with the rinse water.**

Carbon. Deposits remaining from heated motor oil and from fuel are the most difficult type of soil to remove. The entire interior of a high-mileage engine will have a coating of carbon. The heavy carbon deposits can be chipped, scraped, or wire-brushed to loosen them from the surface. There will be more heavy carbon deposits on the valves, pistons, and combustion chambers than on any other engine parts. Typical high-mileage heavy combustion chamber deposits are shown in Figure 15-1. The rest of the light carbon deposits can be removed in chemical cleaning solutions. Any hard deposits that still remain on the surface after chemical cleaning can be removed by mechanical cleaning.

15-2 Cleaning Materials

It is important to use the correct cleaning materials in the proper way to clean the parts thoroughly and rapidly. Generally, *chemical cleaners* are used first to remove all but the hardest deposits. The remaining hard deposits are removed by *mechanical cleaning,* such as scraping and glass bead blasting.

Cleaning Chemicals. Cleaning chemicals are blended together by the manufacturer of the chemicals. Instructions for their use will accompany the chemical. When a service technician follows these instructions, the parts will be satisfactorily cleaned. The following discussions will help you to understand the action of the cleaning materials.

Cleaning chemicals must be used with care. *Safety glasses* should always be used when cleaning engine parts. **Toxic chemicals are poisonous. Caustic chemicals will burn the skin.** In all cases, care should be taken to keep the chemical mixtures from being splashed on clothing or skin. A rubber apron will help keep the chemicals off the clothing. If you are accidentally splashed with the cleaning chemical, immediately flush the chemical off your skin with lots of clear water.

Most chemical cleaners used for cleaning carbon-type deposits are caustic. Acid chemical cleaners are used to clean the cooling system. A scale from 1 to 14, called pH, is used to indicate the amount of chemical activity. Pure water is neutral. On the pH scale, water is pH 7. Caustic materials have pH numbers from 8 through 14. The higher the number, the stronger the caustic action will be. Acid materials have pH numbers from 6 through 1. The lower the number, the stronger the acid action will be. Caustic materials and acid materials neutralize each other. This is what happens when baking soda (a caustic) is used to clean the outside of the battery (an acid surface). The caustic baking soda neutralizes any sulfuric acid that had been spilled or splashed on the outside of the battery.

Caustic cleaning materials have a *soap action.* The chemical molecules surround and coat the soil molecules. This coating on the soil molecules keeps them from combining to form large deposits. The soil molecules will then float in the solution. In this way, the soil is soluble in the chemical solution. **A chemical solution is the chemical dissolved in the liquid.** The soluble soil can readily be flushed or rinsed from the surface of the part.

Before the caustic chemical can act as a soap in the solution, it must neutralize the minerals in the water being used. Minerals cause the water to be hard. The hard-water minerals that float out of a soap solution produce *scum.* Scum frequently appears in the bathtub when soap is used with hard bath water. Once the bath water is softened, the soap can wet the surface and dissolve the soil. This is why more soap is required when bathing with hard water than when bathing with soft water.

Soap must first *wet* the surface before it can begin to clean. Wetting is the result of the soap breaking up the *surface tension* of the water. An example of wetting action can be readily seen when washing a car. Because water has surface tension, clean water will bead up when it is wiped on a polished hood. If soap is put in the water, the soapy solution will spread out over the hood surface as it breaks the surface tension of the water. Cleaning only begins after the soap solution reduces the hardness and surface tension of the water being used. Cleaning begins as the soap encloses the organic soils. Once the soils are enclosed, they can be flushed away to leave the part clean.

Caustic chemical soap cleaners are used in a concentration of pH 10 to 12. They are usually heated above 140 °F (60 °C). The heat causes rapid chemical action and the warm parts will dry rapidly.

Grease and oil are removed by degreasing the parts. Petroleum solvents are convenient but expensive degreasers. Concentrated commercial degreasing chemicals are commonly used in the service shop. They are diluted with water or distillate (kerosene, diesel fuel, or petroleum solvent). The degreasing chemical solution is sprayed or brushed on the soil. Heavy deposits may have to be brushed hard to be loosened. After a soaking period, the loosened soil is flushed off the part with water. As the rinse water mixes with the degreasing chemical, they *emulsify;* that is, they turn into a milky mixture as the surface is flushed with the water.

Acid cleaning chemicals are used to remove rust and scale. They will not penetrate or remove petroleum-based deposits. The cooling system is the usual place for acid chemicals to be used. Acid cleaning chemicals have concentrations of pH 1.5 to 2.0.

Inhibitors are added to the chemical solution when cleaning soft metals. **The inhibitor reduces the cleaning chemical activity so that it will not dissolve the soft metal being cleaned.**

Mechanical Cleaning. Mechanical cleaning involves scraping, brushing, and abrasive blasting. It should, therefore, be used very carefully on soft metals. Heavy deposits that remain after chemical cleaning will have to be removed by mechanical cleaning.

The most frequently used type of scraper is a *putty knife.* The blade of the putty knife is pushed under the deposit to free it from the surface. The blade works best on flat surfaces such as gasket surfaces and the piston head. The broad blade of the putty knife prevents scratching the surface as it is used to clean the parts.

Wire brushes can be used on uneven surfaces. A hand wire brush can be used on the exterior of the block and on the head. A round wire brush used in a hand drill motor does a good job of cleaning the combustion chamber and parts of the head. A wire wheel can be used to clean the heads and fillets of the valves.

In engine work, a *glass-bead blaster* is the best type of abrasive-type blasting cleaner. It only works on hard, *dry* deposits. The deposits are broken into a fine powder by the beads. The powder is then blown from the surface by the air blast that propels the beads. The glass beads used for cleaning have diameters from 0.005 to 0.046 in. (0.13 to 1.17 mm). The beads do not remove metal as sand blasting would do, and they leave no abrasive dust on the parts. Besides cleaning, glass-bead blasting has other benefits. The bead blast *micropeens* the surface, and this work hardens the surface. The work-hardened surface is stronger than it was before micropeening. Micropeening also breaks sharp corners on the parts to reduce stress concentrations. The bead-blasted surface is satin smooth. A partly cleaned head is shown in Figure 15-2.

15-3 Cleaning Machines

There are a large number of cleaning machines. Some are designed for continuous use in a production shop. Others are designed for occasional use.

Figure 15-2. A head partly cleaned by glass bead blasting.

Each of the machines will satisfactorily clean parts when they are operated properly using the correct cleaning materials. Production cleaning equipment cleans fast, so there is little expensive labor time required. On the other hand, the equipment is expensive to buy and operate. The cleaning machines designed for occasional use are less expensive to buy, but they require more expensive labor time to clean the parts satisfactorily. In any case, cleaning is always part of the cost to recondition the engine.

Sprayers. Sprayers are commonly found in automotive service facilities. *Low-pressure sprayers* are powered with compressed air. A typical low-pressure sprayer is shown in Figure 15-3. They are used to spray chemical solutions on parts to be degreased. After soaking, the parts are rinsed with water. The chemical will emulsify in the rinse water as it removes the soil.

High-pressure sprayers are often used to remove the cleaning chemical that has been applied with the low-pressure sprayer. The high-pressure sprayer may use only high-pressure water or it may use a mixture of water and high-pressure air. Figure 15-4 shows a typical high-pressure sprayer. *Safety glasses should always be worn* when cleaning engine parts with sprayers.

Steam cleaners are a special class of sprayers. Steam vapor is mixed with high-pressure water and sprayed on the parts. The heat of the steam and the propellant force of the high-pressure water combine to do the cleaning. Usually, a caustic cleaner is added to the steam and water to aid in the cleaning. This mixture is so active that it will damage, and even remove paint, so painted surfaces must be protected

Figure 15-3. Low pressure sprayer.

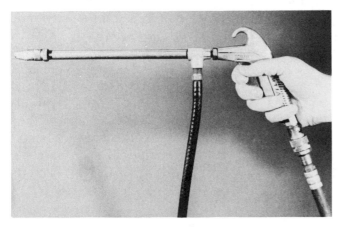

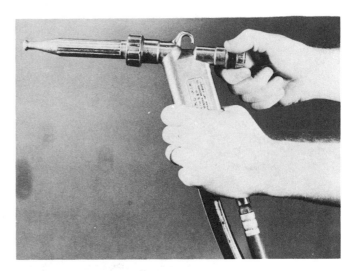

Figure 15-4. High pressure sprayer.

from the spray. Engines are often steam-cleaned before they are removed from the chassis. This makes a clean engine to work on; however, the cleaning mixture may damage water hoses, belts, and electrical wiring. Steam cleaning must be used with extreme care.

The steam cleaning machine has a boiler to heat the water to produce steam. The water pump should be turned on so that a stream of water is flowing through the boiler before the burner is ignited. The water flowing through the boiler keeps the boiler coils from overheating. The steam cleaner is ready for use when steam pressure builds up. The chemical is turned on after the water and steam are flowing freely. Shutting down the steam cleaner properly is just as important as starting it up. The cleaning solution is turned off first, but the water is kept running through the boiler. The burner is turned off next. When the water runs in a clear stream, the pump can be turned off. The boiler must be drained if the steam cleaner is to be stored in a place where the temperature drops below freezing.

Tanks. Most of the engine parts are cleaned in tanks after the engine has been disassembled. It is the usual practice to remove the general soil and the *heavy dirt* and grease deposits before cleaning the parts in the tank. Steam cleaning works well for the first cleaning. Heavy deposits may require scraping. This precleaning reduces the total cleaning cost by minimizing the contamination of the cleaning chemical. The cleaning chemical will work faster and last longer if it is not contaminated.

Figure 15-5. Cold soak tank.

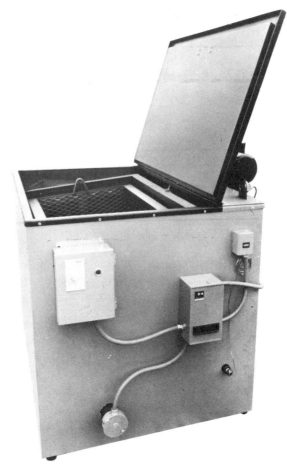

Figure 15-6. Hot soak tank (Courtesy of Geo. Olcott Co.).

Two types of soaking tanks are used. The cold soak tank is used to remove grease and carbon. The disassembled parts are placed in the tank so that they are *completely* covered with the chemical cleaning solution. After a soaking period, the parts are removed and rinsed until the milky appearance of the emulsion is gone. The parts are then dried with compressed air. The clean, dry parts are usually given a very light coating of clean oil to prevent rusting. Carburetor cleaner, purchased with a basket in a bucket, is one of the most common types of cold soak tanks in the automotive shop. An example of this is shown in Figure 15-5. Usually, the chemical will have water over the surface to prevent evaporation of the chemical. This water is called a *hydroseal*.

The hot soak tank (Figure 15-6) is used for cleaning heavy organic deposits and rust from iron and steel parts. Caustic cleaning solution used in the hot soak tank is kept near 200 °F (93 °C) for rapid cleaning action. The solution must be inhibited when aluminum is to be cleaned. After the deposits have been loosened, the parts are removed from the tank and rinsed with hot water or they are steam-cleaned. The hot parts will dry rapidly. They must be given a light coating of oil to prevent rusting.

Parts washers are often used in place of the soaking tanks. The parts are moved back and forth through the cleaning solution or the cleaning solu-

tion is pumped over the parts. This movement, called *agitation,* keeps moving fresh cleaning solution past the soil to help it loosen. The parts washer is usually equipped with a safety cover held open with a low temperature *fusible link*. If a fire occurs, the fusible link will melt and the cover will drop closed to snuff the fire out.

Vapor cleaning is popular in some automotive service shops. The parts to be cleaned are suspended in hot vapors above a perchlorethylene solution. The vapors of the solution loosen the soil from the metal so that it can be blown, wiped, or rinsed from the surface.

Ultrasonic cleaning is used in one of the latest types of cleaning machines. It is used to clean small parts that must be absolutely clean. Hydraulic lifters and diesel injectors are examples of these parts. The disassembled parts are placed in a tank of cleaning solution. The solution is vibrated at ultrasonic speeds to loosen all of the soil from the

Figure 15.7. Bead blasting machine.

Figure 15-8. Cleaning the combustion chamber with a rotary wire brush.

Figure 15-9. Valve guide cleaned with a wire brush.

parts. The soil goes into the solution or falls to the bottom of the tank.

Blasters. Glass-bead abrasive blasting equipment is used for cleaning engine parts. All *soft deposits* should be removed *before* using the blasting machine. The abrasive is fed into a stream of high-pressure air in the nozzle of the blasting machine. This propels the beads at a high velocity. The high-velocity glass-bead abrasive hits and pulverizes the deposit. This breaks them into dust. The air blows the dust deposit away. Blasting is done in an enclosed box. The box has a window for watching the work being done. Long rubber gloves are used to handle the parts while blasting. The used abrasive falls into a hopper in the bottom of the box to be recycled through the blaster. Some blasters have a cyclone filter that separates the used abrasive from the soil dust. This makes the abrasive last longer and keeps it effective. Figure 15-7 shows a typical bead blaster.

15-4 Parts Cleaning

There is no single cleaning process that will satisfactorily clean all engine parts. Several different processes can be used to clean the deposits from each part. It is assumed in the following discussion that the engine parts are *not* being cleaned with equipment that is designed for production cleaning. Cleaning equipment that is designed for occasional use is emphasized.

Heads. The deposits on the heads are primarily hard carbon deposits in the combustion chamber and ports. Some soft deposits may be found on the exterior of the heads and in the rocker arm area. The heads are first cleaned with a degreaser. After they are dried, a glass bead blaster will do the *best* job of removing the remaining deposits. Shops that do not have a bead blaster will use scrapers and a rotary wire brush (Figure 15-8) to remove the hard carbon deposits. In all cases, the valve guides will have to be cleaned with a special spiral brush or scraper on the end of a drill motor. The wire brush cleaner is pictured in Figure 15-9.

Figure 15-10. Chipping heavy carbon off an intake valve with an old valve.

Figure 15-11. Cleaning carbon from a valve with a rotating wire wheel.

Valves. The valves are coated with hard carbon. The heavy carbon on the intake valve is usually chipped off with an old valve, as shown in Figure 15-10. The remaining carbon on both valves is cleaned by holding them against a rotating wire wheel (Figure 15-11). An excellent job of valve cleaning can be done using a bead blast.

Pistons. The top of the piston heads and the interior of the piston can be cleaned with a glass-bead blaster after degreasing. Pistons must be removed from the rod before they are chemically cleaned. The pistons are soaked in an *inhibited* caustic

chemical cleaning tank if they are not blasted. After the soak, they are rinsed and the remaining carbon is scraped off. The piston head can be scraped with a putty knife scraper, as shown in Figure 15-12. Pistons should never be cleaned with a wire brush. The brush would break the square edges of the piston so that it will not do a good job of scraping the oil from the cylinder wall. In addition, the wire brush will remove some of the metal from aluminum pistons. In all cases, the ring grooves must be cleaned with a *ring groove cleaner*. Figure 15-13 shows a typical ring groove cleaner being used. A properly fitting drill bit must be used by hand to open drilled oil return passages in the oil ring groove.

Figure 15-12. Scraping carbon from a piston head.

Figure 15-13. Cleaning piston ring grooves with a typical ring groove cleaner.

Rods and Shafts. Surface carbon and varnish is found on the rods, camshaft, crankshaft, timing gears, rocker arms, pushrods, and oil pump. These deposits are removed in a caustic chemical cleaning tank. In some cases, steam cleaning is effective in cleaning these parts. The hard deposits on these parts do not affect their operation, but the deposits may hide cracks.

Hydraulic Lifters. Hydraulic lifters require special cleaning. They are generally disassembled, cleaned, and reassembled *one at a time.* The disassembled lifter is cleaned in fresh petroleum solvent or in an ultrasonic cleaner (Figure 15-14). The lifter is assembled immediately after cleaning, then it is tested using a special test oil. This oil can remain in lifters that pass the test. The lifters should be protected until they are to be installed in the engine. Wrapping in wax paper or plastic wrap is a good way to do this.

Block and Covers. The engine block, bell housing, timing cover, intake manifold, oil pan, and rocker covers are generally steam-cleaned or soaked in a caustic chemical cleaner. The interior deposits are of the same type as those on the rods and shafts. The exterior of the block and covers should be cleaned enough to allow repainting when the engine is reassembled. If heavy deposits still remain in the block cooling passages, the block may have to be soaked in an acid tank to remove the rust and scale. The soft plugs should be removed before the block

Figure 15-15. A crack in a cam timing sprocket can be seen during a magnetic inspection.

is cleaned so that the cooling passages can be thoroughly cleaned.

15-5 Visual Inspection

After the parts have been thoroughly cleaned, they should be reexamined for defects. A careful examination should be made to locate cracks. A magnifying glass is helpful finding them. Very critical parts of a performance engine should be checked for cracks using specialized magnetic or penetration inspection equipment. The crack in the sprocket pictured in Figure 15-15 was found with magnetic inspection. Internal parts that have cracks should be replaced. Cracks in the block and heads can be repaired. These repair procedures are described in a later section.

Valve stems, rocker arm pivots, pushrod ends, lifters, cam lobes and bearing journals, piston skirts, and crankshaft bearing journals should be checked for scoring and abnormal wear. When these conditions are found, the associated parts should be examined to determine the *cause* of the abnormal wear. The cause should be corrected before the engine is reassembled. All parts that are worn too much to be reconditioned should be set aside.

15-6 Dimensional Inspection

Parts that pass the visual inspection are given a dimensional inspection. In some cases, the *size* of the part is not too important. Generally, however,

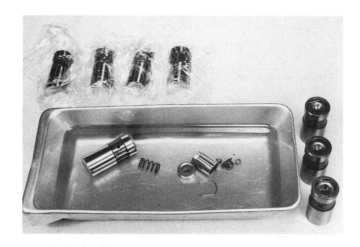

Figure 15-14. Cleaning a hydraulic lifter in clean petroleum solvent.

the *clearance space* between the two parts is of critical importance. For instance, the amount of piston oversize is not as critical as the clearance between the piston and the cylinder wall. Measuring tools are used to check the size of the parts.

Measuring Tools. There are two basic measuring systems used in automotive engines: customary units (inches) and metric units (millimeter). For years, the customary measurements of the inch and thousandths of an inch have been used for domestic engines. Most of the import engines use the metric millimeter as a base measurement. In the late 1970s, metric dimensions gradually started to be used in domestic engines. Many of these engines mix customary and metric measurements. They also used both customary and metric dimensions and fasteners. Use extreme care to make sure the proper specification and correct fastener is used.

The *micrometer* is one of the basic measuring tools for engine service. Micrometers are available with either inch or millimeter scales, as shown in Figure 15-16. An external measuring *outside micrometer* is the type most frequently used in the automotive engine shop. It is used to measure the external diameter of bearing journals (Figure 15-17), pistons, valve stems, and so on. It is used with *small hole gauges* and with *telescopic gauges* to measure the inside diameters of holes. These measuring instruments are shown in Figure 15-18.

Figure 15-17. Measuring a connecting rod journal with a micrometer.

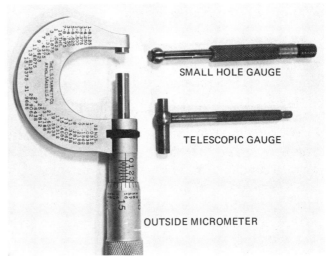

Figure 15-18. Small hole and telescopic gauges used with an outside micrometer to measure the diameter of holes.

The small hole gauge or telescopic gauge is adjusted to the hole diameter (Figure 15-19) and then the gauge is measured with the outside micrometer (Figure 15-20). The clearance space is determined by measuring the diameter of the part fitting the hole. This dimension is subtracted from the diameter of the hole to determine the clearance space. Outside micrometers in customary units are used in 1-in. steps. The micrometer sizes (in customary units) most commonly used for automotive engine service are the 0–1, 1–2, 2–3, 3–4, and 4–5 in. Typical outside micrometers are shown in Figure 15-21.

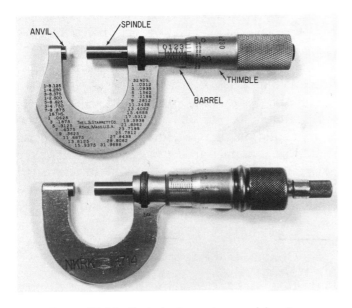

Figure 15-16. Typical micrometers used for dimensional inspection.

Figure 15-19. A small hole gauge set to the size of a hole.

Figure 15-20. Measuring the small hole gauge setting with an outside micrometer.

Before an outside micrometer can be used for exact measurement, it must be checked with a standard gauge. Special *gauge rods* are available for this. Once the micrometer is adjusted to the gauge rod length, it will be accurate throughout its full range. Figure 15-22 shows a micrometer being checked.

Inside micrometers are available to measure the inside diameter of openings. They require a great deal of care. They must be kept clean and they must be assembled correctly to be accurate. This is difficult to do if they are used by several people. This is why the small hole and telescopic gauges are usually used with an outside micrometer to make inside measurements.

A second basic measuring tool used for measurements in the automotive engine shop is the *dial gauge*. The dial gauge is usually used to measure differences in dimensions or to measure slight movements of the parts. The dial gauge is mounted firmly by using a clamp or a magnetic base. The spindle of the dial gauge is placed against the part to be measured. The part is moved and the movement is indicated by the pointer of the dial gauge.

The dial gauge is the only convenient tool to measure shaft runout. The shaft is supported at its ends, in V-blocks or on the end bearings and the dial gauge is set up in the middle. A typical setup is shown in Figure 15-23. The shaft is rotated and the movement of the shaft center is indicated on the dial.

Some small clearances can be measured with a *feeler gauge*. This metal strip is sometimes called a

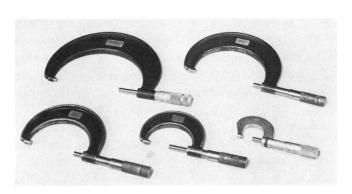

Figure 15-21. Different sizes of typical outside micrometers used to measure the size of engine parts.

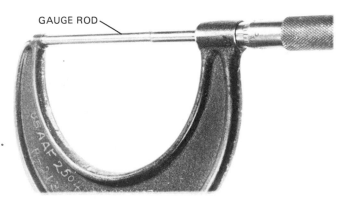

GAUGE ROD

Figure 15-22. A micrometer being checked for accuracy with a gauging rod.

Figure 15-23. A typical dial gauge set up to measure the crankshaft run-out at the center main bearing journal.

Figure 15-24. A group of feeler gauges used for measuring the space between two parts.

Figure 15-25. A feeler gauge used to measure the space between a straight edge and the head to detect head warpage.

thickness gauge. Gauges in customary units are available in sets or as single strips from 0.0015 to 0.025 in. in thickness. They are made in steps of 0.001 in. from 0.002 in. A group of feeler gauges of the type used on engines is shown in Figure 15-24.

Some machinists prefer to use a *vernier caliper*. It can be used to measure both inside and outside dimensions. This tool is not popular with the average automotive service technician. Inside and outside *calipers* are also used in the engine repair shop. They are used for measuring approximate sizes, not for precision measurements.

A *straightedge* approximately 20 in. long is very useful in the engine repair shop. It is usually used with a feeler gauge to check the flatness of a part (Figure 15-25). A square is also needed to make some checks on engine parts.

Valves and Guides. Dimensional inspection of valves includes the valve guides and seats. The valve stem diameter is the only part of the valve that is measured. If the valve has passed both the visual and dimensional inspection, it can be reconditioned.

The valve stem must have the proper clearance

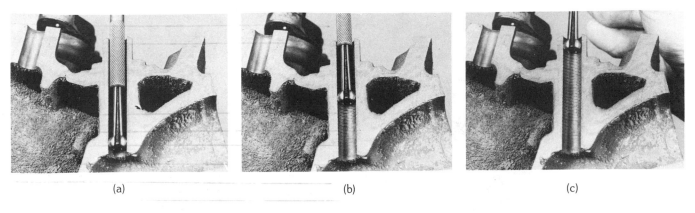

(a)   (b)   (c)

Figure 15-26. A cut away head is used to show how a small hole gauge is used to measure the taper and out-of-round of a valve guide.

in the valve guide. Too little clearance will cause binding, while too much clearance will allow the valve to cock. This will lead to valve burning. Too much clearance will also allow oil to be drawn into the manifold through the valve guide clearance. The valve guide is measured in the middle and with a small hole gauge. The gauge size is checked with a micrometer. The guide is then checked at each end. This is shown in a cutaway valve guide in Figure 15-26. The small hole gauge is being checked with a micrometer in Figure 15-27. The expanded part of the ball should be placed crossways of the engine

where the greatest amount of valve guide wear exists. The dimension of the valve stem diameter is subtracted from the dimension of the valve guide diameter. If the clearance exceeds the specified clearance, the valve guide will have to be reconditioned. Dimension measurements of the valve and the guide have to be made during head reconditioning. It also has to be made during assembly. Use of these measurements is discussed in a later chapter.

Valve Springs. Valve springs close the valves after they have been opened by the cam. They must close squarely to form a tight seal and to prevent valve stem and guide wear. It is necessary, therefore, that the springs must be square and have the proper amount of closing force. The valve springs are checked for squareness by rotating them on a flat surface with a square held against the side. They should be within a $\frac{1}{16}$ in. or 1 mm of being square. This is shown in Figure 15-28. Only the springs

Figure 15-28. Measuring the squareness of a valve spring with a square on a flat surface.

Figure 15-27. After adjusting to the valve guide size, the small hole gauge is measured with an outside micrometer.

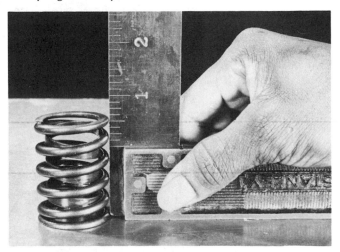

Figure 15-29. One popular type of spring tester used to measure the compressed force of valve springs.

The cylinder should be rechecked to determine the cylinder *taper* and *out-of-round*. This is usually done with a cylinder gauge adjusted to the standard cylinder bore diameter. A cylinder gauge is shown being used in Figure 15-30. The cylinder should be measured across the engine (perpendicular to the crankshaft) where the greatest wear occurs. Most of the wear will be found just below the ridge and the least wear will occur below the lowest ring travel. The amount of cylinder taper and out-of-round is compared to the specifications to determine if the cylinder needs to be reconditioned. Cylinder repair procedures are described in Chapter 17.

The piston skirt is measured to determine the size across the thrust surfaces, as shown in Figure 15-31. This dimension is subtracted from the smallest cylinder bore diameter to determine the minimum piston-to-cylinder clearance. Consideration should be given to slight cylinder enlargement that will be caused by deglazing or honing during reconditioning of the block. The clearance should be compared to the specifications. If the clearance is

that are square should be checked to determine their compressed force. The out-of-square springs will have to be replaced. The surge damper should be *removed* from the valve spring when checking the spring force. A valve spring scale is used to measure the valve spring force. One popular type, shown in Figure 15-29, measures the spring force directly. Another type uses a torque wrench on a lever system to measure the valve spring force. The spring force is usually checked at the valve-closed length and at the valve-open length. Valve spring force specifications are given in the service manual for the engine. Weak valve springs are replaced.

Pistons and Cylinders. The sliding surface between the piston and the cylinder wall is one of the places in an engine that shows a great deal of wear. The pistons wear smaller and the cylinders wear larger. It is unusual when either the piston or cylinder does not require service during engine reconditioning.

Figure 15-30. A cylinder gauge being used to measure the size, taper, and out-of-round of a cylinder.

Figure 15-31. A piston skirt being measured to determine the size across the thrust surfaces.

Figure 15-32. Using a special no-go gauge to check the piston ring groove width.

excessive, the piston skirt can be resized to reduce the piston-to-cylinder clearance.

The piston ring goove should be checked to see how much it has worn in width. A special no-go gauge similar to the one pictured in Figure 15-32 is very convenient for measuring the groove for wear. The groove wear is excessive if the gauge will slip into the groove. It is the general practice to service the upper ring groove when the pistons have been removed from the engine. The ring groove is machined and a ring groove spacer is installed above the new ring to correct the effective groove width.

It has been a common practice to install new piston pins on a routine basis. The clearance be-

tween the piston pin and the piston are only a few ten thousandths of an inch, and measuring to this accuracy is very difficult with standard measuring tools. In addition, the pin hole in the connecting rod is slightly smaller than the piston pin. It forms an interference fit to hold the pin in place. Machining equipment designed to hone piston pin holes in the piston and rod have very special measuring gauges to properly size the holes.

Bearings and Journals. Bearings may not need to be replaced if a low-mileage engine is disassembled. New bearings are almost always installed in a customer's engine during overhaul or rebuilding. For this reason, the used bearings are not measured. During a dimensional inspection, the primary concern is the bearing journals on the crankshaft and camshaft. The journal is measured at four to six diameters close to each end and at the middle to determine the maximum variation in the journal size and shape (Figure 15-33). Differences from end to end of the journal is called *taper*. Differences of the diameter measurements around the journal is called *out-of-round*. Both should be within specifications, usually less than 0.001 in. (0.025 mm).

The crankshaft should be checked for warpage. This can be done with the block inverted by placing one bearing shell in each of the end bearing saddles. One is placed in the front saddle and one in the rear saddle. A dial gauge is placed on the center main

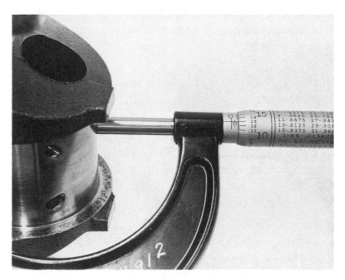

Figure 15-33. Measuring a crankshaft bearing journal with an outside micrometer.

Figure 15-34. Alignment of main bearing saddles using a feeler gauge under a straight edge.

bearing journal of the crankshaft. This was shown in Figure 15-23. The dial reading variation is observed as the crankshaft is rotated. This is called crankshaft *runout*. The runout should be less than the bearing-to-journal oil clearance. The crankshaft should be reground if it has excessive runout or if the journals have too much taper or out-of-round.

The bearing saddles have to have the correct size and they must be in perfect alignment. The saddle can be checked by using a feeler gauge under a straightedge, as shown in Figure 15-34. Misaligned bearing bores are corrected by align boring to straighten them. Precision engine rebuilders always bore or hone the main bearing bores when reconditioning engine blocks. These repairs are described in a later section.

Parting Surfaces. The head and block deck are typical parting surfaces that should be checked. With no gasket, the head is placed on the deck with paper tabs around the edge. Each tab is pulled slightly to determine the location of the loose tab. When the location of the loose tabs is found, the tabs are removed. A feeler gauge is slipped between the head and block at the location of the loose tabs to determine the amount of gap.

Another way to measure the head and block for twist and warp is to place a straightedge on the surface and slide a feeler gauge under the straight edge, as pictured in Figure 15-25. This will show where the twist or warp is located. Engine rebuilders will

resurface the head and block deck as a routine service procedure, so they will not make this check.

This same procedure for measuring the flatness of the parting surface can be used on manifolds and timing cover parting surfaces to determine warp and twist. Parting surfaces found to have excessive twist or warp can be reconditioned by milling or by surface grinding.

Oil Pump. It is very important that the oil pump clearances are within specifications. Usually, oil pump wear can be seen during the visual inspection. A worn oil pump is not given a dimensional inspection. It is always replaced with a new pump. If the oil pump has excessive clearances, it will not pump enough oil and it may lose its *prime*. When a pump loses its prime, air in the pump will go around with the gears so the pump does not pull oil from the pan. This would be called *cavitation*. If this happens, oil will not be supplied to the engine parts and the engine will be ruined. The clearance in the oil pump is usually checked with a feeler gauge. A gear-type pump is checked for tooth clearance in the housing (Figure 15-35), backlash between the gears, and gear-to-cover side clearance (Figure 15-36). The oil pump gears can be replaced, but the housing is usually worn, too. If the oil pump shows *any* indication of wear, it is replaced.

Figure 15-35. A gear-type oil pump being checked for tooth clearance in the housing.

Figure 15-36. Measuring the gear-to-cover clearance of an oil pump.

Additional Measurements. Once the visual and dimensional inspections have been completed, the serviceable engine parts are reconditioned. It is necessary to make many dimensional measurements during the reconditioning operations. This is necessary to make the parts fit properly. During assembly, additional dimensional measurements will be made to make sure the parts are properly adjusted and have the proper clearances. These measurements are described in the following chapters.

REVIEW QUESTIONS

1. When is it economical to use less efficient cleaning methods? [INTRODUCTION]
2. What checking and inspecting must be done to determine if a part should be reconditioned? [INTRODUCTION]
3. Give three examples of exterior and three examples of interior engine soil. [15-1]
4. How does heat affect the cleaning rate of cleaning chemicals? [15-1]
5. How are water-soluble deposits removed from the outside of an engine? [15-1]
6. How does a petroleum-based cleaning solvent remove oil and grease deposits? [15-1]
7. How does steam cleaning remove soil from parts? [15-1]
8. Where are heavy carbon deposits found in the engine? [15-1]
9. How are heavy carbon deposits cleaned from the engine? [15-1]
10. When parts are cleaned mechanically? [15-2]
11. How are toxic chemicals different from caustic chemicals? [15-2]
12. What should you do if you are accidentally splashed with cleaning chemicals? [15-2]
13. What type of chemical cleaners are used for carbon-type deposits? [15-2]
14. Where do acid and caustic materials fit on the pH scale? [15-2]
15. Describe the soap action of a caustic. [15-2]
16. What causes scum to form when cleaning with a soap solution? [15-2]
17. What must the soap solution do before it can clean the soil? [15-2]
18. What parts are cleaned with acid cleaning chemicals? [15-2]
19. Why are inhibitors used in chemical cleaners? [15-2]
20. Name three types of mechanical cleaning. [15-2]
21. What are the advantages and disadvantages of production cleaning equipment over other types? [15-3]
22. When is a high-pressure sprayer used rather than a low-pressure sprayer? [15-3]
23. What are the advantages and disadvantages of steam cleaners? [15-3]
24. What precautions should be observed when starting, using, and turning the steam cleaner off? [15-3]

25. Why should parts be precleaned before being soaked in a tank? [15-3]
26. What are the advantages and disadvantages of a hot soak tank compared to a cold soak tank? [15-3]
27. Why should soft deposits be removed before bead blasting? [15-4]
28. What is the best way to clean heads? [15-4]
29. How should valves be cleaned? [15-4]
30. What precautions must be observed when cleaning pistons? [15-4]
31. What should be done with the varnish-type deposits on the internal engine parts? [15-4]
32. How should hydraulic lifters be cleaned? [15-4]
33. What is checked in a visual inspection? [15-5]
34. Why is the clearance space usually more important than the size of a part? [15-6]
35. Why is it necessary to use care when checking the fastners of engines built since the late 1970s? [15-6]
36. How is the clearance space between a shaft and hole determined? [15-6]
37. What precautions should be observed before using a micrometer? [15-6]
38. What precautions should be observed before using a dial gauge? [15-6]
39. Name three locations in an engine where a feeler gauge can be used to measure a small clearance. [15-6]
40. How is the valve guide clearance checked? [15-6]
41. How are valve springs checked? [15-6]
42. How is the cylinder measured for wear? [15-6]
43. How is the piston-to-cylinder clearance determined? [15-6]
44. How is a bearing journal measured? [15-6]
45. How is the run-out of the crankshaft checked? [15-6]
46. When should the crankshaft be reground? [15-6]
47. How are parting surfaces checked? [15-6]
48. How is a worn oil pump serviced? [15-6]

CHAPTER
16

Valve and Head Service

Valves need to be reconditioned more often than any other engine part. The valves and head should be cleaned and inspected to locate defects, as discussed in previous chapters. Some defects can be repaired; others cannot. Parts that cannot be repaired will have to be replaced. The repairable parts are taken to the automotive machine shop, where the required reconditioning will be done.

Serviceable valves are reconditioned by grinding the valve face and the valve tip. The valve face must be finished at the correct angle with a smooth, bright surface. The valve tip is refinished to make a smooth, flat surface. The required amount of valve-tip grinding is determined after the seat has been reconditioned by the length of the installed valve stem.

Cracked standard automotive engine heads are generally replaced. The crack repair procedure is time-consuming and, therefore, is expensive. Usually, a sound used head is obtained from a salvage yard to replace the cracked standard head. Heads that have special designs and heads for heavy-duty engines are usually expensive. Therefore, if these heads have cracks, the cracks are usually repaired.

A new engine has been machined and assembled within a few hours after the heads and block are cast from melted iron. Newly cast parts have internal stresses within the metal. The stress results from the different thicknesses of the metal sections in the head. Forces from combustion in the engine, plus continued heating and cooling, gradually relieve these stresses. By the time the engine has accumulated 20,000 to 30,000 miles (32,000 to 48,000 km), the stresses have been completely relieved. This is why some engine rebuilders prefer to work with used heads and blocks that are stress-relieved. The head will usually have some warpage when the engine is disassembled. Some tool manufacturers say that the engine has been operating with the warped head. They also say that the head matches a warped block and so the head should not be resurfaced. On the other hand, most equipment manufacturers advise resurfacing the gasket surface of the head to make it flat.

The first operation on a clean, sound, and flat head is to recondition the valve guides. The seats are then serviced using the reconditioned valve guides as pilots. Each seat and guide must be centered and square for the valve to operate correctly. **Centered is called concentric. Square is called perpendicular.** Reconditioning the guide first is necessary to match the seat to the position of the reconditioned valve guide. The valve guide and seat are checked with the valve that is to be used in that valve guide.

The valve seat width and its contact on the valve face is checked using Prussian Blue, sometimes called mechanics blue, or marks from a felt-tip pen. The valve seat is adjusted, as required, using different angle grinding stones or cutters to produce the proper valve-to-seat fit. The valve stem length is then adjusted, when necessary, by grinding the valve tip.

After a thorough cleaning and relubrication, the valves, oil seals, springs, retainers, and locks are assembled to finish the head reconditioning procedure. Rocker arms, rocker arm pivots, pushrods, lifters, and cams are also an important part of the complete valve system reconditioning process. They should not be neglected when doing a complete valve job.

16-1 Valve Service

Each valve grinder operates somewhat differently. The operation manual that comes with the grinder should be followed for lubrication, adjustment, and specific operating procedures that must be followed. The general procedures given in the following paragraphs apply to all valve grinding equipment. Safety glasses should *always* be worn when working on machinery, so they should be worn while doing valve and seat reconditioning.

During grinding operations, fine hot chips fly from the grinding stones. Safety glasses keep these chips from getting into the eyes.

The face of the valve is ground on a *valve grinder*. Before starting, the grinder head is set at the correct valve face *angle* to make the recommended interference fit (Figure 16-1). The grinding stone is *dressed* with a special diamond tool to remove any roughness from the stone surface (Figure 16-2). The valve stem is clamped in the work head as close to the fillet under the valve head as possible to prevent vibrations. The work head motor is turned on to rotate the valve. The wheel head motor is turned on to rotate the grinding wheel. The coolant flow is adjusted to flush the material away, but not so much that it splashes

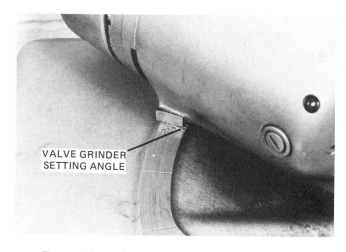

Figure 16-1. The valve grinder is set to the recommended angle.

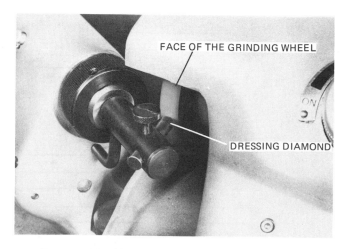

Figure 16-2. Dressing the face of the grinding wheel with a diamond dressing tool.

(Figure 16-3). The rotating grinding wheel is fed slowly to the rotating valve face. *Light* grinding is done as the valve is moved back and forth across the grinding wheel face. The valve is never moved off the edge of the grinding wheel. It is ground only enough to clean the face (Figure 16-4). The margin of the valve should be over 0.060 in. (1.5 mm) when grinding is complete (Figure 16-5). Aluminized valves will lose their corrosion-resistance properties when ground. For satisfactory service, aluminized valves must be replaced if they require refacing.

Slight imperfections on valve tips are removed by grinding in a special fixture (Figure 16-6). Tip grinding is usually completed after the valve seats have been reconditioned. The valve is put in the head and the length of the tip is measured. The tip is

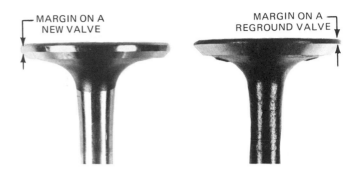

Figure 16-5. The difference in the margin on a new and used valve head.

Figure 16-6. A valve in a fixture to grind the valve tip.

Figure 16-3. Grinding the face of a valve.

Figure 16-4. A finished valve face after grinding.

ground to shorten the valve stem length to compensate for the valve face and seat grinding. The valve will not close if the valve tip extends too far from the valve guide on engines that have hydraulic lifters and nonadjustable rocker arms. If the valve is too long, the tip may be ground as much as 0.020 in. (0.50 mm) to reduce its length. If more grinding is required, the valve must be replaced. If it is too short, the valve face or seat may be reground, within limits, to allow the valve to seat deeper. Where excessive valve face and seat grinding has been done, shims can be placed under the rocker shaft on some engines as a repair to provide correct hydraulic lifter plunger centering. These shims must have the required lubrication holes to allow oil to enter the shaft.

Any sharp edge produced during valve servicing should be removed with a hand stone. This reduces the chance of damaging seals or increasing wear on the sharp edges.

263

16-2 Crack Repair

Cracks in the head will allow coolant to leak into the engine or they will allow combustion gases to leak into the coolant. Cracks across the valve seat cause hot spots on the valves. The hot spots will burn the valve face. A head with a crack will either have to be replaced or the crack will have to be repaired. A cracked head is shown in Figure 16-7. Two methods of crack repair are commonly used: welding and plugging. In each case, the full length of the crack must be determined. Center punch marks can be placed along the crack so that its location can be easily seen. A hole is made at each end of the crack to prevent additional cracking before the crack is repaired. Cracks across the valve seat will require the installation of a replacement valve seat insert.

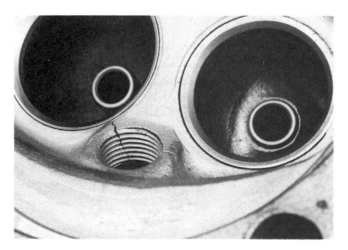

Figure 16-7. A crack through the spark plug hole on a head.

Welding. It takes a great deal of skill to weld cast iron. The cast iron does not puddle or flow as steel does when it is heated. Heavy cast parts, such as the head and block, conduct heat away from the weld so fast that it is difficult to get the part hot enough to melt the iron for welding. When it does melt, a crack will often develop next to the edge of the weld bead. Welding can be satisfactorily done when the entire cast part is heated red hot.

The head or block is placed on fire bricks and supported in a level position so that the crack to be welded will be on top. In this way, the molten metal will not flow away from the area being repaired. The casting is then covered with an asbestos sheet. An opening is made in the sheet over the crack that is to be welded. The hole, in turn, is covered with

another piece of asbestos. The casting is then heated under the asbestos with a fire nozzle until it becomes red hot. This will take several hours. The hole cover is removed and the crack is welded.

First, a hole is burned through the casting at each end of the crack to stop any additional cracking. Then the metal along the crack is melted out to leave a gap where the crack had been. The gap is filled with the molten welding rod to make a bead above the surrounding metal. The hole over the weld is covered and the casting is allowed to cool very slowly. The surface of the weld is smoothed after the casting has cooled.

Plugging. In this process, the crack is closed using interlocking tapered plugs. The ends of the crack are centerpunched and drilled with the proper-size tap drill used for the plugs. The hole is reamed with a tapered reamer (Figure 16-8). The hole is then tapped to give full threads (Figure 16-9). The plug is coated with sealer; then it is tightened into the hole (Figure 16-10). The plug is sawed about one-fourth of the way through; then it is broken off. The saw slot controls the break point (Figure 16-11). If the plug should break below the surface, it will have to be drilled out and a new plug installed. The plug should go to the full depth or thickness of the cast metal. After the first plug is installed on each end, a new hole is drilled with the tap drill so that it cuts in-

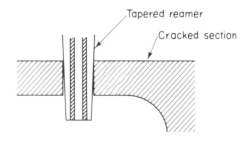

Figure 16-8. Reaming a hole for a tapered plug.

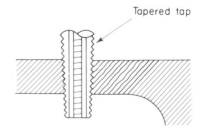

Figure 16-9. Tapping a tapered hole for a plug.

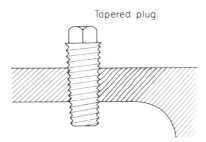

Figure 16-10. Screwing a tapered plug in the hole.

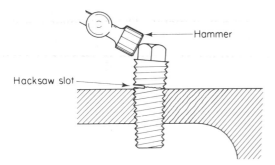

Figure 16-11. Cutting the plug with a hack saw.

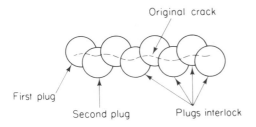

Figure 16-12. Interlocking plugs.

Figure 16-13. Scraping the gasket surface of the head to remove all traces of the gasket or sealer.

Figure 16-14. Finish cleaning the gasket surface of the head by draw filing.

to the edge of the first plug. This new hole is reamed, tapped, and a plug is inserted as before. The plug should fit about one-fourth of the way into the first plug to lock it into place (Figure 16-12). Interlocking plugs are placed along the entire crack, alternating slightly from side to side. The exposed ends of the plugs are peened to help secure them in place. The surface of the plugs is ground or filed down nearly to the gasket surface. The plugs are ground down to the original surface in the combustion chambers and at the ports, using a hand grinder. The gasket surface of the head must be resurfaced after the crack has been repaired.

16-3 Head Resurfacing

The gasket surface of the head should be resurfaced if it does not match the flatness of the block deck by commercial standard reconditioning practice. The surface must be thoroughly clean. It is first scraped (Figure 16-13), then draw-filed (Figure 16-14) to

remove any small burrs. This is determined by trying to slide a 0.006 in. (0.15 mm) feeler gauge under a straightedge held against the head surface, as shown in Figure 16-15. The head should not vary over 0.002 in. (0.05 mm) in any 6 in. (15 cm) length. The head should also be resurfaced if there is any roughness caused by corrosion of the head gasket. This roughness can be felt on the head surface when rubbing your fingernail across it. In precision engine rebuilding, *both* the head and the block deck are resurfaced as a standard practice.

Two common resurfacing methods are used: milling and grinding. A *milling-type* resurfacer uses metal cutting tool bits fastened in a disc. The disc is the rotating work head of the mill. This can be seen in Figure 16-16. The surface *grinder type* uses a large-diameter abrasive wheel. Both types of resurfacing methods are used in table-type surfacers and in precision-type surfacers. The head or block is passed over the cutting head that extends slightly

265

Figure 16-15. Checking the head for warping (a) and twist (b).

Figure 16-16. Milling type resurfacer.

above a work table on the table-type surfacer. This is shown in Figure 16-17. The abrasive wheel is dressed before grinding begins. The wheel head is adjusted to just touch the surface. At this point, the feed is calibrated to zero. This is necessary so that the operator knows exactly the amount of cut being made. Light cuts are taken. The abrasive wheel cuts are limited to 0.005 in. (0.015 mm). The abrasive wheel surface should be wire-brushed after each five passes and it should be redressed after grinding each 0.100 in. (2.50 mm). The mill-type cutting wheel can remove up to 0.030 in. (0.075 mm) on each pass. A special mill cutting tool or a dull grinding wheel is used when resurfacing aluminum heads.

Figure 16-17. A grinder-type resurfacer.

The intake manifold will no longer fit correctly when the gasket surface of the heads of V-type engines are ground. The ports and the assembly bolt holes do not match. The intake manifold surface of the head must be resurfaced to remove enough metal from the gasket surface of the head to rematch the ports and bolt holes. The amount of metal that must be removed depends on the angle between the head gasket surface and the intake manifold gasket surface. Figure 16-18 shows how this is calculated. Automotive machine shops doing head resurfacing have tables that specify the exact amount of metal to be removed. It is usually necessary to remove some metal from the front and the back gasket surface of closed-type intake manifolds used on V-type engines. This is necessary to provide a good gasket seal that will prevent oil leakage from the lifter valley.

16-4 Valve Guide Reconditioning

No matter how good the valve or seats are, they cannot operate properly if the valve guide is not accurate. In use, the valve operating mechanism pushes the valve tip sideways. This is the major cause of valve stem and guide wear. The valve normally rotates a little each time it is opened to keep the stem wear even all around the stem. The valve guide, on the other hand, always has the wear in the same place. This causes both top and bottom ends of the guide to wear in an oval or egg shape. The largest diameter is perpendicular to the camshaft. The valve guide does not have to be measured if the valve feels sloppy in the guide. It obviously requires reconditioning. The guide must be reconditioned to match the valve that is to be used in that valve guide.

The guide is reconditioned to make it round. It

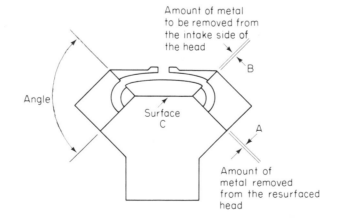

Angle	Amount to be removed from B
90°	A x 1.000
85°	A x 1.100
80°	A x 1.233
75°	A x 1.414
70°	A x 1.673
75°	A x 2.067
70°	A x 2.733
65°	A x 4.072

The amount removed from surface C is 1.4 x A

Figure 16-18. The angle between the intake manifold and head gasket surface is measured (a) and the material that must be removed for a good manifold fit (b).

is then resized to give the correct valve stem clearance. If the stem clearance is too small, oil cannot get in to lubricate the surface. The small clearance will cause the valve to contact the guide, causing a scuffing and galling condition. If the stem clearance is too large, it will allow oil to seep through the clearance into the manifold. This will cause excessive exhaust smoke, combustion chamber deposits, and oil consumption. Large stem clearances will also allow the valve to cock or bend. This, in turn, will either cause poor valve seating or it will lead to valve stem cracks just below the head. Cracks will eventually cause the valve to break. Excessive intake valve guide clearances will allow air as well as oil to enter the intake manifold. This added air will upset the precise air/fuel ratios required for emission-controlled engines. This will make the intake charge excessively lean. The lean charge may also reduce fuel economy by causing the engine to misfire. The less clearance the valve stem has in the valve guide, the more important it is to recondition the valve face and seat to exact dimensions. A valve stem that is loose in the guide will allow the valve to center itself onto a slightly misaligned seat. A closely fit stem requires that the seat is exactly concentric and perpendicular to the guide.

When an engine is designed with replaceable valve guides, their replacement is always recommended when the valve assembly is being reconditioned. The original valve guide height should be measured before removing the guide so that the new guide can be properly positioned.

Valve Guide Replacement. After measuring the valve guide height, the worn guide is pressed from the head with a properly fitting *driver*. The driver has a stem to fit the guide opening and a shoulder that pushes on the end of the guide. If the guide has a flange, care should be taken to make sure that the guide is pushed out from the correct end, usually from the port side toward the rocker arm side. The new guide is pressed into the guide bore using the same driver. Make sure that the guide is pressed to the correct depth. After the guides are replaced, they are reamed or honed to proper inside diameter. Figure 16-19 shows how the driver is used to remove and replace valve guides.

Oversized Valves. All domestic automobile manufacturers that have *integral* valve guides in their engines recommend reaming worn valve guides and installing new valves with *oversized stems.* Figure 16-20 shows a reamer in a valve guide. When a valve guide is worn, the valve stem is also likely to be worn. In this case, new valves are required. If new valves are used, they can just as well have oversized stems as standard stems. The valve guide is reamed or honed to the correct oversize that will fit the oversized stem of the new valve. The resulting clearance of the valve stem in the guide is

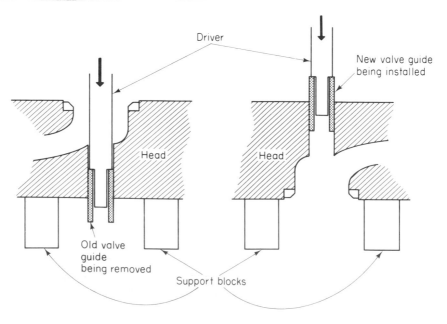

Figure 16-19. Valve guide replacement procedure.

Figure 16-20. Reaming a valve guide.

the same as the original clearance. Original intake valve clearances are 0.001 to 0.003 in. (0.025 to 0.075 mm) and exhaust valves are 0.002 to 0.004 in. (0.05 to 0.10 mm). The oil clearance and the heat-transfer properties of the original valve and guide are not changed when installing new valves with ovesized stems. The service technician will not go wrong when the manufacturer's valve guide reconditioning recommendations are followed.

Valves with oversized stems are not generally used in head reconditioning jobs. These valves are rarely kept in stock to keep the cost low in stock inventory. They can be special-ordered, but this takes

time. In addition, the shop doing the valve work must have special reamers for each oversized valve stem size. As a result, most worn valve guides are *resized* to their original standard dimension using one of several reconditioning procedures. The resized guide is finished with standard sized reamers so that valves with standard stems can be used.

Knurlizing. One of the most popular methods of resizing valves is knurlizing (Figure 16-21). In this process, a tool is rotated as it is driven into the guide. The tool *displaces* the metal to reduce the hole diameter of the guide. In the displacing process, the knurlizing tool pushes a small tapered wheel or dull threading tool into the wall of the guide hole. This makes a groove in the wall of the guide without removing any metal, as pictured in Figure 16-22. The metal piles up along the edge of the groove just like dirt would pile up along the edge of a tire track as it rolls through soft dirt. The dirt is displaced from under the wheel to form a small ridge alongside the tire track.

The knurlizing tool is driven by an electric drill through a speed reducer. The valve tip is placed in the guide after the knurlizing tool is removed. If the valve tip goes into the guide, the guide is reamed and the next-larger-size knurling tool is used. This process is continued until the valve stem tip will no longer go into either end of the guide more than ¼ in. (6 mm). The guide is finally reamed to size. The

Figure 16-21. Knurlizing tool in a valve guide.

Figure 16-22. A sectioned view of a knurled valve guide.

reamers that accompany the knurlizing set will ream just enough oversize to provide the correct valve stem clearance for commercial reconditioning standards. The valve guides are honed to size when precise fits are desired in the precision shop. Clearances of knurled valve guides are intended to be one-half of the new valve guide clearance. This small clearance can be used because knurling leaves so many small oil rings down the length of the guide for lubrication. Only original equipment oil seals should be used with knurlized valve guides. This will allow the proper amount of oil to enter and lubricate the valve guides without having excessive oil consumption. Positive-type oil seals designed for use on sloppy valve guides should never be used with properly sized knurlized valve guides. If they are used, they will keep oil from entering the valve guides, and this will lead to scuffing and galling of the valve stem and guide.

Guide Insert. When the integral valve guide is badly worn, it can be reconditioned using an insert. This repair method is usually preferred in heavy-duty and high-speed engines. Two types of guide inserts are commonly used for guide repair: a thin-walled bronze alloy sleeve bushing and a spiral bronze alloy bushing. The valve guide rebuilding kit used to install each of these bushings includes all the required reamers, installing sleeves, broaches, burnishing tools, and cutoff tools that are needed to install and properly size the bushings.

The valve guide must be bored oversize to accept the thin-walled insert sleeve. The boring tool is held in alignment by a rugged fixture. One type is shown in Figure 16-23. Depending upon the make of the equipment, the boring fixture is aligned with the valve guide hole, the valve seat, or the head gasket surface. First, the boring fixture is properly aligned. The guide is then bored, making a hole somewhat smaller than the insert sleeve that will be used. The bored hole is reamed to make a precise smooth hole that is still slightly smaller than the insert sleeve. The insert sleeve is installed with a press fit that holds it in the guide. The press fit also helps to maintain normal heat transfer from the valve to the head. The thin-walled insert sleeve is held in an installing sleeve. A driver is used to press the insert from the installing sleeve into the guide as illustrated in Figure 16-24. A broach is then pressed through the insert sleeve to firmly seat it in the guide. The broach is designed to put a knurl in the guide to aid in lubrication. The insert sleeve is then trimmed to the valve guide length. Finally, it is reamed or honed to provide the required valve stem clearance. A very close clearance of 0.0005 in. (0.013 mm) is usually used with the bronze thin-walled insert sleeve.

The spiral bronze alloy insert bushing is screwed

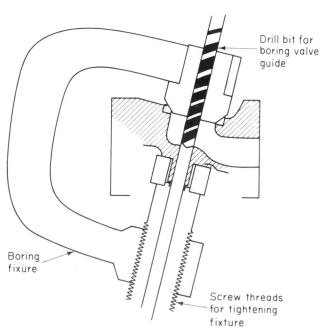

Drill bit for boring valve guide

Boring fixture

Screw threads for tightening fixture

Figure 16-23. A line drawing illustrating the type of fixture required to bore the valve guide oversize to accept a thin wall insert sleeve.

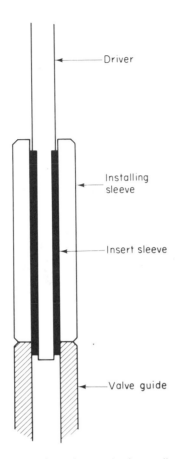

Figure 16-24. The valve guide thin wall insert being pushed into the valve guide from the installing sleeve.

Figure 16-25. An installed spiral bronze insert bushing.

into a thread that is put in the valve guide. The tap used to put cut threads in the valve guide has a long pilot ahead of the thread cutting portion of the tap. This aids in restoring the original guide alignment. The long pilot is placed in the guide from the valve seat end. A power driver is attached to the end of the pilot that extends from the spring end of the valve guide. The threads are cut in the guide from the seat end toward the spring end as the driver turns the tap, pulling it toward the power driver. The tap is stopped before it comes out of the guide and the power driver is removed. The thread is carefully completed by hand to avoid breaking either the end of the guide or the tap. An installed spiral bronze insert bushing can be seen in Figure 16-25.

The spiral bronze bushing is tightened on an inserting tool. This holds it securely in the wound-up position, so that it can be screwed into the spring end of the guide. It is screwed in until the bottom of the bushing is flush with the seat end of the guide. The holding tool is removed and the bushing material is trimmed one coil *above* the spring end of

the guide. The end of the bushing is temporarily secured with a plastic serrated bushing retainer and a worm gear clamp. This holds the bushing in place as a broach is driven through the bushing to firmly seat it in the threads. The bushing is reamed or honed to size before removing the temporary bushing retainer. The final step is to trim the end of the bushing with a special cutoff tool that is included in the bushing installing tool set. This type of spiral bronze bushing can be replaced by using a pick to free the end of the bushing. It can then be stripped out and a new bushing inserted in the original threads in the guide hole. New threads do not have to be put in the guide again. The spiral bushing design has natural spiral grooves to hold oil for lubrication. The valve stem clearances are the same as those used for knurlizing and for the thin wall insert.

16-5 Valve Seat Reconditioning

The valve seats are reconditioned after the head has been resurfaced and the valve guides have been resized. The final valve seat width and position is checked with the valve that is to be used on the seat being reconditioned.

Valve seats will have a normal seat angle of either 45° or 30°. Narrow 45° valve seats will crush lead and carbon deposits to prevent buildup of deposits on the seat. The valve will, therefore, close tightly on the seat. While closed on the seat, the

271

valve heat will transfer to the seat and cylinder head. The 30° valve seat is more likely to burn than a 45° seat because some deposits can build up to keep the valve from seating properly. The 30° valve seat will, however, allow more gas flow than a 45° valve when both are opened the same amount of lift. This is especially true with valve lifts of less than ¼ inch (6 mm). The 30° valve seat is also less likely to have valve seat recession than 45° seats. Generally speaking, when 30° valve seats are used, they are used on the cooler operating intake valves rather than hot exhaust valves.

The valve seats are only resurfaced enough to remove all pits and grooves and to correct any seat runout. As metal is removed from the seat, the seat is lowered into the head (Figure 16-26). This causes the valve to be located farther into the head when it is closed on the seat. The result of this is that the valve tip extends out farther from the valve guide. The valve being low in the head also tends to restrict the amount of valve opening. This will reduce the flow of gases through the opened valve. The reduced flow of gases, in turn, will reduce the maximum power the engine can produce.

Ideally, the valve face and valve seat should have exactly the same angle. This is impossible, especially on exhaust valves, because the valve head becomes much hotter than the seat and so the valve expands more than the seat. This expansion causes the hot valve to contact the seat in a different place on the valve than it did when it was cold.

As a result of its shape, the valve does not expand evenly when heated. This uneven expansion also affects the way the hot valve contacts the seat. In valve and valve seat reconditioning, the valve is often ground with a face angle from 1 to 2° less than the seat to compensate for the change in hot seating. This is illustrated in Figure 16-27. The angle between the valve face and seat is called an *interference angle*. It makes a positive seal at the combustion chamber edge of the seat when the engine is first started after a valve job. As the engine operates, the valve will peen itself on the seat. In a short time, it will make a matched seal. After a few thousand miles, the valve will have formed its own seat, as pictured in Figure 16-28.

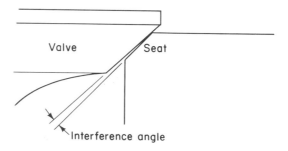

Figure 16-27. A interference angle gives the valve a tight line seal at the combustion chamber edge of the seat.

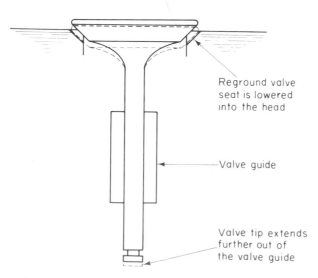

Figure 16-26. The valve seat is lowered when ground. This lets the valve tip extend further from the valve guide.

Figure 16-28. Typical valve-to-seat fit of a valve and seat after use.

The interference angle has another benefit. The valve and seat are reconditioned with different machines. Each must have its angle set before it is used for reconditioning. It is nearly impossible to set the exact same angles on both valve and seat reconditioning machines. Making an interference angle will make sure that any slight angle difference favors a tight seal at the combustion chamber edge of the valve seat when the valve servicing has been completed.

As the valve seats are resurfaced, their width increases. The resurfaced seats must be narrowed to make the correct seat width and to position it properly on the valve face. The normal automotive seat is from 1/16 to 3/32 in. (1.5 to 2.5 mm) wide. There should be at least 1/32 in. (0.8 mm) of the ground valve face extending above the seat. This is called *overhang*. The fit of a typical reconditioned valve and seat are shown in Figure 16-29. Some manufacturers recommend having the valve seat contact the middle of the valve face. In all cases, the valve seat width and the contact with the valve face should comply with the manufacturer's specifications.

Since World War II, most valve seats have been reconditioned with grinding wheels. Valve seat cutters are gradually becoming popular to recondition seats. The cutters will rapidly produce a good commercial-quality valve seat.

Pilots for Seat Reconditioning. Valve seat reconditioning equipment uses a pilot in the valve guide to align the stone holder. Two types of pilots are used: tapered and expandable. Examples of these are pictured in Figure 16-30. *Tapered* pilots locate themselves in the least worn section of the guide. They are made in standard sizes and in oversize increments of 0.001 in., usually up to 0.004 in. oversize. The largest pilot that will fit into the guide is used for valve seat reconditioning. This type of pilot restores the seat as close to the original position as

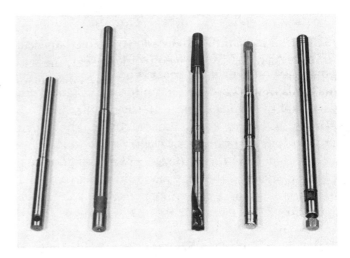

Figure 16-30. The two left pilots are solid tapered. The three right pilots are expandable.

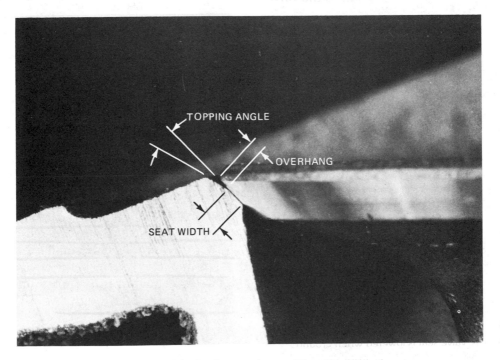

Figure 16-29. The fit of a typical reconditioned valve and seat.

possible when used with worn valve guides. Oversized taper pilots cannot be used to measure the guide wear, because they only fit the *smallest* part of the guide. Wear is indicated by the *largest* part of the guide.

Two types of *expandable* pilots are used with seating equipment. One type expands in the center of the guide to fit like a tapered pilot. Another expands to contact the ends of the guide where there has been the greatest wear. The valve itself will align in the same way as the pilot. If the guide is not reconditioned, the valve will match the seat when an expandable pilot is used. The only time an automotive technician has a choice of piloting methods is when there are several types of seating equipment available for use or when new equipment is to be purchased. The pilot and guide are thoroughly cleaned. A guide cleaner like the one shown in Figure 16-31 that is rotated by a drill motor does a good job in cleaning the guide. The pilot is placed in the guide to act as an aligned support or pilot for the seat reconditioning tools. An expandable pilot is shown in a cutaway valve guide in Figure 16-32.

Valve Seat Grinding Stones. Three basic types of grinding stones are used. All are used dry. A *roughing stone* is used to rapidly remove large amounts of seat metal. This would be necessary on a badly pitted seat or when installing new valve seat

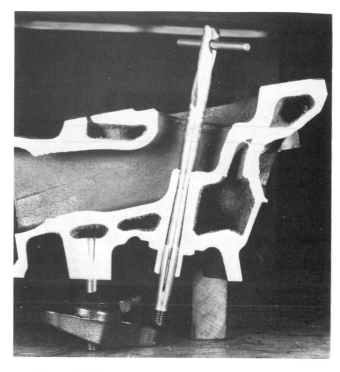

Figure 16-32. An expandable pilot in the valve guide of a sectioned head to show how the pilot fits.

inserts. The roughing stone is sometimes called a seat *forming stone*. After using the seat forming stone, a *finishing stone* is used to put the proper finish on the seat. The finishing stone is also used to recondition cast-iron seats that are only slightly worn. *Hard seat stones* are used on hard Stellite exhaust seat inserts.

The stone diameter and face angle must be correct. The diameter of the stone must be larger than the valve head, but it must be small enough so that it does not contact the edge of the combustion chamber. An oversized grinding stone is shown in Figure 16-33. The angle of the grinding surface of the stone must be correct for the seat. When an interference angle is used with reground valves, it is common practice to use a seat with the standard seat angle. The interference angle is ground on the valve face. In some cases, such as an aluminized valve, the valve has the standard angle and the seat is ground to give the interference angle. This required seat angle must be determined *before* the seat grinding stone is dressed.

Dressing the Seating Stone. The selected grinding stone is installed on the stone holder. A drop of oil is placed on the spindle of the dressing fixture and

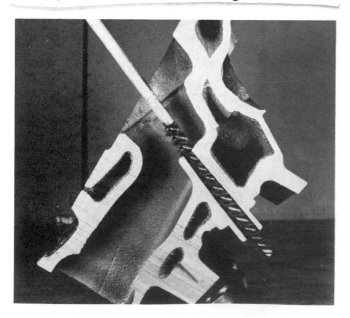

Figure 16-31. A sectioned head showing how a brush valve guide cleaner is used.

Figure 16-33. A valve seat grinding stone that is too large to use.

Figure 16-34. The tip of a diamond dressing tool.

the assembly is placed on the spindle. The dressing tool diamond (Figure 16-34) is adjusted so that it extends ⅛ in. or less from its support. The valve seat angle is adjusted on the fixture. If necessary, the base of the spindle is moved sideways so that the dressing tool just clears the stone face. The driver for the seating tool is placed in the top of the stone holder. This assembly is shown in Figure 16-35. The holder and grinding stone assembly are rotated with the driver. The diamond is adjusted so that it just touches the stone face. If it is too deep, the stone will grind behind the diamond, undercutting it so that it will fall from the holder. This will ruin the dressing tool. The diamond dressing tool is moved slowly across the face of the spinning stone, taking a very light cut. Dressing in this way will give the stone a clean, sharp cutting surface. It is necessary to redress the stone each time a stone is placed on a holder, at the beginning of each valve job, and at any time the stone does not cut smooth and clean while grinding valve seats.

Valve Seat Grinding. It is a good practice to clean each valve seat before grinding. This keeps the soil from filling the grinding stone. The pilot is then placed in the valve guide. A drop of oil is placed on the end of the pilot to lubricate the holder. The holder, with the dressed seating stone, is placed over the pilot with a small piece of emery paper between the stone and seat. The paper is held against the stone and it is rotated by hand so that the emery cleans the seat. The emery is removed so that the seat can be ground.

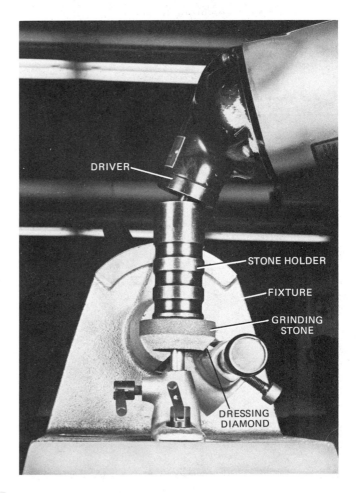

Figure 16-35. A typical assembly for dressing a valve seat grinding stone.

The driver for the stone holder should be held squarely on the holder so that no side loads are applied (Figure 16-36). Side loads will prevent concentric seat grinding. The driver should be supported so that no driver weight is on the holder. This allows the stone abrasive and the metal chips to fly out from between the stone and seat to give fast, smooth grinding. Grinding is done in short bursts, allowing the seating stone to rotate approximately 10 turns. The holder and stone are lifted from the seat between each grinding burst to check the condition of the seat. The finished seat is bright and smooth across the entire surface, with no pits or roughness remaining (Figure 16-37). This is the standard commercial finish. Precision grinding requires that the stone be redressed and the seat be given a final very short grinding burst of three to five turns. This will form an ideal seat finish. Some of the induction hardness from the exhaust valve seat will sometimes extend over into the intake seat.

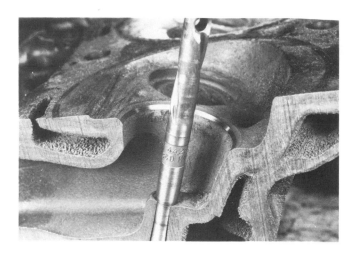

Figure 16-37. A finished valve seat shown on a cutaway head.

It may be necessary to hold a slight pressure on the driver toward the hardened spot to form a concentric seat. The seat is checked with a dial gauge to make sure that it is concentric within 0.002 in. (0.05 mm) before the seat is finished (Figure 16-38). The

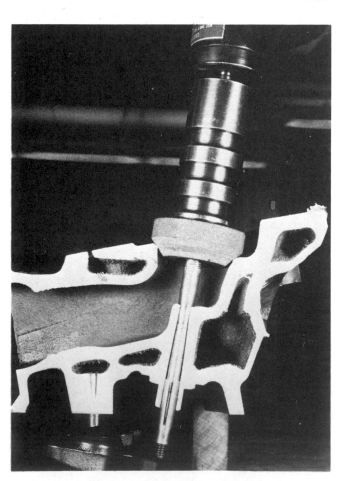

Figure 16-36. A typical set up for grinding a valve seat is shown on a cut-away head.

Figure 16-38. A typical tool for measuring valve seat concentricity.

remaining valve seats of the same size are completed before the stone is removed from the holder.

Narrowing the Valve Seat. The valve seat becomes wider as it is ground. It is therefore necessary to narrow the seat so that it will contact the valve properly. The seat is *topped* with a grinding stone dressed 15° less than the seat angle. Topping lowers the top edge of the seat. The amount of topping required can best be checked by measuring the maximum valve face diameter using dividers (Figure 16-39). The dividers are then adjusted ¹⁄₁₆ in. smaller to give the minimum valve face overhang. The seat is checked with the dividers (Figure 16-40) then topped with short grinding bursts, as required, to equal the diameter set on the dividers. The seat width is then measured (Figure 16-41). If it is too wide, the seat must be *throated* with a 60° stone. This removes metal from the port side of the seat, raising the lower edge of the seat. Throating is done with short grinding bursts until the correct seat width is achieved. Throating and topping angles are illustrated in Figure 16-42.

The completed seat must be checked with the valve that is to be used on the seat. This can be done by marking across the valve face at four or five places with a felt-tip marker. The valve is then inserted in the guide so that the valve face contacts the seat. The valve is rotated 20° to 30° and then removed. The location of the seat contact on the valve is observed where the felt-tip marking has been rubbed off from the valve. Valve seating can

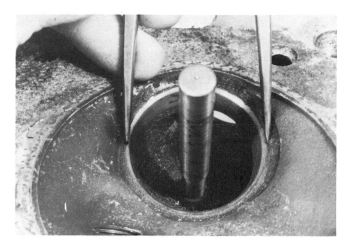

Figure 16-40. Checking the maximum valve seat diameter with the dividers adjusted 1/16 in. less than the maximum valve face diameter.

Figure 16-41. Measuring the valve seat width.

Figure 16-39. Measuring the maximum valve face diameter.

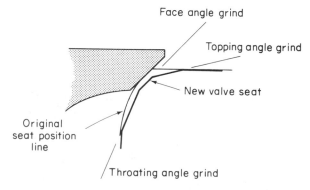

Figure 16-42. Throating and topping angles used to adjust the new valve seat width and contact location on the valve face.

be seen in Figure 16-43. Valve seat grinding is complete when each of the valve seats has been properly ground, topped, and throated.

Valve Seat Cutting. Some automotive service technicians prefer to use valve seat cutters rather than valve seat grinders. Examples of each are shown in Figure 16-44. The valve seats can be reconditioned to commercial standards in much less time when using the cutters rather than the grinders. A number of cutting blades are secured at the correct seat angle in the cutting head of this valve seat reconditioning tool. The cutter angle usually includes the interference angle so that new valves with standard valve face angles can be used without grinding the new valve face. The cutters do not require dressing as stones do. The cutting head assembly is placed on a pilot in the same way the grinding stone holder is used (Figure 16-45). The

Figure 16-45. A valve seat cutter assembly being lowered onto the pilot.

cutter is rotated by hand or by using a special speed-reduction motor. Only metal chips are produced. The finished seat is checked for concentricity and fit against the valve face using the felt-tip-marker method previously described.

16-6 Valve Seat Replacement

Valve seats need to be replaced if they are cracked or if they are burned or eroded too much to be reseated. A badly eroded valve seat is shown in Figure 16-46. It may not be possible to determine this before an attempt is made to recondition the valve seats. Valve seat replacement is piloted in the valve guide. This means that the valve guide must be reconditioned before the seat can be replaced. Damaged *insert valve seats* are removed and the old seat counterbore is cleaned up to accept a new oversized seat insert. Damaged *integral valve seats* must be counterbored to make a place for the new insert seat.

Figure 16-43. On this cut-away head the location of the valve seat is shown where the ink from the felt tip pen has transfered from the seat to the valve face.

The old insert seat is removed by one of several methods. A small *pry bar* can be used to snap the seat from the counterbore. It is sometimes easier to do this if the old seat is *drilled* to weaken it. Be careful not to drill into the head material. Another way to loosen the seat is to run an *arc weld* around the seating surface of the insert. When the weld cools, it will shrink the insert enough to loosen it. Sometimes, an expandable hook-type puller is used to remove the seat insert. The seat counterbore must be cleaned before the new, oversized seat is installed. The replacement inserts have a 0.002 to

Figure 16-44. Examples of a valve seat cutter on the left and a valve seat grinding stone on the right.

Figure 16-46. A badly eroded valve seat.

head loosened, is placed over the tool holder. It is clamped to the cylinder head in a way that will put no loads on the tool holder. The fixture swivel is then clamped in place.

The new insert is placed between the support fixture and the stop ring. The stop ring is adjusted against the new insert so that cutting will stop when the counterbore reaches the depth of the new insert. This setup is pictured in Figure 16-47. The boring tool is turned by hand or with a reduction gear motor drive. It cuts until the stop ring reaches the fixture. Figure 16-48 shows a valve seat counterbore that is nearly finished. The support fixture and the tool holder are removed. The pilot and the correct size adapter are placed on the driving tool. Ideally, the seats should be cooled with dry ice to cause them to shrink. Each insert should be left in the dry ice until it is to be installed. This will allow them to be

0.003 in. (0.05 to 0.07 mm) interference fit in the counterbore. The counterbores are cleaned and properly sized, using the same equipment described in the following paragraph for installing replacement seats in place of faulty integral valve seats.

Cracked or badly burned integral valve seats can often be replaced to salvage the head. All head cracks are repaired *before* the old integral seat is removed. The replacement seat is selected first. It must have the correct inside and outside diameters and it must have the correct thickness. Manufacturers of replacement valve seats supply tables that specify the proper seat insert to be used. If an insert is being replaced, the new insert must be of the same type of material or better than the original insert. Insert exhaust valve seats operate from 100 to 150 °F (56 to 83 °C) hotter than integral seats. Upgraded valve and valve seat materials are required to give the same service life as the original seats.

A counterbore cutting tool is selected that will cut the correct diameter for the outside diameter of the insert. The cutting tool is bottomed securely in the tool holder so that it will cut the counterbore at the correct diameter. The tool holder is attached to the size of pilot that fits the valve guide. The tool-holder feed mechanism is screwed together so that it has enough threads to properly feed the cutter into the head. This assembly is placed in the valve guide so that the cutting tool rests on the seat that is to be removed. The supporting fixture, with the swivel

Figure 16-47. Adjusting the cutting tool stop ring with the new valve seat insert.

Figure 16-48. The seat cutting tool boring out the old eroded valve seat.

counterbore. It serves no purpose to hit the driver after the insert is seated in the bottom of the counterbore. The installed valve seat insert is *peened* in place by running a peening tool around the metal on the outside of the seat. The peened metal is slightly displaced over the edge of the insert to help hold it in place. A fully installed seat is shown in Figure 16-50. Seats are formed on the replacement inserts using the same procedures described to recondition valve seats.

installed with little chance of shearing metal from the counterbore. Sheared chips could become jammed under the insert, keeping it from seating properly. The chilled seat is placed on the counterbore. The driver with a pilot is then quickly placed in the valve guide so that the seat will be driven squarely into the counterbore. The driver is hit with a heavy hammer to seat the insert, as shown in Figure 16-49. Heavy blows are used to start the insert, and lighter blows are used as the seat reaches the bottom of the

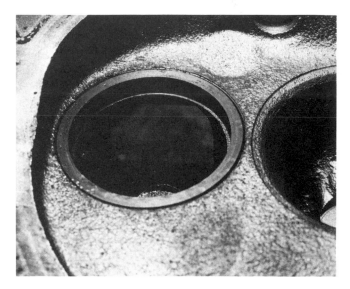

Figure 16-50. A fully installed valve seat insert.

Figure 16-49. Seating the new insert in the counter bore by hitting the driver with a heavy hammer.

16-7 Finishing the Head

All threaded holes should be checked to make sure that the threads are in good condition. Damaged threads should be repaired with threaded inserts. Threaded inserts are available for all standard and metric size threads, including spark plug threads. One type of tool set used to replace threads is shown in Figure 16-51. The damaged threads are drilled out with the proper-size drill. The hole is tapped with a special bottoming tap having the same number of threads as the original hole. Both special drill and special tap are included with the thread repair kit. The thread insert is screwed in (Figure 16-52) and the drive tang of the thread is broken off to complete the repair.

Broken or damaged studs or cap screws must be removed. If two nuts can be fastened to a damaged stud, the nuts are tightened or jammed together, as shown in Figure 16-53. The lower nut is then turned to remove the stud. A damaged stud can also be removed using a locking-type pliers. Broken studs and cap screws must be removed from the casting. This can usually be done by first centerpunching in the center of the exposed broken end of the stud or cap screw (Figure 16-54). The correct size hole is drilled into the center punched end of the

Figure 16-53. Nuts jammed together on a stud for stud removal.

Figure 16-51. Parts of a thread repair kit.

Figure 16-52. A threaded insert being screwed into the retapped hole.

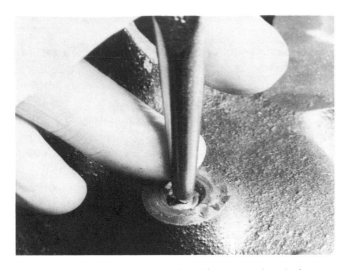

Figure 16-54. Center punching the exposed end of a broken cap screw.

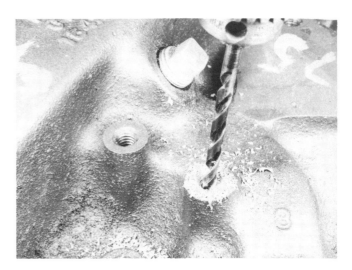

Figure 16-55. Drilling into the center of the broken cap screw.

broken stud or cap screw (Figure 16-55). A stud remover is put into the drilled hole and the broken section is unscrewed, as shown in Figure 16-56. Care must be used to avoid breaking the drill or stud remover in the broken section. The drill and stud remover are made of very hard materials. If they are broken, they will have to be removed with special equipment that will burn them out. New studs are installed in place of the broken stud. Make sure that the new stud has the correct type of thread and that it is screwed in to the correct depth.

Damaged or loose rocker arm studs must be replaced. Pressed-in studs that are damaged must be pulled out. The stud hole is reamed to size and a new, oversized stud is pressed in. Care must be used to press the new stud to the correct depth. This is critical on studs that have a bottoming shoulder for the pivot nut. A special stud installing tool is recommended. If this type of stud were not pushed in far

enough, there could be some looseness in the valve train. Sometimes, the pressed-in replacement studs are locked in place. A pin installed in a hole drilled through the stud boss and stud locks them in place. The damaged pressed-in stud is sometimes replaced with a threaded stud. In this case, the original stud hole must be tapped with a starting tap and finished with a bottoming tap. The replacement stud must be screwed in to the correct depth.

If the soft plugs have been removed from the head for cleaning the cooling passages, they should be reinstalled. The soft plug counterbores should be cleaned. Sealant is placed around the soft plug. It is then put in place with a properly fitting driver.

The head is ready for assembly when all of the reconditioning work is done. The exterior may be painted, but painting is usually done after the engine has been reassembled. Assembly of the valves in the head is described in Section 12-3.

Figure 16-56. Removing the broken section with the stud remover.

VALUE

TRAIN

REVIEW QUESTIONS

1. Why do some engine rebuilders prefer to work on used heads? [INTRODUCTION]
2. Why is valve guide reconditioning the first head servicing operation? [INTRODUCTION]
3. Why is a valve grinding stone dressed before grinding begins? [16-1]
4. Describe the valve grinding procedure. [16-1]
5. What should be done to worn aluminized valves? [16-1]
6. When is the valve tip ground? How do you know how much to remove from the valve tip? [16-1]
7. Why are center punch marks placed along a crack in the head? [16-2]
8. Name two methods used to repair cracked heads. [16-2]
9. Describe the procedure used when welding a cracked head. [16-2]
10. Describe the procedure used to plug a crack in the head. [16-2]
11. What two ways are used to resurface heads? [16-3]
12. Why must part of the manifold for V-type engines be resurfaced when the head is resurfaced? [16-3]
13. How does the type of valve wear differ between the valve stem and guide? [16-4]
14. How do engine problems differ when caused by too little guide clearance as compared to too much guide clearance? [16-4]
15. How does a large valve stem clearance make seating easier? [16-4]
16. When is it recommended to ream a valve guide oversize? [16-4]
17. When is it recommended to replace valve guides? [16-4]
18. Give the reasons why most valve guides are resized during engine reconditioning. [16-4]
19. Describe the process of displacing metal to make the guide hole smaller. [16-4]
20. Why should only standard valve stem seals be used with knurled valve guides? [16-4]
21. Describe the procedure of installing an insert sleeve in a valve guide. [16-4]
22. How is the thread cut in the valve guide for a spiral bronze alloy insert bushing? [16-4]
23. What are the advantages and disadvantages of using valve guide inserts? [16-4]
24. When are valve seats resurfaced? [16-5]
25. What happens to the valve tip when the seat is ground? [16-5]
26. Why is it impossible to have the valve face and seat at the same angle? [16-5]
27. What is an interference angle between the valve and seat? [16-5]
28. Why is an interference angle desirable when seating valves? [16-5]
29. What happens to the seating of the valve after a few thousand miles? [16-5]
30. What are the advantages and disadvantages of tapered pilots compared to expandable pilots for valve seating? [16-5]
31. Why can't pilots be used to measure the size of the valve guide? [16-5]
32. When does the automotive technician have a choice of pilot types when reconditioning the seats? [16-5]
33. What must be done to the guide before installing the pilot? [16-5]

34. Name three types of valve seat grinding stones. When is each one used? [16-5]
35. What three things must be considered when selecting valve seat grinding stones? [16-5]
36. Why is it necessary to dress valve seat grinding stones? [16-5]
37. What precautions should be taken when dressing the valve seat grinding stone? [16-5]
38. When should the valve seat grinding stone be dressed? [16-5]
39. When is it desirable to clean the valve seat before grinding? [16-5]
40. How should the driver for the stone holder be handled while grinding valve seats? [16-5]
41. When are valve seats topped? [16-5]
42. When are valve seats throated? [16-5]
43. Compare valve seat grinding stones with valve seat cutters. [16-5]
44. How are old valve seat inserts removed? [16-6]
45. How is the new valve seat insert selected? [16-6]
46. How is the diameter and depth for the new valve seat insert made to the correct size? [16-6]
47. How is the valve seat insert installed? [16-6]
48. How can damaged threads in a casting be repaired? [16-6]
49. How are studs removed from a casting? [16-6]

CHAPTER
17

Lower Engine Service

It is impossible to say exactly what should be done to recondition the lower engine. When only high-mileage wear exists, the engine *can* be repaired at relative low cost so that the operator can "get by" for awhile. Additional reconditioning and parts replacement can make the engine as good as new. In between these two extremes, the customer and the service technician must agree on the amount of repair that is to be done on the engine.

The combustion chamber is resealed when the engine lacks power. Resealing is done by reconditioning the head, as described in Chapter 16, and by servicing the piston and cylinder wall, then installing new piston rings. When engine power returns to normal, it will put greater stress on the engine bearings. The connecting rod bearings are generally replaced when the piston rings are replaced. Other worn engine parts may be expected to fail before the piston rings need to be replaced again. These other parts can also be reconditioned while the engine is disassembled. This would save money in the long run.

The cost of disassembling, cleaning, inspecting, reassembling, and sealing the engine is the same, regardless of the amount of reconditioning that is done while the engine is disassembled. It costs as much to remove and reinstall the head while doing a valve job as it does when the engine is rebuilt. Once the engine is disassembled, the added cost of reconditioning the parts is only the cost of that reconditioning operation. For example, the cost of removing and reinstalling the crankshaft with new bearings is the same whether the crankshaft journals were reground or not. The added cost of reconditioning a worn crankshaft would only be the cost of crankshaft reconditioning. This would make the clearances the same as the clearances in a new engine. This example illustrates another fact about engine reconditioning. Once the engine is disassembled, many of the components can be reconditioned with no concern given to unrelated parts. An example of this is the crankshaft that can be reconditioned without being concerned with the pistons. In other cases, the reconditioning of parts are related to each other. Reconditioning both piston and cylinder is an example of this interrelationship. Unrelated parts can be reconditioned in any order, while the size of related parts must be checked when either or both of them are reconditioned.

The surface finish of a reconditioned part is as important as the size of the part. The standard measurement for the surface finish is the *microinch root mean square (RMS)*. **A microinch (μin.) is one-millionth part of an inch.** Another way of looking at this is that a microinch is like dividing 0.001 in. into 1000 subdivisions. A scratch in the surface 0.0001 in. deep would be 100 μin. deep.

If we could look at the polished surface under a strong microscope, we would see that the surface was not perfectly smooth. A tool with a very sharp point is moved across the surface to measure and automatically compute the average surface finish in *microinches* RMS. The rougher the surface, the higher the microinch finish will be. A polished crankshaft journal finish will be 15 μin. RMS. A finished cylinder wall will have between 20 and 35 μin. RMS finish.

The size of the abrasive particles in the grinding and honing stones control the surface finish. The size of the abrasive is called the *grit* size. The abrasive is sifted through a screen mesh to sort out the grit size. A course mesh screen has few wires in each square inch, so large pieces can fall through. A fine mesh screen has many wires in each square inch so that only small pieces can fall through. The screen is used to separate the different grit sizes. The grit size is the number of wires in each square inch of the mesh. A low-numbered grit has large pieces of abrasive material; a high-numbered grit has small

pieces of abrasive material. The higher the grit number being used, the smoother the surface finish will be. A given grit size will produce the same finish as long as the cutting pressure is constant. Light cutting pressure produces fine finishes; heavy cutting pressure produces rough finishes with the same grit size. Figure 17-1 shows typical abrasive stones used for cylinder honing.

During the inspection procedure, the nonrepairable parts were put aside. The following discussion describes the typical operations performed on the remaining lower engine parts that are to be continued in service.

Figure 17-1. Typical abrasive grit cylinder honing stones. A fine grit stone is on the left and a course grit stone is on the right.

17-1 Piston and Rod Assembly Service

When engine servicing equipment is not available, the repairable piston and rod assemblies are taken to the automotive machine shop for reconditioning. As a standard commercial practice, the pistons are removed from the rods. After cleaning, the skirts of the used pistons are *resized,* a *spacer* is placed in the top of the upper ring grooves, and the rod is *aligned*. During precision rebuilding, the large end of the connecting rod is also resized to remove any stretch or warp that could affect bearing crush.

Piston Ring Grooves. As the piston goes rapidly up and down in the cylinder, it tosses the rings to the top and to the bottom of the ring grooves. The pounding of each ring in its groove gradually increases the piston ring side clearance. Material is worn from both the ring and the groove. The greater the side clearance, the faster the wear becomes. This results because the ring moves farther before it hits the side of the groove. This is related to a hammer. The greater distance a hammer is swung, the harder the blow will be. Most of the side wear takes place in the top ring groove because this surface is exposed to the most abrasive material from the combustion chamber. It also receives a heavy force from the combustion pressure. A large ring groove side clearance allows the piston ring to twist in the groove. In severe cases, this twist can cause a wedging effect between the piston and the cylinder. In some cases, the rings will wedge so tightly that the starter can hardly crank the engine. Movement of the ring in the groove may, in severe cases, cause rings to break. It should be obvious that ring groove side clearance must be corrected.

The upper ring groove is reconditioned when the groove has worn more than 0.005 in. (0.125 mm). Ring groove wear always occurs on high-mileage engines and it sometimes occurs on low-mileage engines. To correct the ring groove clearance, the top ring groove is machined 0.025 in. (0.625 mm) wider than the standard groove. One type of tool used for this is pictured in Figure 17-2. A steel ring groove spacer is placed above the new piston ring in the reconditioned ring groove to return the ring side clearance to the standard dimension, as shown in Figure 17-3.

Figure 17-2. One type of piston ring groove reconditioner.

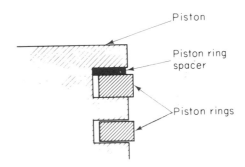

Figure 17-3. Ring groove spacer placed above the new upper ring in the reconditioned ring groove.

Piston Skirt. Most automotive engine pistons are of aluminum slipper skirt design. In operation, the piston supports heavy side loads during the power stroke. This gradually causes a slight collapse of the piston skirt. The aluminum piston skirt also wears away. Piston skirt collapse and wear in time will allow the piston to rock slightly as it moves up and down in the cylinder. This slight rocking action also rocks the piston rings so that they keep changing their contact face against the cylinder wall. New piston rings cannot form a good moving seal if they are not held perpendicular to the cylinder. The best reconditioning procedure is to replace the piston with a new piston. It is necessary to resize the skirt of used pistons when they are to be reused, so that the pistons will not rock as they operate in slightly worn cylinders.

Pistons are resized by knurling to expand their skirts. Knurling interrupts the surface of the piston skirt, displacing the metal outward between the teeth of the knurling tool, as shown in Figure 17-4. This effectively increases the diameter of the piston skirt. The amount the skirt diameter is increased by knurling is controlled by the pressure put on the knurling tool by the service technician. In addition to increasing the piston skirt diameter, the knurled surface carries lubricating oil to help maintain the required thin oil film between the piston and cylinder. It is common practice to fit knurled pistons to clearances one-half the amount specified for piston skirts that are not knurled.

The piston skirt does not operate in the upper inch of the cylinder, where most of the cylinder wear occurs. The bottom of the cylinder has the least amount of cylinder wear. For this reason the piston skirt is knurled to fit the bottom of the

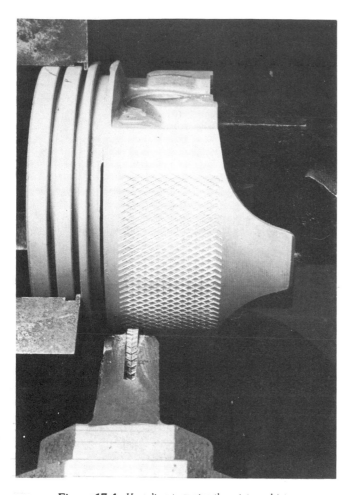

Figure 17-4. Knurling to resize the piston skirt.

cylinder. Fitting the piston must be done *after* the cylinder wall has been reconditioned. Knurling the piston cannot correct for excessive cylinder taper. The cylinders will have to be rebored oversize to correct excessive taper and out-of-round. New pistons will have to be installed in rebored cylinders. Knurling cannot increase the piston size enough.

Piston Pins. Piston pins do not normally become loose enough to cause a knock or tapping sound until the engine has very high mileage. The piston pin is more likely to produce a knock you can hear after the piston skirt has been resized and new piston rings installed. For this reason, automotive machine shops may install oversized piston pins when reconditioning piston and rod assemblies. The majority of domestic automobile manufacturers do not supply oversized piston pins through their parts departments. The manufacturers recommend replacing the used pistons with new pistons when the clearance between the pin and piston becomes excessive (over 0.001 in. or 0.025 mm). The new pistons have prefit-

ted piston pins. This allows the connecting rod to be used without resizing the small eye of the rod. Once the connecting rod eye is honed oversize, a standard-size piston pin can never be used if the piston should require replacement in the future. For most engines, oversized piston pins are available in the parts replacement market. Installing an oversized piston pin in a used piston and rod will help the engine to run quietly for a reasonable length of time. The use of a reconditioned high-mileage piston cannot be expected to give as long a service life as a new replacement piston.

Both the piston and the small eye of the connecting rod are honed with precision equipment that can control the hole surface finish and the hole size, within 0.0001 in. (0.0025 mm), either larger or smaller than the diameter of the piston pin. The piston pin hole in the piston is sized to give a clearance of 0.0002 to 0.0005 in. (0.0006 to 0.0012 mm). A typical pin hone is shown in Figure 17-5. When both the size and the finish of the piston pin hole in the piston are correct, the piston pin will slide through the piston by its own weight when the piston is at room temperature. The small eye of the connecting rod is fitted to a *press* or *interference fit*. In this type of fit, the hole is honed to 0.0008 to 0.0016 in. (0.002 to 0.004 mm) *smaller* than the diameter of the piston pin. This provides the correct press or interference fit. A precision measuring gauge of the type shown in Figure 17-6 is needed to measure the honed hole. When required, the large

Figure 17-5. Honing a piston to fit an oversize piston pin.

Figure 17-6. One type of precision gauge required to check the size of the hole while it is being honed for an oversize piston pin.

Figure 17-7. A heater being used to expand the piston pin eye of the connecting rod.

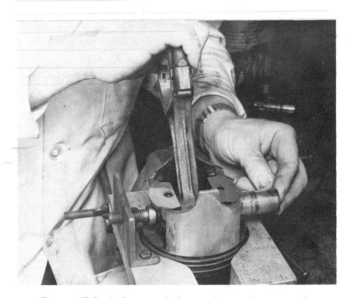

Figure 17-8. A shop made fixture designed to correctly position the piston pin as it is installed in the piston and heated rod.

end of the connecting rod should also be resized before the piston is installed on the rod. This is discussed in a later paragraph. Depending on the type of shop equipment used, the connecting rod may be aligned before assembly with the piston or after.

To assemble the piston and rod, the piston pin is put in one side of the piston. The piston and rod are placed on a press, using adapters and supports. This setup was shown in Figure 12-10. The pin is pressed into the rod until it is centered. The press fit of the pin in the small eye of the rod will hold the pin securely in place during engine operation. This keeps it from sliding out and touching the cylinder wall. In precision engine shops, the small eye of the connecting rod is heated before the pin is installed. One type of heater is shown in Figure 17-7. This causes the rod eye to expand so that the pin can be pushed into place with little force. The pin must be rapidly pushed to the correct center position. The shop-made fixture pictured in Figure 17-8 does a good job of centering the piston pin. There is only one chance to get it in the right place because the rod will quickly seize on the pin as the rod eye is cooled by the pin.

Full-floating piston pins operate in a bushing in the small eye of the connecting rod. The bushing can be replaced. The bushing and the piston are honed to the same diameter. This allows the piston pin to slide freely through both. The full-floating piston pin used in passenger cars is held in place with a lock ring at each end of the piston pin. The lock ring expands into a small groove in the pin hole of the piston. Figure 17-9 shows one type of lock ring in place in the piston. The lock rings should always be replaced with new rings. These must be seated properly in the ring groove. Used rings have

289

Figure 17-9. One type of lock ring used on the end of a full floating piston pin to keep it centered in the piston.

come out of the piston, usually ruining the piston, as was shown in Figure 14-27. They will usually cause heavy cylinder wall scoring as well, as they move with the piston.

Rod Big End. As an engine operates, the forces go through the large eye of the connecting rod. This causes the eye to gradually deform. The large eye of the connecting rod is resized during precision engine service. The parting surfaces of the rod and cap are smoothed up to remove all high spots before resiz-

ing. A couple of thousandths of an inch of metal is removed from the rod cap parting surface. This is done using the same grinder that is used to remove metal from the parting surface of main bearing caps as shown in Figure 17-13. The cap is installed on the rod and the nuts or cap screws are properly torqued. The hole is then bored or honed perfectly round to the size and finish required to give the correct connecting rod bearing crush. Figure 17-10 shows the setup to resize the rod on a typical hone used in engine reconditioning.

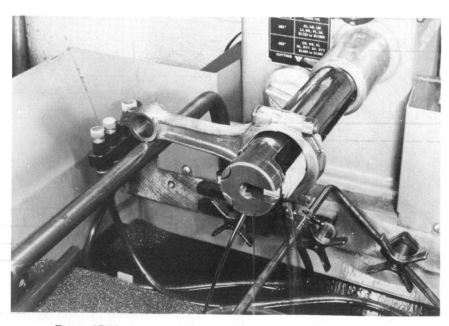

Figure 17-10. Resizing the big end of the connecting rod with a hone.

Rod Alignment. The connecting rod must not be bent or twisted to operate properly. A bent or twisted connecting rod will cause premature rod bearing and piston skirt failure. These failures were described in Section 14-2. A misaligned connecting rod will *not* hold the rings perpendicular to the cylinder wall, so the rings cannot function correctly. Signs of misalignment can first be seen by the wear pattern on the piston skirt and rod bearing as the engine is disassembled as shown in Figure 17-11. Rods can also be bent while they are being handled as they are being serviced. They should be checked as the last piston and rod reconditioning operation.

Figure 17-11. Signs of connecting rod misalignment.

Rod alignment is checked on an alignment fixture. The rod is aligned on some fixtures before it is connected to the piston. The piston and rod are assembled when using other types of alignment fixtures. One of these is illustrated in Figure 17-12. The rod can be straightened either using lever bars or by using a special hydraulic press. The aligning method depends upon the equipment available. The technician will use care in handling connecting rods, once the small force needed to change the alignment is experienced.

Figure 17-12. One type of fixture used to check the alignment of the piston and rod assembly.

17-2 Engine Block Service

The engine block is the foundation of the engine. All parts of the block must have the correct size and they must be aligned. The parts must also have the proper finishes if the engine is to function dependably for a normal service life. At one extreme, the engine block can be reused with little reconditioning. This gives a patch-up, get-by, inexpensive repair if the engine has only normal, high-mileage wear. At the other extreme, all the critical surfaces and dimensions can be reconditioned so that the block is actually more precise than it was when it was new. This process is usually called *blueprinting*.

The following discussion of engine block service will cover the operations in the order that they would usually be done if the block were to be completely reconditioned. In actual practice, reconditioning can start with any service operation and follow the order of the remaining service operations given in this section. The operations to be done on any specific engine will depend on the wishes of the customer and the quality of the engine that is needed when the engine reconditioning is completed. The cost will be higher as more service operations are

291

done. The cost of disassembly and reassembly will be the same, regardless of the type of block reconditioning that is to be done. Extra operations add only the cost of that operation to the total engine reconditioning cost. A discussion of the service operations to be done must consider the cost of these operations compared to the value of each operation to the completely repaired engine.

Main Bearing Bore Alignment. The main bearing journals of a straight crankshaft are in alignment. If the main bearing bores in the block are not in alignment, the crankshaft will bend as it rotates. This will lead to premature bearing failure and it could lead to a broken crankshaft. The original stress in the block casting is gradually relieved as the block is used in service. Some slight warpage may occur as the stress is relieved. In addition, the continued pounding caused by combustion will usually cause some stretch in the main bearing caps. Realigning and resizing the main bearing bores in the block is a procedure called *align boring* or *align honing.*

There are a number of different types of equipment used to align the main bearing bores in the block. Some are simple fixtures that clamp on the block, while others place the block in a large production align boring or align honing machine. The align boring tool is a cutting tool, similar to a lathe tool. The align honing tool is similar to a pin hone. Depending on the skill of the operator, satisfactory alignment to commercial standards can be done with all types of the aligning equipment. Precision automotive machine shops will align hone. Honing produces a finer finish than does align boring. It provides exact control of the main bearing bore size.

The same general steps are followed in align boring, regardless of the type of equipment used. First, a small amount of metal is removed from the main bearing cap parting surfaces. Figure 17-13 shows one method used to do this. It requires that about 0.015 in. (0.38 mm) be removed when using an align boring cutting tool. Only 0.002 in. (0.05 mm) needs to be removed when align honing. The resurfaced main bearing caps are torqued in place on the block. The align boring tool has a large heavy arbor that is placed in the main bearing bores. It extends beyond the block at each end. The arbor is supported at each end, and usually between the main bearing saddles so that it is centered in the

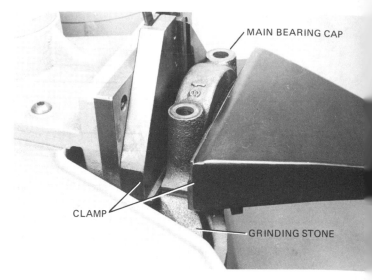

Figure 17-13. One type of fixture used to remove a small amount of metal from a bearing cap.

main bearing bores. The arbor holds the cutting tool. The main bearing bores are checked to determine exactly where metal must be taken from to align the bores. The align boring tool is adjusted at each main bearing bore to cut the correct diameter. A typical align boring fixture is shown in Figure 17-14.

One dimension that is critical in all engines is the spacing between the cam bearing center line and the crankshaft center line. This must be maintained to have the proper cam drive gear mesh or the proper timing chain tension. If this dimension is not correct, it will lead to faulty timing and to premature failure of the cam drive.

The align hone fits through all of the main bearing bores at the same time. The align hone is stroked back and forth through the bearing bores to properly size them, as shown in Figure 17-15. It takes individual instruction and practice to develop a touch and the skill necessary to properly align main bearing bores in the block. The block and oil passages must be thoroughly cleaned after align boring to remove all abrasives and metal chips. The machined surfaces are coated with oil to prevent rusting until the block is finally cleaned for assembly. The aligned main bearing bores are in exact alignment. The diameter and surface finish of the bore provides the proper backing to hold the main bearings in alignment with the proper crush. A crankshaft placed in these bores will rotate freely when the bearings have minimum clearances and the caps are fully torqued.

Figure 17-14. Align boring the main bearing bores.

Figure 17-15. Align honing the main bearing bores.

Decking. An engine should have the same combustion chamber size in each cylinder. For this to occur, each piston must come up to an equal distance from the block deck. The connecting rods are attached to the rod bearing journals of the crankshaft. Pistons are attached to the connecting rods. As the crankshaft rotates, the pistons come to the top of the stroke. When all parts are sized equally, all of the pistons will come up to the same level. This can only happen if the block deck is parallel to the main bearing bores.

The block deck can be *flattened* by moving it across a surfacing table in the same way a head is flattened. This will make a commercially acceptable flat surface to seal the gasket. The block deck must be resurfaced in a surfacing machine that can control the amount of metal removal when it is necessary to match the size of the combustion chambers. The block is set up on a bar located in the main bearing saddles of the block, as shown in Figure 17-16. The bar is parallel to the direction of the cutting head movement. The block is leveled sideways, then the

Figure 17-16. The main bearing saddles of the block is placed on an aligning bar.

deck is resurfaced in the same manner as the head is resurfaced. Figure 17-17 shows a block deck being resurfaced by grinding. Head resurfacing was described in Chapter 16.

Cylinder Boring. Most cylinders are serviceable if they are no more than 0.003 in. (0.076 mm) out-of-round, if they have no more than 0.005 in. (0.127 mm) taper, and if they have no deep scratches in the cylinder wall. Cylinder walls in this condition can be expected to provide a normal service life when they

are resurfaced. Some piston rings are designed to operate with twice as much out-of-round and taper as this, but they cannot be expected to provide a normal service life. The most effective way to correct excessive cylinder out-of-round, taper, or scoring is to *rebore* the cylinder. The rebored cylinder requires the use of a new, oversized piston. Oversized pistons have the same weight as the original pistons, so that a single cylinder on a multicylinder engine can be rebored without upsetting the commercial quality of the engine balance.

Figure 17-17. The block deck being resurfaced with a grinder.

It costs no more to bore the cylinder to maximum size than it does to bore it to the minimum size. Cylinders are often rebored to the size of the largest available piston to give the engine as much horsepower as possible. The maximum oversize is determined by two things: the cylinder wall thickness and the size of the available oversized pistons. Cylinders are rebored to the smallest oversize that will straighten and clean it for commercial reconditioning. This will allow material for another rebore in the future after the cylinder has again become worn in service. The pistons that will be used must always be in hand *before* the cylinders are rebored. The cylinders are then bored and honed to match the size of the piston.

Sometimes, cylinders have a gouge so deep that it will not clean up when reboring the cylinder to the maximum size. This could happen if the piston pin moved endways and rubbed on the cylinder wall. Cylinder blocks with deep gouges can be salvaged by *sleeving* the cylinder. This is done by boring the cylinder greatly oversize to match the outside diameter of the cylinder sleeve. The sleeve is pressed into the rebored block; then the center of the sleeve is bored to the diameter required by the piston. The cylinder can be sized to use a standard-size piston when it is sleeved.

The cylinder must be perpendicular to the crankshaft for normal bearing and piston life. If the block deck has been aligned with the crankshaft, it can be used to align the cylinders. Portable cylinder boring bars are clamped to the block deck. Heavy-duty production boring machines support the block on the main bearing bores. The fixtures that hold the block are perpendicular to the boring head. In this way, the cylinder bores will be perpendicular to the crankshaft, regardless if the deck has been resurfaced or not. Figure 17-18 shows a production cylinder boring machine.

Main bearing caps should be torqued in place when reboring cylinders. In precision boring, a dummy head plate is also bolted in place of the cylinder head while boring cylinders. In this way, distortion is kept to a minimum. The general procedure used to rebore cylinders is to set the boring bar up so that it is perpendicular to the crankshaft. It must be located over the center of the cylinder. The cylinder center is found by installing centering pins in the bar. The bar is lowered so that the center-

Figure 17-18. A production cylinder boring machine set up to begin boring a cylinder of an industrial engine.

ing pins are located and adjusted near the bottom of the cylinder where the least wear has occurred. This locates the boring bar over the original cylinder center. Once the boring bar is centered, the boring machine is clamped in place to hold it securely. This will allow the cylinder to be rebored on the original center line, regardless of the amount of cylinder wear. A sharp, properly ground cutting tool is installed and adjusted to the desired dimension. Some tools can cut deeper than others. Rough cuts remove a great deal of metal on each pass of the cutting tool. The surface of a rough cut is pictured in Figure 17-19. The rough cut is followed by a fine cut that produces a much smoother and more accurate finish, as shown in Figure 17-20. Different-shaped tool bits are used for rough and finish boring. The cutting tools are resharpened before boring each cylinder to accurately control the bore diameter and the surface finish. The last cut is made at least

Figure 17-19. The finish of the cylinder surface after boring a rough cut (Courtesy of Dana Corporation).

Figure 17-20. The finish of the cylinder surface after boring a fine cut (Courtesy of Dana Corporation).

0.0025 in. (0.064 mm) smaller than the required diameter. The cylinder wall is then finished by honing. Honing produces the required cylinder diameter and surface finish. Each cylinder is honed to give the correct clearance for the piston that is to operate in that cylinder.

 Finishing. It is important to have the proper surface finish on the cylinder wall for the rings to seat against. Some ring manufacturers recommend breaking the hard surface glaze on the cylinder wall with a hone before installing new piston rings. When honing is not required, no time is needed for honing or for cleanup. This reduces the reconditioning cost for the engine.

The cylinder wall should be honed to straighten the cylinder when the wall is wavy or scuffed. If honing is being done with the crankshaft remaining in the block, the crankshaft should be protected to keep honing chips from getting on the shaft.

Two types of hones are used for cylinder service. A *deglazing hone* removes the hard surface glaze remaining in the cylinder. It is a flexible hone that follows the shape of the cylinder wall, even when the wall is wavy. It cannot be used to straighten the cylinder. A spring-loaded deglazing hone is shown in Figure 17-21. A brush-type deglazing hone is shown in Figure 17-22. A *sizing hone* can

Figure 17-21. A cut-away engine showing a spring loaded deglazing hone in position for honing.

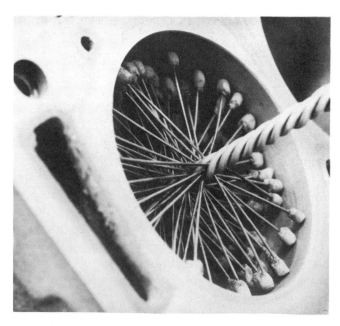

Figure 17-22. A brush type deglazing hone in position for honing a cylinder.

be used to straighten the cylinder. Its honing stones are held in a rigid fixture with an expanding mechanism to control the size of the hone. The sizing hone can be used to straighten the cylinder taper by honing the lower cylinder diameter more than the upper diameter. As it rotates, the sizing hone only cuts the high spots so that cylinder out-of-round is also reduced. The cylinder wall surface finish is about the same when refinished with either type of hone. Figure 17-23 shows a sizing hone. Used pistons will have to be resized when the cylinder taper is straightened by using a sizing honing.

The hone is stroked up and down in the cylinder

Figure 17-24. A typical honed cylinder.

Figure 17-23. A cut-away engine shown with a sizing hone in position for honing.

as it rotates. This produces a *crosshatch finish* on the cylinder wall. A typical honed cylinder is pictured in Figure 17-24. The angle of the crosshatch should be from 20° to 60°. Higher angles are produced when the hone is stroked more rapidly in the cylinder. The roughness of the finish is more important than the crosshatch. A coarse stone with a grit size of 70 is used for removing metal. A 150 grit hone is used to provide normal cylinder finish. If a polished cylinder wall is desired, stones with 280 grit are used in the hone.

Honing oil is wiped or flowed in the cylinders and on the honing stones. The hone is placed in the cylinder. Before the drive motor is turned on, the hone is moved up and down in the cylinder to get the feel of the stroke length needed. The end of the hone should just break out of the cylinder bore on each end. The hone must *not* be pulled from the top of the cylinder while it is rotating. Also, it must not

be pushed so low in the cylinder that it hits the main bearing web or crankshaft. The sizing hone is adjusted to give a solid drag at the lower end of the stroke. The hone drive motor is turned on and stroking begins immediately. Stroking continues until the sound of the drag is reduced. The hone drive motor is turned off while still stroking. Stroking is stopped as the rotation of the hone stops. After rotation stops, the hone is collapsed and removed from the cylinder. The cylinder is examined to check the bore size and finish of the wall. If more honing is needed, the cylinder is again coated with honing oil and the cylinder is honed again. The finished cylinder should be within 0.0005 in. (0.013 mm) on both out-of-round and taper. After honing, the top edge of the cylinder is given a slight chamfer to allow the rings to enter the cylinder during assembly.

Cleaning the honed cylinder wall is an important part of the honing process. If any grit remains on the cylinder wall, it will rapidly wear the piston rings. This wear will cause premature failure of the reconditioning job. Degreasing and decarbonizing procedures will only remove the honing oil. They will *not* remove the abrasive. The *best* way to clean the honed cylinders is to scrub the cylinder wall with a brush using a mixture of soap or detergent in water. The block is scrubbed until it is absolutely clean. This can be determined by wiping the cylinder wall with a clean cloth. The cloth should pick up no soil when the cylinder wall is clean.

The cylinder walls *can* be cleaned with oil when the crankshaft has not been removed. Cleaning with oil takes more time and uses a great many more shop towels than scrubbing. When oil cleaning, the honed cylinder wall is given a heavy coat of motor oil. This oil is wiped off. Some of the honing grit will come off with the oil. Coating the wall with oil and wiping is repeated until the shop towel picks up no more soil. It will be necessary to put oil on the cylinder wall a dozen or more times to get each cylinder wall clean. The clean cylinder wall should be given a final coat of oil to protect it from rust until the block is recleaned for assembly.

17-3 Crankshaft Service

When required, the bearing journals on the crankshaft are reconditioned. In some cases, the crankshaft is rebalanced. Crankshafts may require straightening before grinding. The bearing journals of the used crankshaft should have less than 0.001 in. (0.025 mm) out-of-round or taper from one end to the other. The finish on the bearing journal should be no rougher than 16 μin. This can be checked by rubbing the edge of a penny across the journal. If the shaft picks up copper from the penny, the shaft is too rough. If the shaft is normal and bearings can be selected to give the proper operating clearances, the crankshaft can be reused without reconditioning.

Before assembly, the crankshaft journal surfaces should be examined for nicks similar to the ones pictured in Figure 17-25. If they exist, the raised portion of the nick can be removed with a fine hand stone so that it will not scratch the bearing. Nothing is done with the small pit or nick that remains in the bearing journal.

Crankshaft journals that have excessive out-of-round or taper should be reground. The crankshaft will also have to be reground if the journals are badly scored. The journal will have to be ground if bearings are not available to compensate for wear of the journal.

Both crankshaft ends are placed in rotating heads on one style of crankshaft grinder. The main bearing journals are ground on the centerline of the crankshaft. The crankshaft is then offset in the two rotating heads just enough to make the crankshaft main bearing journal center line rotate around the center line of the crank pin. The crankshaft will

Figure 17-25. Nicks on a crankshaft caused by the threads of the connecting rod bolt being allowed to hit the crankshaft during assembly.

then be rotating around the crank pin center line. The journal on the crank pin is reground in this position. The crankshaft must be repositioned for each different crank pin center.

In another type of crankshaft grinder, the crankshaft always turns on the main bearing center line. The grinding head is programmed to move in and out as the crankshaft turns to grind the crank pin bearing journals. The setup time is reduced when this type of grinder is used. Figure 17-26 shows a crankshaft being ground.

It takes a skilled machinist to operate a crankshaft grinder. The machinist must keep the grinding wheel properly dressed so that it will produce a smooth finish grind. The finished fillet radius must be the same size and shape as the original fillet. The journal is polished, using 320 grit polishing cloth and oil, to remove the fine metal *fuzz* remaining on the journal from grinding. The crankshaft is rotated in its normal direction as the polishing cloth pulls the fuzz away from the direction of rotation. This leaves a smooth shaft with the proper surface finish. The oil hole chamfer in the journal should be smoothed so that no sharp edge remains to cut the bearing. Finally, the crankshaft oil passages are thoroughly cleaned. The reground journals are coated with oil to keep them from rusting until they are to be cleaned for assembly.

Sometimes, it is desirable to salvage a

Figure 17-26. A crankshaft being ground.

crankshaft by building up a bearing journal, then grinding it to the original journal size. This is usually done by either electric arc welding or by metal spray. Sometimes the journal is chromeplated. Chromeplating makes an excellent bearing surface when the chrome is well bonded. If the bonding loosens, it will cause an immediate bearing failure.

It is desirable to balance an engine that is to be operated at high rpm. Balancing will also improve the durability of a low-speed engine. It is reported that 10 g of unbalance on a standard automotive crankshaft can cause an unbalance effect of 60 lb at 6000 rpm.

When the crankshaft is balanced, bob weights are selected to equal the total weight of the piston and rod assembly. The bob weights are installed on the crank throws. Externally balanced crankshafts must also have the flywheel and damper installed. The balancer spins the crankshaft, usually at low rpm. The unbalance readout is similar to that of modern wheel balancers. Counterweights that are too heavy are drilled to reduce their weight. Metal is welded to light counterweights to add the required weight for balance.

17-4 Camshaft Service

The production automotive camshaft is usually replaced if it has become worn beyond acceptable standards. Camshaft replacement may be less expensive than reconditioning. Camshafts with special grinds and heavy-duty camshafts are usually reconditioned. The cam bearing journals can be ground to a standard undersize in the same way in which crankshaft main bearing journals are ground. Worn cam lobes can also be ground undersize. When this is done, longer pushrods are required with hydraulic lifter valve trains to properly position the rocker arms. Engines with solid lifters can be adjusted to compensate for cam lobe grinding. An alternative means of reconditioning the cam lobes is to build them back up with weld or metal spray, then grind them to the original size and shape. The lobe must be polished to obtain the proper μin. finish.

Many engine rebuilders retreat the cam lobe with the original-type phosphate coating process. This forms a wear-resistant iron phosphate surface that duplicates the original camshaft finish.

REVIEW QUESTIONS

1. What part of the engine reconditioning cost is the same regardless of the amount of service done? [INTRODUCTION]
2. How much added cost would there be to engine reconditioning if the crankshaft were reground? [INTRODUCTION]

3. What two things must be considered when reconditioning a bearing journal? [INTRODUCTION]
4. How is the required abrasive grit size determined? [INTRODUCTION]
5. How does the grit size affect the surface smoothness? [INTRODUCTION]
6. What is done to recondition the piston and connecting rod as a standard commercial practice? [17-1]
7. What causes the piston ring groove clearance to widen in service? [17-1]
8. How is the piston ring groove clearance corrected? [17-1]
9. What two conditions will allow the piston to rock slightly in the groove? [17-1]
10. How are piston skirts resized? [17-1]
11. In what part of the cylinder is the fit of the resized piston checked? [17-1]
12. When should the piston be fit to the cylinder? [17-1]
13. What should be done when the piston pin has too much clearance in the piston? [17-1]
14. What is an interference fit of the piston pin? Why is this fit necessary? [17-1]
15. How is the piston pin installed in the piston and rod assembly? [17-1]
16. How is the large eye of the connecting rod reconditioned? [17-1]
17. What is the last piston and rod reconditioning operation? [17-1]
18. What determines the service operations that will be done on the block? [17-2]
19. Why are the main bearing bores aligned during reconditioning? [17-2]
20. How do align boring and align honing differ? [17-2]
21. Why is decking done on a block? [17-2]
22. What is the difference between commercial standards and precision standards of block decking? [17-2]
23. When should a cylinder be rebored? [17-2]
24. What determines the maximum size that the cylinder can be rebored? [17-2]
25. Why should the oversize pistons be on hand before reboring the cylinders? [17-2]
26. When is a cylinder reconditioned by sleeving? [17-2]
27. What are the procedures that must be followed to properly align the boring bar before boring a cylinder? [17-2]
28. How is the cylinder boring done? [17-2]
29. What is done to give the correct surface finish to the cylinder wall? [17-2]
30. When should the cylinders be honed? [17-2]
31. What does a sizing hone do that is different from a deglazing hone? [17-2]
32. What conditions are important on a honed cylinder wall? [17-2]
33. Describe the process of honing a cylinder. [17-2]
34. What is the best way to clean a cylinder after honing? [17-2]
35. When should a crankshaft be reground? [17-3]
36. What do you do when small nicks are found on the crankshaft bearing journals? [17-3]
37. Describe the crankshaft grinding procedure. [17-3]
38. What is done to recondition the camshaft? [17-4]

CHAPTER
18

Assembly For Operation

Engine performance, operating economy, and durability are all affected by the way the engine is assembled and prepared for operation. Before assembly is started, all the parts to be reused should have been reconditioned. The required new parts should also be on hand.

The subassembly parts can be reassembled right after reconditioning or the entire engine can be reassembled after all the parts have been reconditioned. Reassembled subassemblies should be protected until they are required for assembly in the engine so that they are not exposed to dust or moisture. Placing them in plastic bags will work well for this.

The fit and clearances of the parts are rechecked before they are reassembled. Parts that have relative motion between them are coated with oil or assembly lubricant to provide initial lubrication during the first engine start.

Some new assembly fasteners are often used. They must have at least the same strength as the original parts. Care must be taken to make sure that the seals and gaskets are positioned correctly to prevent leaks. The assembly bolts should be tightened in the proper sequence and they should be tightened to the specified torque. Experienced automotive service technicians first tighten the assembly bolts to about half-torque. Then they are retightened to the specified torque.

The reassembled engine is usually repainted so that it looks good. The paint also protects the engine from rusting. After the paint has dried, the engine accessories can be installed. Oil pressure should be built up in the engine before any attempt is made to start the engine.

18-1 Assembly Fasteners

Most of the assembly fasteners used on engines are cap screws. Cap screws are called *cap screws* when they are threaded into a casting. Automotive service technicians usually refer to these fasteners as bolts, regardless of how they are used. In this chapter, they are called bolts. Sometimes, studs are used for assembly fasteners. A *stud* is a short rod with threads on both ends. Often, a stud will have coarse threads on one end and fine threads on the other end. The end of the stud with coarse threads is screwed into the casting. A nut is used on the opposite end to hold the engine parts together. These fasteners are pictured in Figure 18-1.

The assembly fastener threads must match the threads in the casting or nut. The threads may be either customary or metric. First, the threads must be the same size. The size is measured across the outside of the threads. This is called the *crest* of the thread.

Customary threads are either coarse or fine. Coarse threads are called UNC (Unified National Coarse) and the fine threads are called UNF (Unified National Fine). Standard combinations of sizes and number of threads per inch (called pitch) are used. Pitch can be measured with a thread gauge (Figure 18-2). Customary threads are specified by

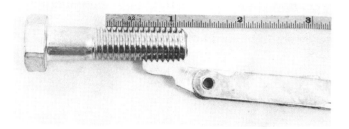

Figure 18-2. A thread gauge used to measure the pitch of the thread.

giving the diameter in fractions of an inch and the number of threads per inch. Typical UNC thread sizes would be 5/16-18 and 1/2-13. Similar UNF thread sizes would be 5/16-24 and 1/2-20.

Metric threads used in automotive engines are coarse threads. The size of a metric bolt is specified by the letter M followed by the diameter in millimeters across the outside (crest) of the threads. Typical metric sizes would be M8 and M12. Fine metric threads are specified by the thread diameter followed by an × and the pitch in millimeters (M8 × 1.5).

Both customary and metric threads have tolerance. A loose-fitting thread has a lot of tolerance. A very close fitting thread has very little tolerance. Threaded fasteners used in automotive engines work are designed with the correct tolerance. The threads of the right fastener will always fit properly.

One more important property must be considered when selecting assembly bolts. Bolts are made from many different types of steel. For this reason, some are stronger than others. Their strength is called the *grade* or classification of the bolt. The bolt heads are marked to indicate their grade strength. The customary bolts have lines on the head to indicate the grade, as shown in Figure 18-3. The actual grade of

Figure 18-1. Typical engine fasteners. A bolt and nut on the left, a cap screw in the middle, and a stud on the right.

Figure 18-3. Head markings indicating the grade of bolts with customary threads.

these bolts is two more than the number of lines on the bolt head. Metric bolts have a decimal number to indicate the grade. More lines or a higher grade number indicates a stronger bolt. In some cases, nuts and machine screws have similar grade markings. Figure 18-4 shows head markings for common metric bolts used on engines.

Figure 18-4. Common metric bolt and cap screw grade markings. The 10.9 is stronger than the 9.8 bolt.

Figure 18-5. Typical automotive washers.

When installing or replacing threaded fasteners, technicians must make sure that all of the following are correct:

1. Customary or metric threads
2. Diameter
3. Thread pitch
4. Bolt length
5. Length of the threads on the bolt
6. Grade of the bolt

Often, washers are used with an assembly bolt. Plain washers are used to spread the holding force over a wide area to reduce the chance of stress concentration that will lead to cracks. Lock washers are used to keep the bolt or nut from loosening. Two types of lock washers are in common use: the helical spring (split washer) and the tooth (star washer) lock washers. Conical spring washers are used where it is important to maintain a specific tension on the bolt. Typical automotive engine washers are shown in Figure 18-5.

On critical assemblies, such as connecting rod caps, any loosening will lead to destruction of the engine. Self-locking nuts are usually used in modern engines. Some older engines used cotter keys in the connecting rod nuts to prevent accidental loosening. In these older engines, the main bearing cap screws were often wired together to prevent the screws from loosening. When bolts or nuts use cotter

pins or wire to prevent accidental loosening, they are said to be safetied.

The connecting rod bolts tend to stretch and fatigue while in operation. A rod bolt broken from fatigue is pictured in Figure 18-6. As the nuts are reinstalled, they tend to lose their self-locking properties. Because of this, it is a good repair practice to replace all the connecting rod bolts and nuts with new ones when an engine is rebuilt.

Sometimes, threads in bolt holes of some of the engine castings are damaged. This is especially true of aluminum castings. The threads can be repaired in the same way they are repaired on heads. The procedure was discussed in Section 16-7.

Assembly bolts must be tightened to the proper torque. If the torque is too low, the joint will move and the bolt will loosen or break. If the torque is too great, the assembled parts will warp, the threads will strip out, or the bolt will break. In either case, leaks will develop. Figure 18-7 shows the normal assembly torque used on automotive engine fasteners.

The required clamping force is what determines

Figure 18-6. A fatigued connecting rod bolt that broke as it was being torqued during assembly.

Bolt torque

Size	Grade 5		Grade 8	
	in-lb	N·m	in-lb	N·m
1/4 - 20	95	10.7	125	14.1
1/4 - 28	95	10.7	150	16.9
5/16 - 18	200	22.6	270	31.2
	ft-lb	N·m	ft-lb	N·m
5/16 - 24	20	27.1	25	33.9
3/8 - 16	30	40.7	40	54.2
3/8 - 24	35	47.5	45	61.0
7/16 - 14	50	67.8	65	88.1
7/16 - 20	55	74.6	70	95.0
1/2 - 13	75	101.7	100	135.6
1/2 - 20	85	115.2	110	149.1
9/16 - 12	105	142.4	135	183.4
9/16 - 18	150	156.0	150	203.4

Figure 18-7. Normal assembly torque on grade-5 and grade-8 bolts and cap screws.

what fastener is to be used. Larger fasteners, in the same grade, can have greater tightening torques and, therefore, greater clamping forces. Higher-grade bolts are used where the space is limited and still high clamping forces are required. High-grade bolts are not used throughout the engine because they are expensive.

18-2 Subassemblies

During engine reconditioning, it is the usual practice to reassemble each of the small subassemblies after it has been serviced. These subassemblies include items such as the oil pump, coolant pump, piston and rods, heads, and block. Each reassembled subassembly is covered to protect it from dust and moisture until it is to be installed in the engine.

The following assembly discussion will cover the assembly of a typical internal combustion engine. The engine assembly procedure is similar for all engines but they are not identical. The specific assembly procedure given here may have to be modified for the engine being worked on.

Block. All of the parts are attached to the engine block, so it must be ready to accept them. In pushrod engines, the cam bearings are installed in the block. The best rule of thumb to follow is to replace the cam bearings whenever the main bearings are replaced. The replacement cam bearings

must have the correct outside diameter to fit snugly in the cam bearing bores of the block. They must have the correct oil holes. Cam bearings must also have the proper inside diameter to fit the camshaft bearing journals.

In many engines, each cam bearing is a different size. The largest is in the front and the smallest is in the rear. The cam bearing journal size must be checked and each bearing identified *before* assembly is begun. The location of each new cam bearing can be marked on the outside of the bearing with a felt-tip marker. This will help avoid mixing bearings. Marking in this way will not affect the bearing size or damage the bearing in any way.

A cam bearing installing tool is required to insert the new cam bearing without damage to the bearing. There are a number of tool manufacturers that design and sell cam bearing installing tools. The feature they have in common is a shoulder on a bushing that fits inside the cam bearing with a means of keeping the bearing aligned as it is installed. Figure 18-8 shows a camshaft bearing on the removing and installing tool. The bearing is placed on the bushing of the tool and rotated to properly align the oil hole. The bearing is then forced into the bearing bore of the block by either a pulling screw or a slide hammer. A pulling screw-type is illustrated in Figure 18-9. The installed bearing should be checked to make sure that it has the correct depth and the oil hole is indexed with the oil passage in the block. No additional service is required on the cam bearings that have been properly installed.

The opening at the back of the camshaft is closed

Figure 18-8. A cam bearing tool being used to remove a used cam bearing.

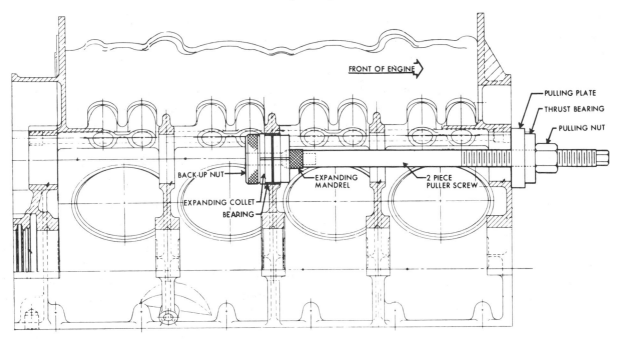

Figure 18-9. A screw-type puller being used to install a new cam bearing (Courtesy of Buick Motor Division, General Motors Corporation).

with an expansion plug (Figure 18-10). At the same time, expansion plugs (core plugs) are installed in all the core openings in the cooling system. The openings are wiped to make sure that no foreign object will keep the expansion plug from properly seating. A properly sized expansion plug is given a light coat of sealing compound around its outer edge and it is then placed in the opening. The expansion plug is forced into place using a properly fitting driver. Sometimes, a large socket can be used for

Figure 18-10. The opening in the back of the camshaft is closed with a cup-type expansion plug. Screw type plugs are installed in the oil passages.

this. Cup-type expansion plugs are driven flush with the casting surface. Convex plugs are pushed against their counterbore seats. They are then upset slightly to secure them in place. The excess sealing compound sheared from the expansion plug is wiped from the surface.

Any oil passage plugs that have been removed should be replaced. It is a good practice to give the oil passages a good final cleaning before the plugs are installed.

Some engine blocks use replaceable bushings to help support the distributor-oil pump drive. When required, these bushings should be replaced with new bushings before other parts are assembled on the block. The new bushings are reamed to size if reaming is required.

Oil Pump. When an engine is rebuilt, the oil pump is replaced with a new pump. This assures positive lubrication and long pump life. Sometimes, oil pumps are reconditioned. Reconditioning is only the replacement of the pump gears and a cover gasket, when used. The cover is removed to check the condition of the oil pump. The gears and housing are examined for scoring. If the gears and housing are heavily scored, the entire pump should be replaced. If they are lightly scored, the clearances in the pump should be measured. These clearances include the space between the gears and housing, the space between the teeth of the two gears, and the space between the side of the gear and the pump cover. Usually, a feeler gauge is used to make these measurements, as discussed in Section 16-6. Gauging plastic can be used to measure the space between the side of the gears and the cover. The oil pump should be replaced when excessive clearance or scoring is found.

The space between the oil pump gears is filled with assembly lubricant before the cover is put on the pump. This provides initial lubrication and it primes the pump so that it will draw the oil from the pan when the lubrication system is first operated.

Coolant Pump. The coolant pump is usually serviced as an assembly. This means that it is replaced with a new or rebuilt pump. The coolant pump is normally replaced when the engine is reconditioned. The pump bearings and seal both wear during use. These can be identified in Figure 18-11. If the coolant pump is reused, it is likely to have a short service life on the reconditioned engine. Coolant pump failure can cause overheating, which, in turn, could cause the reconditioned engine to fail prematurely.

If the pump is to be reconditioned, the parts must be obtained *first*. These may have to be special

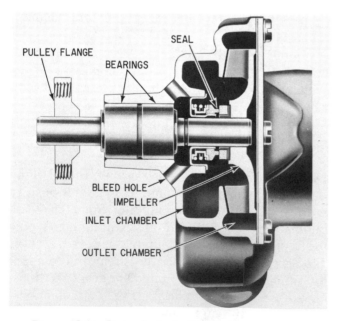

Figure 18-11. Parts of a typical coolant pump (Courtesy of Chevrolet Motor Division, General Motors Corporation).

ordered. The shaft is pushed from the pump. The pulley flange may need to be pulled from the shaft before it is removed from the pump. New bearings and seal are installed. It is a good practice to also install a new impeller. The parts are pressed together in the proper order. Extreme care must be taken with the seal to make sure that it is *absolutely* clean and that no damage occurs to the face of the seal. It should not even be touched because soil from your hands could cause a leak to start.

Timing Gears and Sprockets. The camshaft timing gear or sprocket and chain are replaced. The thrust plate is put on the shaft, then the timing gears and sprockets are pressed on the shaft in some engines. In other engines, they have a slip fit and they are held in place with bolts. Press-on gears and sprockets are installed on the shaft *before* the shaft is put in the block. Pressing requires proper holding fixtures. It is done with a special clamping fixture or on an arbor press. The gear or sprocket is aligned with the drive key. It is pressed on until it is fully seated.

Piston. The piston and connecting rod are reassembled right after the piston pin has been fit, as discussed in Section 17-1. Then the assembly alignment is checked and corrected if necessary. The piston should be rechecked to make sure that it properly fits the cylinder in which it is to operate. The fit can be checked by determining the difference in the measured size of the piston and cylinder. Usually, a service technician will measure the piston-to-cylinder clearance with a strip feeler gauge placed between the piston and the cylinder.

The cylinder and piston, without rings, are wiped thoroughly clean to remove any excess protective lubricant and dust that has accumulated on the surface. The strip feeler gauge is placed in the cylinder along the thrust side. The piston is inserted in the cylinder upside down with the piston thrust surface against the feeler gauge. The piston is held in the cylinder with the connecting rod as the feeler gauge strip is withdrawn. A moderate pull (from 5 to 10 pounds) on the feeler gauge indicates that the clearance is the same as the gauge thickness. A light pull indicates that the clearance is greater than the gauge thickness, while a heavy pull indicates the clearance is smaller than the gauge thickness. This measuring method is shown in Figure 18-12.

Figure 18-12. Measuring the clearance between the piston and cylinder wall with a strip feeler gauge. There are no rings on the piston when making this measurement.

When the clearance is too small, the size of a knurled piston can be reduced by lightly filing the knurled surface. If the clearance is too great, the piston will have to be either knurled deeper or it will have to be replaced with an oversize piston.

Rings. The bottom of the combustion chamber is sealed by the piston rings. They have to fit correctly in order to seal properly. Piston rings are checked for both side clearance and for gap as discussed in section 12-2.

If the ring gap is too large, the ring should be replaced with a ring having the next oversize diameter. If the ring gap is too small, the ring should be removed and the gap filed larger. This is done by fastening a file in a vise. While holding both ends of the ring against the file, the ring is drawn along the file, as shown in Figure 18-13. In this way, metal is removed to increase the gap while keeping the ring ends parallel to maintain a square gap. The edges of the ring are stoned after filing to

remove any sharp points that could scratch the piston or cylinder. The ring is frequently checked in the cylinder to bring the gap to the correct dimension.

Figure 18-13. Method used to file the but ends of a ring gap when the gap is too small.

307

Heads. Like the block, all openings that require expansion plugs should have them replaced. The replacement procedure of the expansion plugs in the head is the same as for those in the engine block openings. The valves are assembled as described in Section 12-3.

Shims are placed under the valve spring (Figure 18-14) as required, and the valve spring, retainer, and appropriate umbrellas are put in place and compressed with a valve spring compressor. Where used, an "O" ring seal is placed in the lower valve stem groove and then the split locks are set in place, as pictured in Figure 18-15.

Figure 18-14. A shim under the valve spring may be required to reset the correct valve spring length.

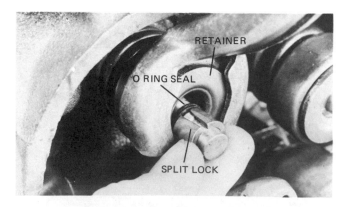

Figure 18-15. Using the correct assembly order the retainer is installed, the spring compressed, the "O" ring is placed in the lower groove, and the split locks are put in place.

18-3 Seals and Gaskets

In automotive engines, gaskets are used to seal the joints between adjacent parts. Oil seals are used to seal a joint where one part is turning and the other is stationary. Both are designed to keep liquids and gases in the engine and the dirt and moisture out. Gasket types have been discussed in Section 10-3 and seals in Section 10-4.

Gaskets and seals are packaged individually and in commonly used sets. Typical gasket sets would include all the gaskets and seals required for a specific engine repair job. Typical repair jobs are: valve grind, timing gear replacement, oil pan replacement, and complete overhaul. In addition, manufacturers often package different-quality gasket sets to fit one specific model engine. The gaskets in the less-expensive sets generally do not hold their shape or seal as long as those in the more expensive sets. The automobile service technician will have to decide which quality is to be used during the engine assembly.

Gaskets. Gasket sets are sometimes packaged to fit several different but similar model engines. When this is done, there will be several gaskets in the package for the same purpose. For example, there may be several different timing cover gaskets. The service technician will have to make sure that the proper gasket is selected for use on the engine being reconditioned. It is checked by putting the gasket in place to see that the exterior fits, all the holes are in the correct place, and the holes have the correct size and shape.

Sometimes, the proper gasket does not fit. It may be slightly small or slightly large. This is especially true of inexpensive gasket materials that have been stored for a long time. Gaskets that are too small should be placed in water until they soften and expand to the proper size. Gaskets that are too large should be placed in a warm, dry place until they shrink to size.

Some gaskets require sealers, whereas others do not. In some cases, the use of sealers on gaskets will cause an adverse reaction to the gasket so that the gasket will not properly seal. It is important to follow the sealing recommendations of the gasket manufacturer.

Sealers are often used to hold the gasket in place while the parts are being assembled. Nonhardening sealers will let the gasket slip as the parts are being tightened. This will usually result in fluid leakage. A general rule to follow when installing modern gaskets is to use no sealer if the parts can be assembled without it. Only hardening-type sealers should be used to hold the gasket in place for assembly. The sealer should only be put on one part, not on the gasket itself.

Following the assembly steps below will provide the longest gasket life:

1. The parts must fit properly before assembly. If covers such as rocker covers, timing covers, or oil pans have been distorted, they can be straightened using a hammer and a piece of flat bar stock (Figure 18-16).
2. All gasket surfaces must be clean and dry.
3. The gasket must align properly and fit all the mating openings.
4. The assembly bolts must be evenly tightened to the specific torque using the proper tightening sequence. They are tightened to half-torque, then retightened to the full torque. Too little torque will not compress the gasket enough to seal the joint. Excessive tightening will cause the gasket to either break or deform. This will allow the gasket to leak.

Starting in the late 1970s, *formed-in-place* (FIP) gasket materials came into more common use for some gasket applications. FIP materials lend themselves to high-volume assembly processes and they reduce the gasket inventory needed by the service technician. Most FIP materials used for gaskets are *room-temperature vulcanizing* (RTV) silicone materials. These materials cure when they are contacted by the moisture that is present in air. Curing takes place from the outside inward. Sometimes, *anaerobic* sealing materials are used for gaskets. These materials cure in the absence of air. Anaerobic materials are also used to seal bolt threads and to minimize the chance of loosening. FIP materials are used by spreading a ⅛ in. (3-mm) diameter bead on one of the parting surfaces. The bead is always placed on the inboard side of the attaching bolt holes, as shown in Figure 18-17. The formed-in-place RTV can be used in place of gaskets on most of the lower engine assemblies. The hot upper engine parts still require gaskets for sealing. Sometimes, RTV is recommended for use in the corner of the intake manifold gasket on V-type engines where uneven clamping pressure occurs.

It is a good practice to coat the assembly bolt threads with an antiseize thread compound, where thread sealing is not required. This allows good assembly torque measurements and prevents thread seizing so that the parts can be easily disassembled at a later time.

Seals. Seals are always used at the front and rear of the crankshaft. Overhead-cam engines may also have a seal at the front end of the camshaft and at the front of an auxiliary accessory shaft. Either a lip seal or rope seal is used in these locations. The rear

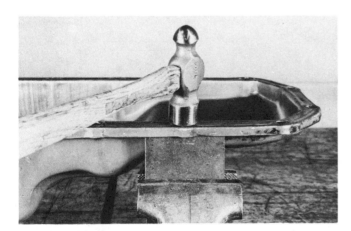

Figure 18-16. Flattening the gasket surface of an oil pan rail.

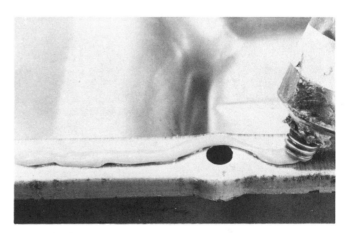

Figure 18-17. A bead of RTV placed on a gasket surface.

crankshaft oil seal is installed *after* the main bearings have been properly fit.

The lip seal may be molded in a steel case or it may be molded around a steel stiffener. The counterbore or guide that supports the seal must be thoroughly clean. In most cases, the back of lip seal is dry when it is installed. Occasionally, a manufacturer will recommend the use of sealants behind the seal. RTV may be used to seal the mating surface of the rear main bearing cap on the block. The engine service manual should be consulted for specific sealing instructions. The lip of the seal should be well lubricated before the shaft and cap are installed.

Rope-type seals (braided fabric seals) are most often used as rear crankshaft oil seals. Engines manufactured by Buick use rope-type seals at both front and rear of the crankshaft. Rope-type oil seals must be compressed tightly into the groove so that no oil can leak behind them. With the crankshaft removed, the upper half of the rope seal is put in a clean groove and compressed by rolling a round object against the seal to force it tightly into the groove. A piece of pipe, a large socket, even a hammer handle can be used for this, as shown in Figure 18-18. When the seal is fully seated in the groove, the ends that extend above are cut flush with the parting surface using a sharp single-edge razor blade (Figure 18-19) or a sharp tool specially designed to cut the seal. The same procedures are used to install the lower half of the rope seal in the rear main bearing cap or seal retainer. The use of a small strip of RTV is often recommended to help seal the main bearing cap and block parting surface joint as the cap is installed.

Figure 18-19. Trimming the rope-type seal to length.

The front rope-type oil seal packing is held in place with a retainer called a *shredder*. The old seal and shredder are driven from the timing cover case. The seal groove is cleaned and a new rope-type seal packing is fitted in place with the ends of the seal at the top. The shredder is installed and staked in place. A staked shredder is pictured in Figure 18-20. Staking is done by upsetting the timing cover metal over the edge of the seal in several places around the seal. The rope seal packing is sized for the damper hub by rolling. The same techniques are used to seat the front seal as was used to seat the rear rope-type seal.

Engines that have the oil pan rail extended below the crankshaft center line have an additional rear seal requirement. The small gap between the side main bearing cap or seal retainer and the block must be sealed. There is no way to tighten the parts

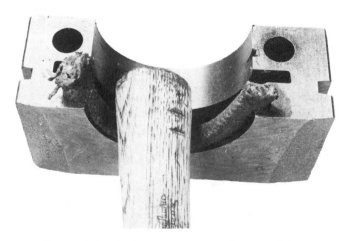

Figure 18-18. Rolling a rope-type seal in a main bearing cap.

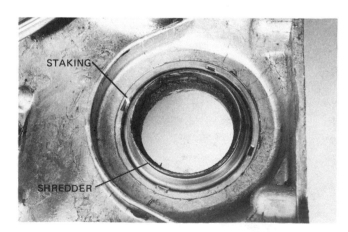

Figure 18-20. The shredder is staked in place to hold the rope-type seal in a timing cover.

to close this space, so it must be *filled* with some type of a seal. Most of these spaces are sealed with a swelling-type seal strip. The sealing strip is put in a groove in the cap or retainer after it has been torqued in place. The sealing strip is first soaked several minutes in petroleum solvent, then inserted in a groove, as shown in Figure 18-21. It may be necessary to use a blunt tool to firmly seat the sealing strip in the groove. The solvent and engine oil cause the sealing strip to swell in place to seal the gap. Sometimes, the seal will seep slightly when the engine is first run. The seepage will stop after a little use.

FIP gasket materials have found application as rear main bearing cap and retainer side seals. They

Figure 18-21. A sealing strip being placed in the side groove of the main bearing cap.

are used in place of the expanding sealing strips. A special installing tool enables the service technician to squeeze the FIP material in the gap until the space is full. The joint is properly sealed when the FIP gasket material cures.

In the following engine assembly procedure, it is assumed that the gasket and seal installing procedures discussed in this section will be followed. They should be modified when other specific directions are given in the engine service manual.

18-4 Engine Assembly

All parts of the engine must be thoroughly clean before they are assembled. It is necessary to reclean surface soil and protective lubricant from the recon-ditioned and new parts going into the assembly. Do not overlook the oil passages. A *thin* coating of assembly lubricant, such as white grease, should be put on all bearing journals as the shafts are installed. Cam lobes may require special lubricants to provide proper lubrication during startup. These will be specified in the service manual for the engine. Cylinders and pistons are coated with motor oil. Oil or grease should not be put on any of the gasket parting surfaces.

The engine is assembled from the inside out. This method allows the technician to support the inner parts as they are assembled. Checks are made during assembly to assure correct fits and proper assembly of the parts. When this is done, the reconditioned engine should be able to be started on the vehicle battery without the use of any booster battery.

The main bearings are properly fit *before* the crankshaft is lubricated or turned. The oil clearance of both main and connecting rod bearings is set by *selectively fitting* the bearings. In this way, the oil clearance can be adjusted within 0.0005 in. of the desired clearance.

Standard bearings are available in 0.001, 0.002, and 0.003 in. undersize to compensate for worn bearing journals. Bearings are also made in 0.010 and 0.012 in. undersize for use on reground journals.

The crankshaft bearing journals should be measured with a micrometer to select the required bearing size. Remember that the main bearings caps will only fit one location and the cap must be positioned correctly. The correct-size bearings should be placed in the block and cap, making sure that the bearing tang locks into its slot. The upper main bearing has an oil feed hole. Carefully rest the clean crankshaft in the block on the upper main bearings. Lower it squarely so that it does not damage the thrust bearing, as shown in Figure 18-22. Place a strip of gauging plastic on each main bearing journal. Install the main bearing caps and tighten the bolts. Remove the cap and check the width of the gauging plastic with the markings on the gauge envelope, as shown in Figure 18-23. This will indicate the oil clearance. If the shaft is out-of-round, the oil clearance should be checked at the point that has the *least* oil clearance. The oil clearance can be reduced 0.001 in. by replacing both bearing shells

Figure 18-22. The crankshaft being carefully lowered in place.

Figure 18-23. Checking the width of the plastic gauging strip to determine the oil clearance of the main bearing.

with bearing shells 0.001 in. undersize. The clearance can be reduced by 0.0005 in. by replacing only one of the bearing shells with a 0.001 in. smaller bearing shell. This smaller bearing shell should be placed in the engine block side of the bearing (the upper shell). Oil clearance can be adjusted accurately using this procedure. Never mismatch the bearing shells by more than 0.001 in. difference in size. Oil clearances normally run from 0.0005 to 0.002 inches.

The crankshaft is removed once the correct oil clearance has been established. The rear oil seal is installed in the block and cap, then the crankshaft journals are lubricated with assembly lubricant. The crankshaft is then installed following the procedure discussed in Section 12-1.

Cam Timing. On pushrod-type engines, the cam timing should be engaged next as described in Section 12-1. The backlash between the timing gears should be checked. This can be done by fitting the proper size feeler gauge between the engaged teeth of the two gears (Figure 18-24). The backlash can also be checked with a dial gauge as one of the gears is rotated forward and backward (Figure 18-25). When used, the chain damper must be properly positioned.

In precision engine service, it is important to know the exact position of the cam. This can be checked using a degree wheel on the crankshaft and a dial gauge against a lifter that is temporarily installed in the intake lifter bore of number one

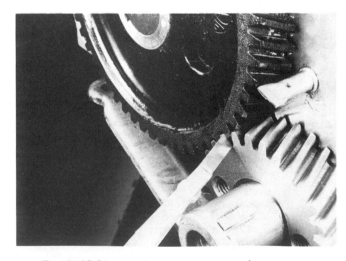

Figure 18-24. A feeler gauge being used to measure the clearance between the timing gears.

312

Figure 18-25. Checking the backlash on timing gears with a dial gauge.

cylinder, as pictured in Figure 18-26. The centerline or point of maximum lift is always one half of the valve duration. Rotate the crankshaft until the dial gauge indicates maximum lift, then note the reading. Rotate the crankshaft backward until the dial gauge reads over 0.100 in. Now slowly and carefully rotate the crankshaft forward. The lifter will be raising. Stop rotating the crankshaft when the dial gauge reads exactly 0.025 in. less than the maximum lift. Note the degree wheel position at this point. Again, carefully rotate the crankshaft in the forward direction, going past maximum lift. Stop rotation as the dial gauge again reads exactly 0.025 in. as the lifter is going down. Note the degree wheel position at this point. The center of the cam is halfway between the two degree wheel positions. Offset cam keys, eccentric bushings for the cam bolt holes, and other devices are available to properly center the camshaft. The degree wheel reading when the camshaft is centered should be exactly in the middle of the cam duration shown on the engine specifications.

Timing Cover. The timing cover of a pushrod engine is installed as described in Section 12-1. Most timing covers are installed with a gasket, but some use RTV sealer in place of the gasket. In either case, it is very important that any oil passages between the cover and the block are properly sealed. An air leak between the oil screen in the block and the inlet passage going to an oil pump in the timing cover will keep the oil pump from building up oil pressure. Air will be pulled into the pump, rather than pulling oil from the oil pan. Leaks in the pressure passage from the oil pump will reduce the amount of oil available to the lubricating system. If the leak is external, it will show up on the outside of the reconditioned engines.

Figure 18-26. The exact center of the cam is being determined with a dial gauge and a degree wheel.

313

The gasket or RTV sealant is put on the gasket surface of the timing cover. A light coat of assembly lubricant is wiped on the oil seal and damper hub for initial lubrication. The timing cover is put in place and the assembly bolts are installed loosely until after the vibration dampener has been installed. The vibration damper is also called a harmonic balancer or a pulley hub. The hub of the vibration dampener will properly position the timing cover by centering the front oil seal. The damper is placed on the crankshaft, making sure that the drive key is aligned with the slot in the hub. After the hub is on the crankshaft, the timing cover assembly bolts can be tightened to the specified torque.

Vibration dampeners are seated in place by one of three methods. First, the dampener hub of some engines is pulled into place using the hub attaching bolt (Figure 18-27). The second method uses a special installing tool that screws into the attaching bolt hole to pull the hub into place. The tool is removed and the attaching bolt installed and torqued. The last method is used on engines that have no attaching bolt. These hubs depend on a press fit to hold the hub on the crankshaft. The hub is seated using a hammer and a special tube-type driver.

The crankshaft and camshaft rotating force of the partial engine assembly should be rechecked. The assembly of the shafts should rotate freely. Only the drag of the seals should be noted. The rotating torque should be less than five ft lbs when measured with a torque wrench.

Piston and Rod Assemblies. The piston and rod were reassembled as the pin was fit. They should be cleaned and installed as described in Section 12-2. Adjustment of the connecting rod oil clearance follows the same procedure used to adjust the oil clearance of the main bearings. The rotating torque of the crankshaft with all connecting rod cap bolts fully torqued should be *no more than* 20 ft lb for a four-cylinder engine or *no more than* 30 ft lb for an eight-cylinder engine.

The connecting rods should be checked to make sure that they still have the correct side clearance. This is measured by fitting the correct thickness feeler gauge between the connecting rod and the crankshaft cheek of the bearing journal (Figure 18-28). A dial gauge can also be set up to measure the connecting rod side clearance.

Top dead center of the number 1 piston can be accurately located after the pistons have been installed. Attach a dial gauge so that it can measure the movement of the number 1 piston as it goes past top center. A degree wheel is attached to the crankshaft. The crankshaft is rotated forward until the dial gauge reaches some selected value less than top center. The position of the degree wheel is noted at this point. The crankshaft is again carefully rotated forward until the dial gauge again reaches the same selected value it had before. Note the new position of the degree wheel. Top center is exactly in the middle of the two degree wheel positions (Figure 18-29).

Figure 18-27. The attaching bolt in the center of a vibration damper.

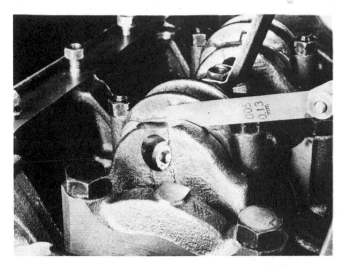

Figure 18-28. The connecting rod side clearance is measured with a feeler gauge.

Oil Pan. The oil pump and pan are installed as described in Section 12-2. Make sure it is attached with the proper gaskets or seals so there will be no air leak that would keep the oil from being pulled into the oil pump or oil leaks from the oil pump.

The oil pan gaskets and seals should be properly positioned on the block. Some manufacturers recommend putting a small bead of RTV sealer in difficult-to-seal corners of the oil pan gasket. The pan is carefully installed so the gaskets are not moved. The attaching bolts are all started in their threads, then they are tightened to the correct torque. Avoid excessive tightening that could damage the gasket and cause an oil leak.

Head. The engine block can be turned upright to complete the assembly. On some overhead cam

Figure 18-29. A degree wheel used to locate top dead center of number one piston.

engines, it may be necessary to put the cam in the head before the head is installed on the block. It may be necessary to install the valve lifters before the head is installed. The head installation was discussed in Section 12-4.

Overhead Camshaft. The assembly of overhead camshafts were discussed in Section 12-2.

The cam sprocket is attached. It is necessary to properly align the timing marks on both crankshaft and camshaft sprockets and engage the sprocket teeth in the chain before the cam sprocket is attached to the camshaft. The belt-type sprocket is attached to the cam before engaging the timing belt. The crankshaft sprocket and the camshaft sprocket timing marks are aligned; then the timing belt is positioned in the slots of the sprockets. The timing belt tension is adjusted to properly tighten the timing belts.

Overhead valve drive trains that use hydraulic pivots require no further service. Those with mechanical adjustments must have the lash checked following the same procedure described for pushrod-type engines.

Rocker Covers and Intake Manifolds. The assembly of rocker covers and intake manifolds were discussed in Section 12-6.

Water Pump. A reconditioned, rebuilt, or new water pump should be used. Gaskets are fitted in place. Sealers are used where they are recommended by either the gasket manufacturer or the engine manufacturer. The pump is secured with assembly bolts tightened to the correct torque.

A new thermostat is usually installed at this time. It is put in place, taking care to place the correct side of the thermostat toward the engine. The thermostat gasket is put in place. Sealers are used on the gasket where they are required. The thermostat housing is installed and the retaining bolts are tightened to the proper torque.

Engine Painting. Engine painting is optional. It does not improve the engine's performance in any way, but can make an engine look reconditioned. If the engine is to be painted, this point in the engine assembly is the best time to paint it.

Standard engine paints with original engine colors are usually available at the parts stores. Engine paints should be used rather than other types of paints. Engine paints are compounded to stay on the metal as the engine temperatures change. Normal engine fluids will not remove them. These paints are usually purchased in pressure cans so that they can be sprayed from the can directly on the engine.

All parts that should not be painted must be covered before spraying painting. This can be done with old parts, such as old spark plugs and old gaskets. They can also be covered by taping paper over the areas. If the intake manifold of an inline engine is to be painted, it can be painted separately. Engine assembly can continue after the paint has dried.

REVIEW QUESTIONS

1. Name the three common types of fasteners used on engines. [18-1]
2. How are the customary thread sizes of fasteners specified? [18-1]
3. How are metric thread sizes of fasteners specified? [18-1]
4. What is meant by the grade of a bolt? [18-1]
5. How do the bolt head grade markings differ between customary bolts and metric bolts? [18-1]
6. What information is necessary to correctly specify a bolt? [18-1]
7. What is meant when a fastener is safetied? [18-1]
8. Why is it a good practice to install new rod bolts and nuts? [18-1]
9. Why is it important to tighten the fasteners to the specified torque? [18-1]
10. When should cam bearings be replaced? [18-2]
11. How are cam bearings installed? [18-2]
12. Where are expansion plugs used in the engine block? [18-2]

13. What is done to recondition an oil pump? [18-2]
14. How is the coolant pump serviced? [18-2]
15. What can be done if the ring gap is too large or too small? [18-2]
16. What is done to the ring gap after filing? [18-2]
17. Describe the valve assembly procedure. [18-2]
18. How is an "O" ring seal properly placed in the lower valve groove? [18-2]
19. What is the basis for selecting gaskets for an engine? [18-3]
20. How can you make composition gaskets shrink or expand slightly for a good fit? [18-3]
21. When are sealers used on gaskets? How should they be used? [18-3]
22. What are the gasket assembly steps? [18-3]
23. What are the advantages of FIP gasket materials? [18-3]
24. When is the rear main bearing seal installed? [18-3]
25. Why is the back of a seal installed dry and the lip oiled? [18-3]
26. How is a braided fabric rope-type oil seal installed? [18-3]
27. How is the side of the rear main cap and retainer sealed on an engine with an extended skirt block? [18-3]
28. What is used to lubricate the parts as the engine is assembled? [18-4]
29. How are the main bearings selectively fit? [18-4]
30. How is the oil clearance checked with gauging plastic? [18-4]
31. How is the center of the cam lift determined? [18-4]
32. Why should the damper be installed before the timing cover bolts are tightened? [18-4]
33. How are vibration dampers seated in place on the front of the crankshaft? [18-4]
34. What precautions should be taken when installing an intake manifold on a V-type engine? [18-4]
35. When should the exterior of an engine be painted? [18-4]

CHAPTER
19

Engine Installation and Repair

Major engine service is required when an engine is worn out or there has been premature failure of the lower end. The engine can be reconditioned or it can be replaced with a used engine, a remanufactured short or long block, or a remanufactured engine. It could even be replaced with a new engine. These options have been discussed in Section 13-3. No matter which option is selected, accessories will have to be put on the engine and it will have to be put in the automobile. The entire engine installation will have to be thoroughly checked to make sure that it is in condition to give the customer dependable operation for a long time.

19-1 The Power Package

All of the operating accessories have to be reinstalled on the engine. They have to be adjusted so that the engine will operate correctly. Some of the accessories can be checked as they are assembled, whereas others will have to be checked after the engine is running.

Oil Pressure. This is a good time in the engine assembly process to establish engine oil pressure. If there is an oil pressure problem, it can be easily corrected before the engine is put in the chassis. Put an oil pressure test gauge in the oil pressure tap of the engine. Install a new oil filter and put in at least 3 quarts of fresh motor oil. Use the same type and grade of oil that will be used in the engine. The oil pump is rotated with an adapter, as described in Section 12-7, until oil pressure is established. The test oil pressure gauge is then removed. The rest of the engine assembly and installation can be completed without concern about having to disassemble part of the engine for low oil pressure.

Transmission—Standard. If the engine was removed with the transmission attached, the transmission should be reinstalled on the engine before other accessories are added. The flywheel is installed on the back of the crankshaft. Often, the attaching bolt holes are unevenly spaced so that the flywheel will fit in only one way to maintain engine balance. The pilot bearing or bushing in the rear of the crankshaft is usually replaced with a new one. This minimizes the possibility of premature failure of this part.

The bell housing is installed next if it has an opening in the bottom through which the clutch can be installed (Figure 19-1). The bell housing alignment is checked by mounting a dial gauge on the flywheel. The anvil of the dial gauge is placed on the transmission opening of the bell housing. The crankshaft is rotated and the dial gauge observed to determine the indicated runout. It is then placed on the transmission mounting flange and the runout is rechecked. The runout can be corrected using offset alignment dowels and shims between the engine block and bell housing.

The clutch is installed next. The parts of a typical clutch are identified in Figure 19-2. Usually,

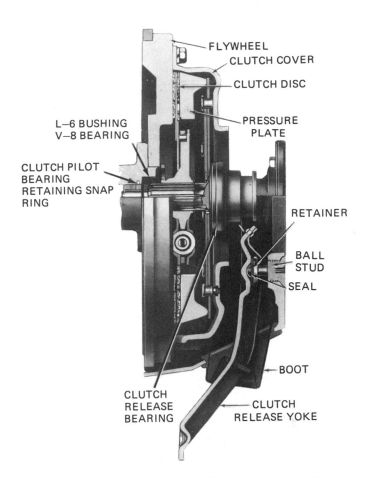

Figure 19-2. Identification of the parts of a typical clutch.

a new clutch is used; at least a new clutch friction disc is installed. The clutch friction disc must be held in position using an alignment tool (sometimes called a dummy shaft) that is secured in the pilot bearing. This holds the disc in position while the pressure plate is being installed. Finally, the engine bell housing is put on the engine, if it was not installed before. A closed bell housing is shown in Figure 19-3. The alignment of this type of bell housing is then checked.

Figure 19-1. A bell housing designed so the clutch can be installed through the bottom.

Figure 19-3. A closed-type bell housing that must be removed to install the clutch.

319

The clutch release yoke should be checked for free movement. Usually, the clutch release bearing is replaced with a new one, making sure that it is securely attached to the clutch release yoke. The transmission can then be installed.

The transmission clutch shaft must be guided straight into the clutch disc and pilot bearing. Often, two guide pins are made that will screw into two of the transmission mounting bolt holes. The transmission clutch shaft is rotated, as required, to engage in the splines of the clutch disc. The clutch shaft splines are shown in Figure 19-4. The assembly bolts are secured when the transmission fully mates with the bell housing.

Transmission—Automatic. The drive plate is attached to the back of the crankshaft, as shown in Figure 19-5. Its assembly bolts are tightened to the specified torque. The bell housing is part of the transmission case on most automatic transmissions. Usually, the torque converter (Figure 19-6) will be installed on the transmission before the transmission is put on the engine. The torque converter should be rotated as it is pushed onto the transmission shafts until the splines of all the shafts are engaged in the torque converter. Typical automatic transmission shaft splines are pictured in Figure 19-7. The torque converter is held against the transmission as the transmission is fitted on the back of the engine (Figure 19-8). The transmission mounting bolts are attached finger-tight. The torque converter should be rotated to make sure there is no binding. The bell housing is secured to the block, then the torque converter is fastened to the

Figure 19-5. A typical automatic transmission drive plate.

drive plate. The engine should be rotated. Any binding should be corrected before any further assembly is done.

Starter. It is generally easier to install the starter before the engine is put in the chassis. The starter should be checked to make sure that the starter drive pinion does not bind on the ring gear. Shims can be installed between the starter mounting pad

Figure 19-4. The clutch splines on a transmission shaft.

Figure 19-6. A typical automatic transmission torque converter.

Figure 19-7. Shaft splines on the front of an automatic transmission.

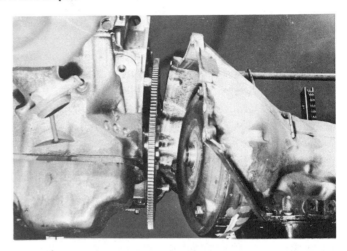

Figure 19-8. The automatic transmission, with the torque converter installed, is attached to the back of the engine.

and the starter to adjust the pinion to ring gear clearance on the GM-type mounting. The starter mounting bolts are then tightened to the specified torque.

Front Accessories. All of the belt-driven engine accessories are mounted on the front of the engine. Some engines drive all of these accessories with one belt. Other engines may use as many as four belts. The service manual should be checked to determine the specific belt routing for the accessories used on the engine being built up. On some engines it is more convenient to install the front accessories before the engine is installed; on other engines, it is easier to put the engine in the chassis before installing the front accessories.

This is a good time to install the secondary ignition cables that connect the distributor cap to the spark plugs. The cables are threaded through their supports. The service manual should be checked for the proper routing. Cross fire can occur if the spark plug cables are misrouted. The cables should not be either too long or too short. Long cables could contact and wear on parts of the engine, causing an electrical ground. Short cables will stretch and break, causing an electrical open. It may be desirable to install new spark plugs at this time because it is easy to reach them. Leave the spark plug out of the number 1 cylinder so that the compression stroke of this cylinder can be identified for setting the initial timing of the distributor.

Engine Installation. A sling, either a chain or lift cable, is attached to the manifold or head bolts on the top of the engine. A hoist is attached to the sling and snugged up to take the weight and to make sure the engine is supported and balanced properly. A typical sling can be seen holding the engine in Figure 19-5.

The engine is then lifted over the radiator support. It must be tipped as it was during removal to let the transmission go into the engine compartment first. The transmission is worked under the floor pan as the engine is lowered into the engine compartment. The front engine mounts are aligned; then the rear cross member and rear engine mount are installed. The engine mount bolts are installed and the nuts are torqued. Then the hoist is removed. Controls are connected to the transmission under the automobile. This is also a good time to connect the electrical cables and wires to the starter. The exhaust system is then attached to the exhaust manifolds. If any of the steering linkage was disconnected, it can be reattached while working under the vehicle. After the engine is in place, the front engine accessories can all be installed, if they were not installed before the engine was put in the chassis. The air conditioning compressor is reattached to the engine, being careful to avoid damaging the air conditioning hoses and lines.

Cooling System. The radiator is installed and secured in place. This is followed by the cooling fan and shroud shown in Figure 19-9. The fan and new drive belts are installed and adjusted. New cooling hoses, including new heater hoses, are installed. Coolant, a 50:50 mixture of antifreeze and water, is put in the cooling system after making sure that the radiator petcock (Figure 19-10) is closed and the block drain plugs (Figure 19-11) are in place.

Carburetor and Emission Controls. The carburetor is installed with a new gasket. If all lines, hoses, and wires were marked as recommended in Section 13-4, they will be easy to reattach. If they are not marked, the engine emission decal and service manual will have to be followed to properly attach them. The fuel line, fuel pump, fuel separator, and a new fuel filter should be connected to the carburetor. The hoses to the carbon canister should be attached. Follow through the EGR system to make sure that all the hoses are connected properly. Posi-

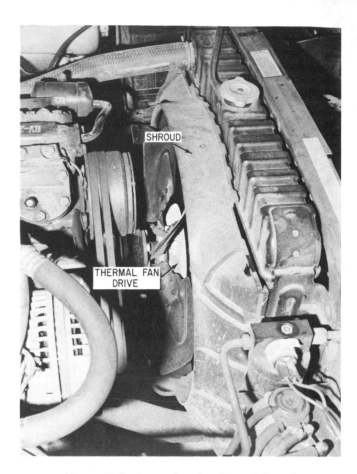

Figure 19-9. A typical cooling fan and shroud.

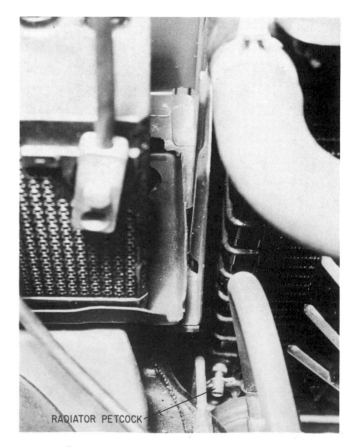

Figure 19-10. A typical radiator petcock drain.

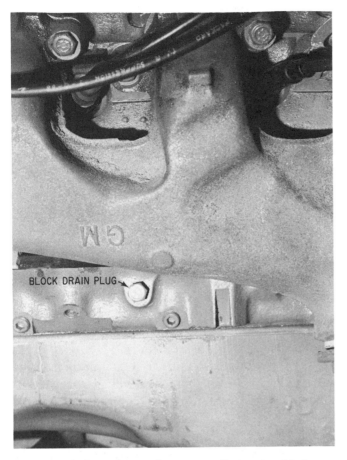

Figure 19-11. A typical engine cooling system block drain plug.

tion and attach the hoses and emission control devices in the distributor vacuum advance system. This is one of the major emission control systems. Attach the carburetor controls, including the choke, throttle, transmission TV linkage, and speed control. Figure 19-12 shows the top of a typical engine assembly with the air cleaner tipped for visibility.

Figure 19-12. The top of a typical V-8 engine assembly with the air cleaner lifted to show the engine lines, hoses, and controls.

Electrical System. The wires are connected to the alternator or generator. If the carburetor has an electrical solenoid, it should be connected. Connect the instrument wires to the electrical sending units on the engine. This includes a cooling temperature indicator and an oil pressure indicator. Make sure the ignition primary wiring is not contacting any metal. Recheck all wiring to make sure it is all connected properly, then install a fully charged battery. Secure the battery hold-down clamps. Clean the surfaces of the battery terminals and cables. Attach the positive cable first, then the ground cable. Check to make sure the starter will crank the engine. Install and time the distributor, then connect the ignition cables.

19-2 Engine Break-In

The engine oil and coolant levels are checked. Both oil and coolant should be brought up to the normally full level. The level of the oil should be checked in the standard transmission. If the automatic transmission fluid has been drained, it must be filled with about 7 quarts of the proper type ATF fluid before the engine is started. Additional ATF fluid will have to be added to bring the fluid level up to the full mark after the engine has been run and the transmission has been shifted through all the gears.

The engine installation should be given one last inspection to see that everything has been put together correctly before the engine is started. If the engine overhaul and installation are done properly, the engine should crank and start on its own fully charged battery without the use of a fast charger or jumper battery. As soon as the engine starts and shows oil pressure, it should be brought up to a fast idle speed and *kept there.* This is necessary to make sure that the engine gets proper lubrication. The fast-running oil pump develops full pressure and the fast-turning crankshaft throws plenty oil on the cam and cylinder walls.

Just as soon as you can tell that there are no serious leaks and the engine is running reasonably well, the automobile should be driven to a road having minimum traffic. Here, the automobile should be accelerated, full throttle, from 30 to 50 mph (48 to 80 km/h). Then the throttle is fully closed while the automobile is allowed to return to 30 mph (48 km/h). This sequence is repeated from 10 to 12 times. The acceleration sequence puts a high load on the piston rings to properly seat them against the cylinder walls. The piston rings are the only part of the modern engine that need to be broken in. Good ring seating is indicated by a dry coating inside the tailpipe at the completion of the ring seating drive.

The automobile is returned to the service area, where the basic ignition timing is set and the idle speed is properly adjusted. The engine is again checked for visible fluid leaks. If the engine is dry, it is ready to be turned over to the customer.

The customer should be instructed to drive the automobile in a normal fashion, neither babying it at slow speeds nor beating it at high speeds for the first 100 miles (160 km). The oil and filter should be changed at 500 miles (800 km) to remove any dirt that may have been trapped in the engine during assembly and to remove the material that has worn from the surfaces during the break-in period.

A well-designed engine that has been correctly reconditioned and assembled using the techniques described should give reliable and durable service for many miles. Premature failure will result from problems that have been overlooked or from operational abuse.

19-3 In-The-Car Repairs

Occasionally, some parts of the engine will fail prematurely while the rest of the engine is in good condition. These parts can often be repaired with minimum engine disassembly. Details of the repairs have been described when discussing the complete engine reconditioning. It will not be necessary to discuss them again in this section. Only the heads need to be removed to do a valve job. The pan will also have to be removed to replace the piston rings and connecting rod bearings. The main bearings can also be replaced with the pan off and the crankshaft in place. It is even possible to replace the camshaft without removing the engine. The engine would have to be pulled from the chassis to remove the crankshaft.

The battery should be disconnected any time repairs are being made on the engine. The battery ground cable is always removed first. The battery should be removed from the vehicle while repairs are being made on the front of the engine. This will give the service technician more working room. There will be less chance of damage to the battery or from electrical grounds.

Valve Seals and Springs. Occasionally, the valve seals leak. This would cause excessive oil consumption and a smoky exhaust. The valve seals are under the valve springs, but they can be replaced without removing the heads on most engines. Valve springs sometimes break. They can be replaced using the same technique used to replace valve seals.

An air-pressure fitting that fits the spark plug opening is required. This can be made by combining an air line fitting and an old spark plug shell. These can be brazed together. Air pressure is put in the cylinder to push on the valve head, holding the valve closed. A valve head 1.5 in. in diameter would have a closing force of 176 lb when 100 lb/in.^{2} of air pressure is put in the cylinder. Both valves must be closed when using this procedure.

The rocker covers and the rocker arms must be removed. A special lever-type valve spring compressing tool is required to compress the valve springs. One end of the tool fits under the rocker arm pivot or it will have a special adapter to connect to the cylinder head. With air pressure in the cylinder, the lever is moved to compress the valve spring. The valve locks, retainer, and spring are removed. The valve seal can be replaced while air pressure remains in the cylinder to hold the valve closed. The spring, retainer, and valve locks are then replaced and the air pressure is released. The rocker arm and pushrod assemblies can then be put back on the engine. The rocker arms are adjusted, when required, and the rocker covers are installed with new gaskets.

Front Oil Seal. The front oil seal will have to be replaced when it leaks. It can be removed from outside some engines. On most engines, the timing cover must be removed to replace the front oil seal. In either case, the battery, fan, shroud, alternator, power steering pump, air conditioning compressor, hoses, and water pump must be removed to have space to work on the engine. The coolant will have to be drained for this. It is a good practice to at least cover the radiator with a fender cover to protect it while working on the front of the engine. It would be even better to remove the radiator. It may still be necessary to cover the fins of the air conditioner condenser.

The vibration dampener must be removed. An impact wrench works well to remove the damper attaching bolt. If an impact wrench is not available, some means must be found to hold the crankshaft as the dampener attaching bolt is loosened. It is often possible to hold it through the starter opening in the bell housing or by removing the flywheel inspection plate at the front of the torque converter or under the clutch. The damper puller is positioned on the damper and the puller bolts are attached. The damper is then pulled from the engine.

Some front seals can be replaced with no further disassembly of the engine. These seals may require special removing and installing tools. The service manual will have to be consulted. The timing cover must be removed on the rest of the engine to replace the front oil seal.

The timing cover bolts are removed. The timing cover is then carefully loosened from the engine. This is important because the lower edge of the timing cover is sealed at the front of the oil pan. It may be necessary to *loosen* all the pan bolts to free the lower edge of the timing cover. A sharp knife should be slid between the block and the timing cover to cut the front of the oil pan side rail gasket at the front edge of the block. The timing cover is carefully lifted from the engine to avoid dropping dirt or parts of the gasket into the front of the oil pan. This opening on the front of the pan should be closed with a shop towel. The old gasket material is scraped from the block. Make sure that the pan side rail gasket is cut flush with the front of the block. The front oil seal is removed from the timing cover. All parts are thoroughly cleaned.

The timing cover gasket set has a new oil seal and all the required gaskets, including the small front section of the oil pan side rail gasket. These will have to be trimmed to fit the space on the pan that mates with the timing cover. The shop towel is removed from the pan opening. The new seal is placed in the timing cover. The seal lip is coated with assembly lubricant. The new front oil pan gasket pieces are fitted and the timing cover gasket is put in place. A bead of RTV silicone sealer is placed across the corner between the oil pan side rail gasket and the timing cover gasket to seal the joint. The timing cover is forced downward against the oil pan until it fits the aligning dowels on the front of the block. Sealant is put on the threads of the timing cover bolts that go into the cooling system. The timing cover assembly and front oil pan bolts are screwed

in finger-tight. The vibration dampener is installed, then the timing cover and oil pan bolts are torqued.

The rest of the parts are replaced on the engine. Belts and hoses should be replaced with new ones if they show any sign of wear. It is a good practice to replace the cooling system thermostat at this time because it is easy to get to it. The belts are properly tensioned and the coolant replaced. The cooling system should be topped off after the engine has become fully warmed up.

Timing Gears. Occasionally, the timing gears or chain will slip or break. Figure 19-13 shows a timing chain and sprocket that failed. The belt that drives an overhead camshaft can be replaced after removing an external cover on the front of the engine. Pushrod-type engines must have the timing cover removed to replace the timing gear. The timing cover is removed in the same way just described for replacing the front oil seal. The timing gears or chain are removed and replaced following the procedures described in Section 18-4. Some engines with timing gears must have the camshaft removed to install the new timing gear on the camshaft. The radiator and grill will have to be removed to slide the camshaft out.

Small particles that break off the timing gears and sprockets will fall to the bottom of the oil pan. It is recommended that the oil pan be removed to remove the pieces from the engine. Some technicians only change the oil and filter. They claim that the oil pump pickup screen will keep large pieces out of the oil system and the filter will remove the small particles.

Figure 19-13. Timing gear teeth have worn down. This allowed the timing chain to slip.

The engine is reassembled following the same procedures described for the replacement of the front oil seal.

Camshaft. The camshaft lobes may wear down prematurely before the rest of the engine is worn out. The camshaft can be replaced without removing the engine from the vehicle. This is easy on an overhead camshaft engine because the camshaft is right on the top of the engine. In most cases, the head will not have to be removed. The camshaft on overhead cam engines is removed using the same tools used for assembly described in Sections 18-2 and 18-3. Camshaft removal is more difficult on pushrod-type engines.

The timing cover must be removed on pushrod-type engines as described for front oil seal replacement. The rocker cover is removed; then the rocker arms are loosened or removed so that the pushrods can be removed. The lifters have to be taken out of the engine to clear the camshaft lobes so that the shaft will slide out. This will require the removal of the intake manifold on V-type engines. The thrust plate must be loosened when it is used on the engine. In order to have room to slide the camshaft out of the front of the block, the radiator must be removed and the air conditioning condenser must be loosened and lifted high enough to clear the camshaft. In most cases, this can be done without discharging and disconnecting the air conditioner lines. In some vehicles, the grill may even have to be removed to make room for the camshaft to be pulled from the engine. The distributor, fuel pump, and the drive chain are removed. Support the camshaft as you carefully slide it out of the block so that the camshaft bearings are not damaged.

If the camshaft is faulty, a new camshaft gear or sprocket is installed with the new camshaft. If only the camshaft drive is faulty, a new gear or sprocket can be installed on the used camshaft.

On some engines, the camshaft gear must be pressed from the camshaft. The camshaft is placed in a press using a press plate and an adapter. The adapter will prevent damage to the thrust plate. The timing gear hub is pressed from the shaft. Both the drive end of the camshaft and the thrust plate should be examined to make sure they are in good enough condition to be used again.

When the camshaft timing gear is pressed on, the thrust plate is first placed on the drive end of the camshaft. A spacer ring may also be used behind the timing gear. The new gear hub slot is aligned with the camshaft drive key. The gear is pressed in place on the camshaft until it bottoms against the spacer ring. A feeler gauge is used between the gear hub and thrust surface of the front camshaft bearing, as shown in Figure 19-14. If the clearance is too great, the thrust plate should be replaced. If the clearance is too small, the spacer ring should be replaced.

Figure 19-14. Measuring the cam thrust plate clearance with a feeler gauge.

The camshaft is supported to prevent damage to the cam bearing as it is carefully slid back into the block or head. The timing marks on the camshaft and crankshaft timing gears must be aligned as the gear teeth engage. The sprocket marks are aligned as the sprockets and timing chain are attached. The thrust plate is then secured in place. Both crankshaft and camshaft sprockets are aligned before the overhead cam drive belt is installed. The belt is then tightened to the correct belt tension.

With the camshaft in place, the engine is reassembled as described in the discussion on the front oil seal and timing gear replacement.

Oil Pan. The oil pan has to be removed to inspect and repair any problem occurring in the lower engine. Most engines have the oil pump located inside the oil pan. The oil pan will have to be removed to determine the cause of oil pressure problems. It will also have to be removed to determine the cause of any bearing noise.

A few automobiles have been built in such a way that the oil pan cannot be removed with the engine remaining in the chassis. There are a few other designs that allow the oil pan to be removed without disconnecting any other parts. Most automobiles require the removal of some other parts, such as the exhaust pipe, loosening the steering linkage, or removing the starter and the flywheel inspection cover.

The oil is drained from the engine. Examine the oil for contaminants that can verify the problem suspected. Disconnect and remove all vehicle parts that are in the way of the oil pan removal. Where necessary, loosen at least one of the front engine mounts, then jack the engine up an inch or two. Place a block between the engine and frame to hold the engine up. When the pan bolts are removed, the pan can be loosened from the gaskets. It may be necessary to rotate the crankshaft to provide connecting rod and counterweight clearance for the pan as it is removed. Examine the deposits remaining in the oil pan for evidence of the suspected problem. Visually inspect the oil pump pickup screen and all parts on the bottom of the engine. It is always a good policy to clean the oil pump pickup screen when the oil pan has been removed.

The required repairs should be made. All gaskets are removed from the pan rail of the block and from the oil pan. The oil pan should be thoroughly cleaned and the pan gasket surface flattened if it is warped.

New gaskets are put in place. Front and rear pan gaskets usually fit into grooves in the timing cover and in the rear main bearing cap. The oil pan side gaskets are placed on the pan. Hardening-type gasket sealing compound will help to hold the gasket parts in place. Often, a ribbon of RTV sealant is placed in the corners between the side and end gaskets. The oil pan is carefully put in place, making sure that the gasket parts remain in place. All the oil pan side rail bolts are installed loosely. They should be gradually tightened in sequence to the specified torque. The oil pan-to-timing cover bolts are then put in place and torqued.

A check is made to see that the gaskets and all the bolts are in place. The engine is lowered and the engine mounts secured. The rest of the parts that had to be removed are reattached. A new oil filter is installed and the engine is filled with fresh motor oil. The engine is started and run at idle until oil pressure is established.

Rear Oil Seal. The rear oil seal can be replaced when it leaks without removing the crankshaft. Rear oil seals are made in two different ways. In one design, the oil seal is a large ring fully enclosing the lip seal. When this type is used, the transmission, clutch, and flywheel or drive plate have to be removed. The seal can be pulled out of the back of the engine and a new seal installed using special tools without removing the pan or the rear main bearing cap. The second seal design is made of two half-seals. An upper half is in the block behind the rear main bearing. A lower half is in the rear main bearing cap. Replacement of these seals requires the removal of the oil pan and rear main bearing cap.

The oil pan is removed in the manner described under the discussion of the oil pan. The rear main bearing cap is removed to expose the rear crankshaft oil seal. Either of two types of rear oil seals will be found: a lip seal or a rope seal. Care must be taken to prevent damage to the rear bearing journal of the crankshaft while working on the rear seal.

All the main bearing caps are loosened to allow the crankshaft to drop slightly when replacing the upper half of a lip-type rear main bearing seal. This provides enough clearance for seal replacement. One end of the seal is driven inward with a soft drift until the other end of the seal extends far enough to be pulled from the block (Figure 19-16). Another removal method suggests turning a screw into the end of the seal, then using the screw as an attachment, pulling the seal from the block. It is sometimes helpful to rotate the crankshaft as the seal is being removed. The other half of the oil seal can be easily pulled from the main bearing cap.

The oil seal grooves are thoroughly cleaned. Use a bottle brush to clean the retaining groove for the upper seal. The sealing surface of the crankshaft should be polished with a fine abrasive to remove accumulated deposits.

With the sealing lip facing the inside of the engine, the new lip-type seal is slipped into the block retaining groove above the crankshaft. It is usually helpful to rotate the crankshaft as the upper half of the seal is being put in place. Care must be taken to make sure that no material is sheared from the seal by the sharp edge of the block. A thin piece of shim stock can be cut to fit behind the edge of the seal. It is positioned between the block and the seal to "shoe-horn" the seal into position. The shim stock tool is then removed. The other half of the seal is installed in the main bearing cap. The seal lip and surface of the shaft are lubricated. The required gaskets and seals are installed on the main bearing cap. Anaerobic sealant can be put on the mating surface of the main bearing cap to help seal the joint. The crankshaft is forced forward and backward to center the thrust bearing; then the main bearing cap bolts are tightened to the specified torque.

The rope-type main bearing oil seal is repaired in a different way. The rear main bearing cap is removed. The rest of the main bearing caps must remain tight. The old upper seal is *not* removed. Instead, each end of the old seal is driven into its groove behind the crankshaft with a soft drift as shown in Figure 19-15. Care must be taken to avoid damage to the crankshaft bearing journal. Sections are cut off from the *old* lower seal that was removed from the cap. These sections are driven into the upper seal groove. When the upper seal groove is fully packed, the seal ends are cut flush with the parting surface. A new rope seal is packed into the cap or seal retainer by rolling it into place with a round tool (Figure 18-18). This seal is also cut flush with the parting surface. Anaerobic sealant is placed

Figure 19-16. Rolling a lip-type seal out of the block while the crankshaft remains in place.

on the parting surface, when required, and the rear main bearing cap is torqued in place. If the rear main bearing is also the thrust bearing, the shaft must be forced forward and backward before torquing the cap bolts.

In some engines, the original equipment rope-type oil seal is replaced by a service-type lip seal. The old upper oil seal is pulled from the groove above the crankshaft. This is done by using a screw turned into one end of the seal. Be careful to avoid crankshaft journal damage. It is usually helpful to loosen all the main bearing caps to allow space for removal of the upper seal. The shaft and grooves for the seal are cleaned using a bottle brush. The replacement lip seal is installed as described in the paragraph on lip seal replacement.

After the seal is replaced, the oil pan is reinstalled as described in an earlier paragraph.

Main Bearings. Main bearing wear is one reason that an oil seal begins to leak. The worn bearing allows the shaft to move up and down to loosen the oil seal. It is a good practice to check the rear main bearing clearance before replacing a leaking rear main bearing oil seal. The engine is opened up in the manner just described for oil seal replacement.

The simplest way to check main bearing oil clearances with the crankshaft still in the block is to use gauging plastic. It only works to measure clearances when there is no weight on the gauging plastic strip. Because of this, the crankshaft must be raised and supported while the main bearing oil clearance is being checked with the gauging plastic.

Figure 19-15. Driving the end of the rope-type seal into the groove.

One way this can be done is to jack the crankshaft up against the upper main bearings. Another way is to remove the main bearing caps, one at a time, and carefully place a strip of shim stock across the bearing in the cap. The cap is put back on and tightened only enough to lift the crankshaft. The bearing being checked will have a strip of gauging plastic put in place of the shim stock. The cap is torqued, then removed, and the width of the compressed gauging plastic is checked with the markings on the envelope that the gauging plastic came in to indicate the oil clearance.

When the bearing shell is to be removed, a special main bearing removing tool is used. It fits into the oil hole in the main bearing journal of the crankshaft. The tool extends above the journal surface less than the thickness of the bearing shell. All the main bearing caps should be *loosened*. The cap is removed from the bearing to be replaced. With the bearing shell removing tool in place in the crankshaft oil hole, the crankshaft is rotated so that the tool contacts the shell on the *opposite* end from the bearing tang. Continued crankshaft rotation will roll the bearing shell from its seat. When removing the rear main bearing shell, hold the upper oil seal as the crankshaft is rotated to keep the seal from coming out of the block, too.

The bearing is replaced by coating the bearing surface of the shell with assembly lubricant, then placing the bearing shell on the shaft. Position the bearing removing tool on the *tang* end of the bearing shell. Rotate the crankshaft slowly to roll the shell into its seat above the crankshaft. Hold the rear seal to keep it in place as the crankshaft is rotated. After replacing the upper rear bearing shell, the lower main bearing shell is put in the main bearing cap. The bearing is then coated with assembly lubricant. The rear main seal must be properly installed. The caps are put on with the cap bolts finger-tight. The crankshaft is forced endways to align the thrust bearing surfaces; then the caps are tightened to the proper torque. The engine is assembled as described in the section discussing the oil pan service.

REVIEW QUESTIONS

1. Why is oil pressure established before the engine accessories or transmission is installed? [19-1]
2. How are the clutch and bell housing installed? [19-1]
3. How are the torque converter and automatic transmission installed on the engine? [19-1]
4. Why is it important to have the secondary ignition cables placed properly in their holders? [19-1]
5. Describe the procedure used to put the engine and transmission assembly in the chassis. [19-1]
6. What parts are attached when working under the vehicle after securing the engine mounts? [19-1]
7. What accessories are attached to the front of the engine? [19-1]
8. What accessories are attached to the top of the engine? [19-1]
9. How is the engine first started? [19-2]
10. How are the rings seated after the engine is started? [19-2]
11. How should the engine be driven after it has been turned over to the customer? [19-2]
12. Why should the battery be removed any time repairs are made on the engine? [19-3]
13. How can the valve spring be replaced without removing the head? [19-3]
14. Describe the procedure used to replace a front oil seal and timing chain. [19-3]

15. What must be done to remove the camshaft on a pushrod engine without removing the head or pan? [19-3]
16. Describe the procedure used to remove the oil pan from an engine installed in an automobile. [19-3]
17. How is a leaking rear lip-type oil seal replaced? [19-3]
18. How is a leaking rear rope-type oil seal repaired? [19-3]
19. How is the main bearing oil clearance measured with the crankshaft remaining in the engine? [19-3]
20. How can the upper main bearing be removed without removing the crankshaft? [19-3]

PART
IV

ENGINE PERIODIC SERVICE

The last part of this book deals with the engine systems that require service during the life of the engine. A brief description of the system operation is followed by a discussion of ways to test it. This is necessary to positively identify faulty parts of the system. Routine maintenance and repair of the parts are discussed. This type of routine maintenance is often classified as tune-up.

Service to the systems discussed in this part is usually considered as preventive maintenance. It is done as periodic service to prevent premature failure or breakdown. Procedures to repair the systems are not covered. They are beyond the scope of this book.

Lubricating System Service

An engine must have a good supply of lubricating oil. The lubricating oil minimizes metal-to-metal contact within the engine. This allows the engine to have a long service life. The oil must be periodically replaced to remove contaminants from the engine and to replace the additives that are used up. New balanced additives are contained in the new oil so that the oil can again lubricate effectively. The useful service life of an engine is longer with the use of good-quality motor oil and frequent oil and filter changes.

20-1 Lubrication Principles

Lubrication between two moving surfaces results from an oil film that separates the surfaces and supports the load. To understand the lubrication principle, consider how slippery a floor seems to be when a liquid is spilled on it. The liquid, either water or oil, supports a person's weight until it is squeezed out from under the shoes. If oil were put on a flat surface and a heavy block pushed across the surface, the block would slide more easily than if it were pushed across a dry surface. The reason for this is that a wedge-shaped oil film is built up between the moving block and the surface. This wedge-shaped film is thicker at the front or leading edge than at the rear or trailing edge. If the block were to be held still, the oil would be gradually squeezed out from under it. The block would then settle down on the surface. As soon as the block is moved again, the wedge-shaped oil film will be reestablished. This is illustrated in Figure 20-1.

The force required to push the block across a surface is dependent upon the weight of the block, how fast it moves, and *viscosity* of the oil. **Viscosity is the thickness or fluid body of the oil.** If the block is heavy, it will quickly squeeze the oil from under the surface. A lightweight block will squeeze it out slowly. As the block is moved faster there is less time for oil to be squeezed out. Because of this, a heavier load can be supported as the speed of the block is increased. This principle is used in water skiing.

The other factor in the ability of the oil film to support a load is the oil's viscosity or thickness. Thin oil would squeeze out faster than thick oil; therefore, the thick oil can support a greater load. The oil can be too thick. If the oil becomes too thick, it becomes sluggish, so that it will require a great deal of effort to move the block over the oil. If the oil is too thin, it will not completely support the block and the block will drag slightly on the surface. For any given block weight and moving speed, there is one oil thickness which requires the least effort to move the block. This is called the lowest *coefficient of friction*. **The force required to move the block divided by the block weight is the coefficient of friction.**

The principle just described is called *hydrodynamic* lubrication. *Hydro* refers to liquids, as in hydraulics, and *dynamic* refers to moving materials. Hydrodynamic lubrication occurs when a wedge-shaped film of lubricating oil develops in between two surfaces that have relative motion between them. When this film becomes so thin that the surface high spots touch, it is called *boundary lubrication.*

The engine oil pressure system feeds a continuous supply of oil into the *lightly loaded* part of bearing oil clearance. Hydrodynamic lubrication takes over as the shaft rotates in the bearing to produce a wedge-shaped hydrodynamic oil film that is curved around the bearing. This film supports the bearing and reduces the turning effort to a minimum when oil of the correct viscosity is used. Changes in the oil viscosity, shaft speed, or load affect the bearing lubrication in the same way they do with the block moving on a flat surface just described.

A crankshaft main bearing will be used to describe the lubrication of a typical journal bearing. When the engine is not running, the crankshaft pushes much of the oil from around it as it settles to the bottom of the bearing. As the engine is cranked, the crankshaft tries to roll up the side of the bearing. If some surface oil remains on the bearing, the shaft will slide back to the bottom of the bearing when it rolls onto this oil. Continued crankshaft rotation will repeat this sequence of climbing and sliding back. This sequence continues until the oil pump supplies fresh oil to the bearing. The shaft continues to try to climb up the side of the bearing; however, it now grabs oil instead of the bearing surface. The rotation pulls oil around the shaft, forming a curved wedge-shaped oil film that supports the crankshaft in the bearing as shown in Figure 20-2. Most bearing wear occurs during the initial start. Wear continues until a hydrodynamic film is established.

A continuous supply of new oil is required to maintain the oil film because oil will leak from the side of the bearing. This oil leakage flushes con-

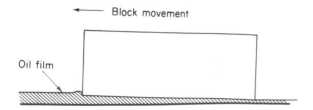

Figure 20-1. A wedge shaped oil film developed below a moving block.

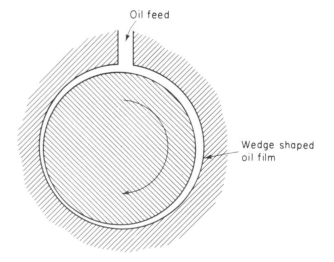

Figure 20-2. A wedge-shaped oil film curved around a bearing journal.

taminants from the bearing and removes heat that is generated in the bearing.

The primary function of the engine lubrication system is to maintain a positive and continuous oil supply to the bearings. Engine oil pressure must be high enough to get the oil to bearings with enough force to cause the required oil flow for proper cooling. Normal engine oil pressure range is from 30 to 60 lb/in.² (200 to 400 kPa). On the other hand, hydrodynamic film pressures developed in the high-pressure areas of the engine bearings may be over 1000 lb/in.² (6900 kPa). The relatively low engine oil pressures, obviously, could not support these high bearing loads without hydrodynamic lubrication.

20-2 Engine Lubrication Requirements

The greatest lubrication demand in engines is usually considered to be the bearings. This is a misconception. Their lubrication is necessary for the maximum service life of the engine; however, bearing lubrication is quite simple and it is easily met with properly designed bearings using oil having the correct viscosity.

The highest unit pressures and the most difficult place to lubricate is between the cam lobes and valve lifters. Much of modern motor oil formulation is based on the oil's ability to minimize scuffing and wear of the lifters. Cam lobes on pushrod engines are not lubricated with positive pressure but rely on oil thrown from the connecting rods and on oil that drains back from the rocker arm and lifter chambers. Overhead cam engines usually have an oil hole in the lightly loaded part of the cam lobe.

Valve assemblies, pistons, piston pins, oil pump-distributor drive, and cam drives require only a film of oil on their surfaces. The loads are relatively light, so that the parts are properly lubricated by the oil splashed on them. Oil under slight pressure is usually directed to the rocker arms. The amount of pressure is not important, only that the oil is *positively* delivered to the moving surface of the valve train that needs lubrication. Many engines direct a positive flow of oil to the cam drive gears or chain. This oil lubricates these surfaces and it helps to cushion the drive and reduce noise.

Automobile engines also use engine oil to operate hydraulic valve lifters. This places another and different kind of requirement on the engine oil. Hydraulic lifters are manufactured with extremely close fitting parts, to minimize leakage. Small foreign particles that get into these clearances could cause the lifter to malfunction. The engine oil must keep the lifter clean and limit deposit formation that would cause the lifter plunger to stick.

20-3 Properties of Motor Oil

The most important motor oil property is its thickness or viscosity. As an oil cools, it thickens. As the oil heats up, it gets thinner. Therefore, its viscosity changes with temperature. The oil viscosity must not be too thick at low temperatures to allow the engine to start. Thick, low temperature oil causes a very high coefficient of friction. If the coefficient of friction becomes too great, the cold engine will not have enough energy to carry the crankshaft over from one firing impulse to the next. When this happens, the engine will not start. There is a maximum oil viscosity with the engine cold that will keep an engine from starting below a specified temperature. On the other end of the scale, with the engine hot, the oil thins and its viscosity lowers. If the viscosity of the oil becomes too low, boundary lubrication will occur and the coefficient of friction will increase. Motor oil viscosity must be between these two extremes. The oil must be thin enough to allow the engine to start when cold. It must be able to flow to the oil pickup and flow through the screen to the pump. It must still have enough body or viscosity to develop the correct hydrodynamic lubrication film when the engine reaches its normal operating temperature. An index of the change in viscosity between the cold and hot viscosity is called the *viscosity index*. All oils with a high viscosity index thin less with heat than oils with a low viscosity index.

Motor oils are sold with an SAE grade number stamped on the top of the oil can. SAE stands for Society of Automotive Engineers. This grade number indicates the viscosity range in which the oil fits. The oil grade numbers are taken from a range of oil viscosities. Oils tested at 212°F (100°C) have a number with no letter following. For example, SAE 30 indicates that the oil has only been checked at 212°F (100°C). It falls within the SAE 30 grade number *when hot*. SAE 20W-20 indicates that the oil has been tested at both 0°F (−18°C) and 212°F (100°C). It falls into the SAE 20W grade number when cold and into the 20 grade number when hot. An SAE 10W-40 multigrade oil is one that meets the SAE 10W number when cooled to 0°F (−18°C) and it meets the SAE 40 number when tested at 212°F (100°C). Multigrade oils must have a higher viscosity index than do straight-grade oils. SAE viscosity numbers are shown in Figure 20-3.

Even though viscosity is the most important oil property, motor oil has other important properties. The lowest temperature at which oil will pour is called its *pour point*. Below this temperature, the oil will become plastic. A plastic will not produce hydrodynamic lubrication; therefore, the oil cannot be used below this temperature.

As oil is heated, the light, volatile parts of the oil boil off. When there are enough volatile vapors, they can be ignited with a small flame. The oil is at its *flash point* temperature when the vapors will ignite. The temperature at which the vapor continues to burn is called the *fire point*. Oil cannot function above the flash point temperature because it changes characteristics at this temperature. The maximum useful temperature range of motor oil is between the pour point and the flash point temperature. The flash point test is also used to determine if used motor oil contains gasoline contaminants. Gasoline is more volatile than the motor oil, so the gasoline vapors will evaporate and burn at a much lower temperature than the oil itself.

Motor oil must also resist oxidation. Oxidation is a form of oil breakdown as the oil combines with oxygen from the air. The motor oil must not bubble or foam, which would upset the hydrodynamic film. Neither the oil nor the additives in the oil should break down and form acids that will cause corrosion, scuffing, or rusting of engine parts. The additives help to disperse and hold contaminants in suspension in the oil. The contaminants will be drained with the oil to keep the engine clean.

The American Petroleum Institute (API), working with the engine manufacturers and oil companies, has established an engine oil performance classification. Oils are tested and rated in full-size multicylinder engines. The oil can is marked (usually printed) with the API classification that the oil meets. The API performance or service classification and SAE grade markings are the only information available to determine that the oil properties are satisfactory for use in the customer's engine. Typical oil can markings are shown in Figure 20-4.

SAE Viscosity Number	Centipoise cP at −18°C	Centistoke cSt at 100°C	
	max	min	max
5 W	1250	3.8	
10 W	2500	4.1	
20 W[a]	10,000	5.6	
20		5.6	9.3
30		9.3	12.5
40		12.5	16.3
50		16.3	21.9

a. SAE 15 W may be used to identify SAE 20 W oils which have a maximum viscosity at −18°C of 5000 cP.

Figure 20-3. Table comparing the oil viscosity to SAE viscosity numbers (Courtesy of Society of Automotive Engineers).

Figure 20-4. Typical oil can markings.

Oil performance classifications available in the service stations have the prefix "S". These are designed for spark-ignited engines. Straight mineral oil without additives has a type SA classification. It should only be used in engines that operate under mild conditions requiring no additional protection. Oil for mild engine operation requiring some extra additive protection has an SB classification. Engines requiring protection from high- and low-temperature deposits, wear, rust, and corrosion require oil classification SC. The SD classification was designed for the most severe type of engine service. This includes low deposit buildup developed during stop-and-go driving and oxidation from high speed. In use, SD oil would thicken at high temperatures. Within 2 years, the SE oil was developed. Automobile manufacturers recommend only SE oil for use in automobiles built after 1971. The service classification of the motor oil is open-ended. When required, an SF classification will be placed on the market.

Additional service classifications are used for diesel engines. These classifications are CA, CB, CC, and CD. The *C* oils are commercial oils developed for compression ignition engines. Oils may be rated for several performance service classifications at the same time. This means that the oil passed all the tests required for each classification. If the oil has several classifications, it may be used for any one of the service types.

Synthetic motor oils have recently been developed for use in automotive engines. They are produced in oil refineries by rearranging the molecular structure of the oil to form synthesized hydrocarbons. In this way, the molecules formed in the oil include only those that have the desired properties for use in automotive engines.

The synthesized hydrocarbon motor oil must meet the same SAE oil grade and API performance classification as standard motor oils do. The synthetic oil does not break down as fast as standard motor oils, so it lasts longer. Its most important property, however, is its very high viscosity index. This means that it does not thin with heat as fast as the standard motor oils do. Because of this, a lower-viscosity synthetic oil can be used and still provide protection for the engine. This leads to lower friction and lower fuel consumption. Synthetic motor oils cost several times more than standard oils.

Their long service life and improved fuel economy is expected to offset the higher price of the oil.

20-4 Motor Oil Deterioration

As oil is used, it deteriorates and requires changing. Two general types of oil deterioration exist—contamination and breakdown.

Oil may be contaminated with dirt or coolant. The most common type of contamination, however, is from the blow-by gases that work their way past the piston and rings. If crankcase ventilation is poor and the engine is cool, blow-by gases remain in the crankcase and mix with the oil. The undesirable blow-by gases consist primarily of partially burned fuel and water vapor. Unburned fuel dilutes the oil. The highly acidic water from the blow-by gases causes rusting and corrosion. Both combine with polymerized and oxidized hydrocarbons, produced during combustion, to undergo further change in the oil. These changes form "binders" that hold organic solids, inorganic salts, wear particles, and fuel soot together. When these combined particles get large enough, they drop out of the oil and are deposited in the engine as a cold-engine sludge having a consistency of mayonnaise. These contaminants will be found in both standard and synthetic motor oils. Less oil contaminants are produced when using no-lead gasoline than when using leaded gasoline.

When an engine is run at normal operating temperatures for some time, the oil gets hot and breaks down. Oil breakdown is the result of hot oil combining with oxygen, and it is called *oxidation*. Continued oxidation will eventually form hard carbon and varnish deposits on engine parts when it is allowed to build up over a long period of time. The higher the temperature, the more the deposits build up. Two hundred and fifty degrees Fahrenheit (121 °C) is the normal maximum engine oil temperature. Temperatures above this causes rapid deposit formation. Synthetic motor oil does not break down as rapidly as the standard motor oil, so it forms fewer high temperature deposits.

Engine oil additives tend to deteriorate and to be used up as the oil is used. When the additives can no longer do their designed job, the oil loses some of its required properties. Engine oils should be changed before sludge develops, before oxidized deposits form, and before the additives lose their effectiveness. Recommend oil change periods when using synthetic motor oils is about twice as long as those recommended for standard motor oils.

20-5 Motor Oil Additives

Additives are used in motor oils for three different reasons: (1) to replace some properties removed during refining, (2) to reinforce some of the oil's natural properties, and (3) to provide the oil with new properties it did not originally have. Oils from some petroleum oil fields require more and different additives than oils from other fields. Additives are usually classified according to the property they add to the oil.

Antioxidants reduce the high-temperature contaminants. They prevent the formation of varnish on the parts, reduce bearing corrosion, and minimize particle formation.

Corrosion preventatives reduce acid formation that cause bearing corrosion.

Detergents and *dispersants* prevent low-temperature sludge binders from forming and keep the sludge-forming particles finely divided. The finely divided particles will stay in suspension in the oil to be removed from the engine as the oil is removed at the next drain period.

Extreme pressure and *antiwear additives* form a chemical film that prevents metal-to-metal seizure any time boundary lubrication exists.

Viscosity index improvers are used to reduce viscosity change as the oil temperature changes.

Pour point depressants coat the wax crystals in the oil so they will not stick together. The oil will then be able to flow at lower temperatures.

A number of other oil additives may be used to modify the oil to function better in the engine. These include rust preventatives, metal deactivators, water repellants, emulsifiers, dyes, color stabilizers, odor control agents, and foam inhibitors.

Oil producers are careful to check the compatibility of the oil additives they use. A number of chemicals that will help each other can be used for each of the additive requirements. However, with improper additive selection, the additives may oppose each other and lose their benefit to the oil. Each oil producer balances the oil additives to provide a motor oil with desirable properties that meet the engine's needs. The balanced additives are called an additive package.

Additives available at service stations are called *proprietary additives*. They generally cannot add any needed desirable property to the oil that it does not already have. It is even possible that these proprietary additives may neutralize some of the additives already in the oil, thus degrading the oil instead of improving it. The procedure usually recommended by the engine manufacturer is to use only motor oil with an SE service classification without additional proprietary additives. When adding oil between changes, it is a good practice to add the same brand and grade of oil that is already in the engine, thus minimizing the chance of having conflicting additives. Changing oil brands at an oil and filter change will cause no problem.

20-6 Lubrication System Maintenance

During the service life of an engine, most of the lubrication system service involves oil and filter changes. Oil selection is based on the manufacturer's or petroleum company's recommendations. These recommendations, in turn, are based on extensive laboratory and fleet testing to check the compatibility of the oil with the engine. One consideration that the oil recommendation must include is the lowest expected temperature. A viscosity grade is recommended that will allow easy engine starting. A maximum starting viscosity is about 5000 centipoise. The engine will not carry over to the next combustion with the viscosity any higher. Most automobile engine manufacturers recommend a multigrade oil that will allow easy cold starts and still provide adequate protection when hot. A second consideration is the type of service that is to be encountered. Automobile manufacturers recommend an oil having an SE performance classification.

All automobile companies recommend using an original-equipment oil filter. These are made to meet the engine requirements. High-quality replacement filters are available that will do an acceptable job of filtering the lubricating oil. They must, however, be designed for the specific engine application. When required, the filter must have either or both a *check valve* and a *bypass valve*.

There is often a debate about the most desirable

oil change period. The automobile manufacturer's recommendations usually differ from the petroleum company's recommendations. The oil drain period is established by considering the oil's contamination, the additive's continued effectiveness, and the original oil performance classification. Running a laboratory analysis on the used oil drained from an engine is the only way to determine the *actual condition* of the oil. This, of course, is only possible in research, engineering laboratories, and in fleet operation. Manufacturers use used-oil analysis, along with extensive service tests by both the engine manufacturers and the petroleum companies, to determine safe oil drain periods. The results of these tests are passed along to their respective service organizations as the recommended oil change period.

Low-speed stop-and-go traffic driving causes the engine to run cool, so sludge will start to build up. Stop-and-go vehicle operation puts a very few miles on the vehicle each month; however, the contamination builds up rapidly for each mile traveled. A time period in terms of months is specified to take care of vehicles that are operated in this manner.

High-speed, high-temperature engine operation builds up varnish and carbon-type deposits. This type of operation adds mileage rapidly over a short period of time. An oil change period based on mileage or hours of operation is recommended to take care of this type of vehicle operation.

Most automobiles are operated with a mixture of low-speed and high-speed driving. The usual oil change recommendations are made on *both* a time and mileage basis to take care of all "normal" operation, either time or mileage. The oil should be changed at the change interval that occurs first. It is further recommended that if the vehicle is operated under abnormal conditions, such as dusty roads or pulling a trailer, the oil should be changed more frequently.

If the oil change periods are extended beyond the manufacturer's recommendations, contaminants may become so large and heavy that they will fall out of the oil and deposit in the engine. When this happens, they cannot be removed from the engine when draining the used oil. These deposits continue to build until the engine is disassembled and cleaned.

Engine designs have reduced contamination buildup and have increased the normal oil change interval. Some of these engine design features are: splash pans under the intake manifold exhaust crossover to keep the oil away from this hot surface so that the oil will not be oxidized; rocker assembly oil return passages have been enlarged to allow the oil to rapidly return from hot head areas to the oil pan to keep the oil from oxidizing; and high-temperature thermostats quickly warm the oil to reduce cold sludge buildup. Extended oil drain periods will not satisfactorily protect older engines that do not include the engine features designed for long drain periods. The drain period should not extend beyond the manufacturer's time or mileage recommendation for each specific engine model.

Superpremium engine oils have been developed to operate for extended oil drain periods. These oils are made from selected grades of lubricating oil stock. Improved high-quality additives are blended into the oil to give it the required long-life properties. Synthetic motor oils have recommended drain periods that are still longer.

The difference between the automobile engine manufacturer and the petroleum company recommended oil drain interval is the result of different philosophies. The automobile manufacturer's recommendations are to the original automobile owner. If these recommendations are followed, the average original automobile owner will have minimum maintenance expense. The petroleum company recommendations, on the other hand, are based on conservative views that would provide the longest service period between engine overhauls. This would give minimum engine lifetime costs if one drove enough miles to "use up" the engine's useful life. In general, it can be stated that the petroleum companies recommend oil changes twice as often as the automobile manufacturers. The reverse is true of synthetic oils. The engine manufacturer recommends the same drain interval for all oils. The synthetic oil manufacturer recommends twice-as-long drain intervals for synthetic oils than for standard motor oils.

Engine oil consumption is another maintenance requirement. The average automobile owners have no idea how much oil is being consumed by the engine until they find that it is necessary to add oil between oil changes. The owners are seldom concerned about oil consumption until it becomes excessive. The only way they can tell if the consumption is excessive is by the amount of oil added to the engine between oil changes, regardless of how the oil gets out of the engine.

Motor oil can get out of the engine either through the combustion chamber or through leaks. Leaks generally result from loose, poor fitting, or damaged gaskets, or from damaged oil seals. In some cases, leaks may result from cracked engine parts. Oil leaks are usually repairable without requiring an engine overhaul.

Oil can get into the combustion chamber past the piston rings, through the valve guides, through some intake gasket leaks, and through the PCV system. Of these leaks, the most expensive one to repair is the oil leak past the piston rings.

Unless excessive oil consumption affects engine operation, such as hard starting or sluggish performance, its correction will be based on economics, pride, or law. Most automobile manufacturers do not consider oil consumption to be excessive and requiring repair of new engines unless the engine uses more than 1 quart each 500 to 700 miles (800 to 1120 km). Fortunately, the oil mileage is seldom this low. As the engine mileage builds up, its oil consumption will increase.

The economic consideration in oil consumption relates the cost of the oil to the cost of the repair job. If oil consumption is a result of a leaking gasket, faulty seal, or cracked part, the repair cost is generally low. If it requires overhaul, such as re-ringing, the repair cost will far exceed the cost of the additional oil consumed.

Many people have so much pride in their automobiles that they are willing to pay the possible high cost to have all leaks repaired, to eliminate exhaust smoke, and have minimum oil consumption. These people will have the oil leaks repaired without considering economics. Some states have pollution laws that will not allow vehicles to be operated with visible exhaust emissions. In these states, the oil consumption will have to be kept at a low level, regardless of cost.

Excessive oil consumption, then, is a compromise between the value of the car, the cost of repair, the cost of the extra oil, the owner's pride, and the operating laws.

In addition to motor oil, many petroleum and chemical companies package *proprietary additives* that the customer may add to motor oil. This means they make the additive using their own private for-

mula to work properly in engines. Some of these materials are solvents that dissolve deposits, some are detergents and dispersants that keep contaminants in suspension, and others are oil thickeners. These products are similar to the products in the additive package the oil manufacturers have already blended into the oil. The additive package gives the oil the required properties at a minimum cost to the consumer. The proprietary products are high-priced additives that increase the cost of the oil when they are used. They are of doubtful value in a "normal" engine that is using recommended oils and oil change periods. Solvents and detergents added to the oil may be useful for freeing sticking hydraulic lifters and valves. When they free up, the oil with the additive should be drained and replaced with a fresh change of the proper engine oil. If a thicker oil is desired, purchasing a heavier grade for a refill is less expensive than using a lighter grade with an oil thickener. As the engine runs, the additive thickener will break down and the oil will thin out, thus losing the desired thickening characteristic.

When one considers the great amount of money spent by the automobile manufacturers and petroleum companies to match the engine and the oil, it seems wise to follow their recommendations. The manufacturer's warranty is based on these recommendations. Following these recommendations will provide satisfactory engine lubrication for the normal service life of the engine.

20-7 Oil Drain

The oil can be drained more rapidly from a warm engine than from a cold one. In addition, the contaminants are more likely to be suspended in the oil immediately after running. Position a drain pan under the drain plug, then remove the plug with care to avoid contact with hot oil. Allow the oil to drain freely so that the contaminants come out with the oil. It is not critically important to get every last drop of oil from the engine oil pan because a quantity of used oil still remains in the engine oil passages and oil pump. While the engine oil is draining, the oil plug gasket should be examined. If it appears to be damaged, it should be replaced. When the oil stops running and starts to drip, reinstall and tighten the drain plug. Replace the oil filter if it is to be done during this oil change. Refill the engine with the proper type, grade, and quantity of oil. Restart the engine and allow the engine to idle until it develops oil pressure; then check the engine for leaks, especially at the oil filter.

20-8 Oil Filters

The oil within the engine is pumped from the oil pan through the filter before it goes into the engine lubricating system passages. The filter is made from either closely packed cloth fibers or a porous paper. Large particles are trapped by the filter. Microscopic paticles will flow through the filter pores. These particles are so small that they can flow through the bearing oil film and not touch the surfaces, so they do no damage.

Either the engine or the filter is provided with a *bypass* that will allow the oil to go around the filter element. This is necessary when an operator neglects to have the filter changed at the proper time. The bypass allows the engine to be lubricated with dirty oil rather than having no lubrication if the filter becomes plugged. The oil also goes through the bypass when the oil is cold and thick. Most engine manufacturers recommend filter changes at every other oil change period. Correct oil filter selection includes the use of a filter with an internal bypass when the engine is not equipped with one. A typical internal bypass is shown in Figure 20-5.

Some oil filters have a replaceable element. A long central tube holds the cover over the element. The cover seals against the engine flange. This tube is removable to enable the technician to replace the filter element inside the cover. The filter housing is thoroughly cleaned and the new element installed with new seals. The tube should be tightened to the recommended torque to prevent leakage.

Most automobile engines use a sealed spin-on filter. The spin-on filter has the new filtering element sealed in a can that is screwed onto an engine fitting. The spin-on type filter is more expensive, but it is much faster and cleaner to use than the replaceable-element type.

If the spin-on filter has been installed properly, it can be removed by hand. Filter wrenches are available and should be used in cases where the filter cannot be loosened by hand. A drain pan placed

Figure 20-5. A typical internal by-pass valve used inside an oil filter.

below the filter will catch the oil that drips as the old filter is removed. After the filter is removed, the sealing surface on the engine should be examined to make sure it is clean and smooth so that the new filter will seal properly. The oil seal of the new filter is coated with motor oil and the filter is screwed on by hand. It should be rotated an additional one-half to three-fourths of a turn after the oil seal first contacts the engine sealing surface. The engine is restarted and allowed to idle until the filter fills with oil and the instrument panel indicator shows oil pressure. With the engine running, a check is made for leaks around the filter. A final check of the oil level should be made before the filter replacement job is complete.

Periodic replacement of the oil and the filter with the recommended oil and the correct-type filter is the best preventive measure to ensure a long engine life that is free from costly repair bills.

REVIEW QUESTIONS

1. What causes a wedge shaped oil film to form in the oil? [20-1]
2. What happens in the oil film when the oil is too thick? [20-1]
3. What happens in the oil film when the oil is too thin? [20-1]
4. When does hydrodynamic lubrication occur? [20-1]
5. Where does oil feed into the bearing oil clearance? [20-1]
6. Why does most bearing wear occur during engine start up? [20-1]
7. Why is it good to have some oil leakage from the edge of a bearing? [20-1]
8. What is the primary function of the engine lubricating system? [20-1]
9. What place is the most difficult to lubricate in an engine? [20-2]
10. How are the cam lobes lubricated on pushrod engines? [20-2]
11. Why is splash lubrication satisfactory for some places in the engine? [20-2]
12. What determines the minimum and maximum viscosities of the motor oil? [20-3]
13. How do the tests for a single grade oil differ from those for multigrade oil? [20-3]
14. Why is the pour point test important? [20-3]
15. What is the value of the flash and fire point tests? [20-3]
16. What information about the oil is available to the customer? [20-3]
17. What information do the letters SE marked on the oil can indicate to the vehicle operator? [20-3]
18. What is indicated by several performance service classifications marked on the same can of oil? [20-3]
19. How do synthesized hydrocarbon motor oils differ from the standard motor oils? [20-3]
20. Name two types of oil deterioration. [20-4]
21. What causes contamination in the motor oil? [20-4]
22. List three reasons additives are put in motor oil. [20-5]
23. List the properties of additives used in motor oils. [20-5]
24. What are the advantages of using a multigrade motor oil? [20-6]
25. How is the recommended oil change period determined? [20-6]
26. Why is it necessary to specify the oil change period on both a time and mileage basis? [20-6]
27. How do the oil change periods recommended by the automobile manufacturer differ from those recommended by the petroleum company? [20-6]
28. How does an operator know that the engine is using oil? [20-6]
29. How can oil get out of an engine? [20-6]
30. When should an overhaul be recommended as a result of oil consumtion? [20-6]
31. When should proprietary additives be used in engine oil? [20-6]
32. Describe the oil drain procedure. [20-7]
33. When does oil go through the filter bypass? [20-7]
34. Describe oil filter replacement. [20-7]

Cooling System Service

Satisfactory cooling system operation depends upon the design and operating conditions of the system. The design is based on heat output of the engine, radiator size, type of coolant, size of coolant pump, type of fan, thermostat, and system pressure. Operating conditions change, for example, when driving in traffic, when changing engine speeds and engine loads by driving up and down hills, and when towing a trailer. Correct operation of all cooling system parts becomes critical when the engine is operated under heavy loads in a hot climate. Unfortunately, the cooling system is usually neglected until there is a problem. Correct routine maintenance can prevent problems. This will minimize the overall cost and the inconvenience that a cooling system problem causes.

21-1 Cooling System Requirements

The cooling system must allow the engine to warm up to the required operating temperature as rapidly as possible, then maintain that temperature. It must be able to do this when the outside air temperature is as low as $-30\,°F$ $(-35\,°C)$ and as high as $110\,°F$ $(45\,°C)$. Quick engine warm-up aids proper fuel vaporization, provides the correct oil viscosity, and helps to make the part clearances normal within the engine.

Peak combustion temperatures in the engine cycle run from 4000 to $6000\,°F$ (2220 to $3330\,°C$). The combustion temperatures will *average* from 1200 to 1700°F (650 to 925°C). Continued temperatures as high as this would weaken engine parts, so heat must be removed from the engine. The cooling system keeps the head and cylinder walls at a temperature that is within the limits of their physical strength.

Low-Temperature Requirements. Engine operating temperatures must be above a minimum temperature for proper engine operation. When the temperature is too low, there is not enough heat to properly vaporize the fuel in the intake charge. As a result, extra fuel must be added to supply more volatile fuel to make a combustible mixture. The heavy, less-volatile portion of the gasoline does not vaporize and so it remains as unburned liquid fuel. In addition, cool engine surfaces quench part of the combustion gases, leaving partially burned fuel as soot. The combustion chamber surface also cools the burned combustion by-products, condensing the moisture that is produced during combustion. The unburned fuel, soot, and moisture go past the piston rings as blow-by gases. They wash oil from the cylinder wall and dilute the oil in the pan. Where there is not enough lubrication, the cylinder wall and piston rings will show excessive scuffing and wear.

Gasoline combustion is a rapid oxidation process in which heat is released as the hydrocarbon fuel chemically combines with oxygen from the air. For each gallon of fuel used, a moisture equivalent of a gallon of water is produced. It is a part of this moisture that condenses and gets into the oil pan, along with unburned fuel and soot, and causes sludge formation. The condensed moisture combines with unburned hydrocarbons and additives to form carbonic acid, sulfuric acid, nitric acid, hydrobromic acid, and hydrochloric acid. These acids are responsible for engine wear by causing corrosion and rust within the engine. Rust occurs rapidly when the coolant temperature is below 130°F (55°C). Below 110°F (45°C), water from the combustion process will actually accumulate in the oil. High cylinder wall wear rates occur whenever the coolant temperature is below 150°F (65°C).

High-Temperature Requirements. Maximum temperature limits are required to protect the engine. High temperatures will oxidize the engine oil. This breaks the oil down, producing hard car-

bon and varnish. If high temperatures are allowed to continue, the carbon that is produced will plug piston rings. The varnish will cause the hydraulic valve lifter plungers to stick. High temperatures always thin the oil. Metal-to-metal contact within the engine will occur when the oil is too thin. This will cause high friction, loss of power, and rapid wear of the parts. Thinned oil will also get into the combustion chamber by going past the piston rings and through valve guides to cause excessive oil consumption.

The combustion process is very sensitive to temperature. High coolant temperatures raise the combustion temperatures to a point that detonation and preignition may occur. These are common forms of abnormal combustion. If they are allowed to continue for any period of time, the engine will be damaged.

Normal Temperatures. There is a normal operating temperature range between low-temperature and high-temperature extremes. The thermostat controls the minimum normal temperature. This temperature has been gradually increased from 160°F to 190°F (72°F to 87°C). Some engines have a minimum temperature as high as 200°F (95°C). The maximum operating temperature on liquid-cooled engines is limited by the coolant's boiling point and the heat-transfer capacity of the radiator. Engine operating temperatures should be kept between normal minimum and maximum temperature extremes, usually close to the minimum temperature. This will provide proper engine operation and the maximum service life of the engine.

21-2 Coolant

The majority of automobiles use liquid-cooled systems. Coolant flows through the engine, where it picks up heat. It then flows to the radiator, where the heat is given up to the outside air. The coolant continually recirculates through the cooling system, as illustrated in Figure 21-1. Its temperature rises as much as 15°F (8°C) as it goes through the engine; then it cools back down as it goes through the radiator. The coolant flow rate may be as high as 1 gallon or 4 liter per minute for each horsepower the engine produces.

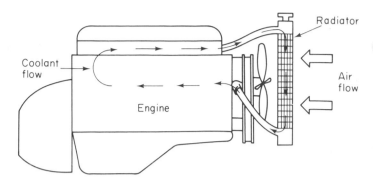

Figure 21-1. Coolant flow through an engine cooling system.

Coolant Types. Water is able to absorb more heat per gallon than any other liquid coolant used in automobiles. Water, however, has both a high and a low usable temperature limit. Under standard conditions, water boils at 212 °F (100 °C) and freezes at 32 °F (0 °C). There are very few places in the continental United States where the temperature does not at some time drop below the freezing point of water. Antifreeze protection is required when these low temperatures are expected. Low-temperature protection is also required on some factory-installed air conditioned cars to keep the heater core from freezing. All manufacturers recommend the use of *ethylene-glycol-based antifreeze* mixtures for this protection. The freezing point curve compared to the percent antifreeze mixture is shown in Figure 21-2. It should be noted that the freezing point in-

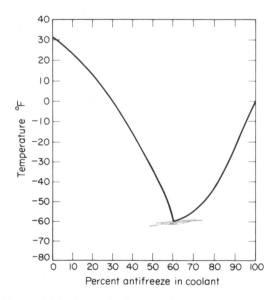

Figure 21-2. A graph showing the freezing point of the coolant in relationship to the percentage of antifreeze used in the coolant.

creases as the antifreeze concentration is increased above 60%. The normal mixture is 50% antifreeze and 50% water. This type of antifreeze is sometimes called permanent type, even though its replacement is recommended each year or two. Ethylene-glycol antifreezes have anticorrosion additives, inhibitors, and water pump lubricants blended into them. When antifreezes are not used, these required additives may be purchased separately by the operator and added to the cooling system water.

Only the minimum required amount of ethylene-glycol-based antifreeze should be used. It is expensive, and therefore it is economical to use the minimum required amount. At the maximum protection, an ethylene-glycol concentration of 60% will absorb about 85% as much heat as water. This can be seen in Figure 21-3. An added advantage in using ethylene-glycol-based antifreeze is the fact that its boiling point is higher than water's boiling point, as shown in Figure 21-4. Ethylene-glycol antifreeze also puts more corrosion inhibitors in the cooling system. Its use will allow the cooling system to run at a higher temperature level, so that a smaller radiator may be used on the vehicle. This coolant mixture is also helpful in transferring heat from the engine to increase the boiling point of the coolant on cars equipped with air conditioning.

21-3 Cooling System Design

Hot coolant comes out of the thermostat housing on the top front of the engine. The engine coolant outlet is connected to the top of the radiator by the upper hose and clamps. The coolant in the radiator is cooled by air flowing through the radiator. As it cools, it moves from the top to the bottom of the radiator. Cool coolant leaves the lower radiator through an outlet and lower hose, going into the inlet side of the water pump, where it is recirculated through the engine.

Much of the cooling capacity of the cooling system is based on the function of the radiator. Radiators are designed for the maximum rate of heat transfer using minimum material and size. This will help keep cost as low as possible. Vehicle designs also limit radiator space, and this affects the radiator designs. Cooling air flow through the radiator is aided by a belt- or electric motor-driven cooling fan.

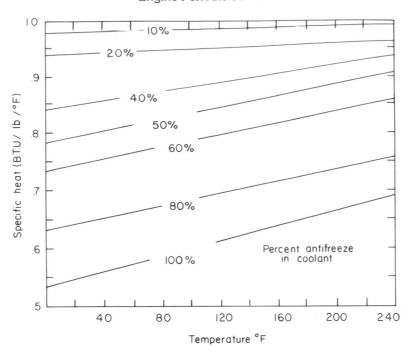

Figure 21-3. A graph showing that the amount of heat carried by the coolant is decreased as the percentage of antifreeze in the coolant increases.

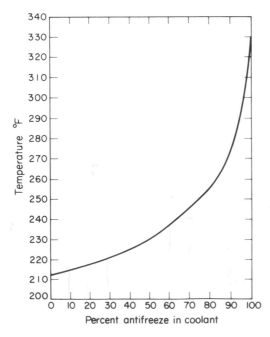

Figure 21-4. A graph showing how the boiling point of the coolant increases as the percentage of antifreeze in the coolant increases.

Radiator. Two types of radiator cores are in common use in domestic automobiles, the serpentine fin core, shown in Figure 21-5, and the plate fin core, shown in Figure 21-6. In each of these types, the coolant flows through oval-shaped *core tubes.*

Heat transfers through the tube wall and soldered joint to *fins.* The fins are exposed to an air flow, which removes heat from the radiator and carries it away by moving air through it.

Most automobile radiators are made from yellow brass or copper. These materials are corrosion-resistant, have good heat-transfer ability, are easily formed, and are easily repaired by soldering. Some heavy-duty applications have used corrosion-protected steel; however, steel is not used in automobile radiators. Aluminum is used for radiators in applications where weight is critical.

Core tubes are made from 0.0045 to 0.012 in. (0.1 to 0.3 mm) sheet brass, using the thinnest possible materials for each application. The metal is rolled into round tubes and the joints are sealed with a locking seam. The tubes are then coated with solder, compressed into an oval shape, and cut to length. Fins are formed from 0.003 to 0.005 in. (0.1 to 0.2 mm) copper or brass. Again the thinnest possible material is used to save weight and cost.

Serpentine fins are formed from solder-coated sheets, stacked between the tubes. The core assembly of fins and tubes is held in a fixture for soldering. The assembly is heated in an oven to fuse the joints and then the entire radiator assembly is submerged in liquid solder. Capillary action pulls solder into

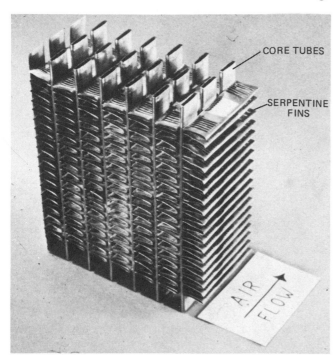

Figure 21-5. A section from a serpentine core radiator (Courtesy of Modine Manufacturing Company).

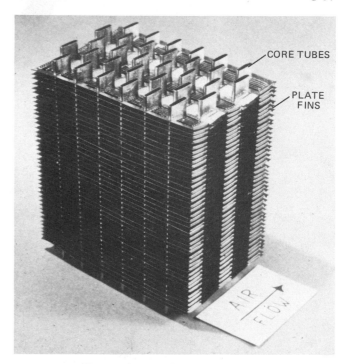

Figure 21-6. A section from a plate fin core radiator (Courtesy of Modine Manufacturing Company.)

the joints, assuring a tight joint that will readily transfer heat. The serpentine-type radiator core is usually used in automobiles. It is the least expensive of the two types and cools as well as the plate type. The serpentine-type core is held together by the soldered joint alone, while the plate type is mechanically held by the plate fins and the solder. Plate-type cores are, therefore, stronger then the serpentine cores.

The main limitation of heat transfer in a cooling system is from the radiator to the air. Heat transfers from the water to the fins as much as seven times faster than the heat transfers from the fins to the air, assuming equal surface exposure. The radiator's *heat-transfer capacity* is the result of the number of fins per inch, height, width, thickness, and the number of coolant tubes. The radiator must be capable of removing the amount of heat energy approximately equal to the heat energy of the power produced by the engine. Each horsepower is equivalent to 42.4 Btu (10,800 cal) per minute. As the engine power is increased, the heat removing requirement of the cooling system is also increased.

Coolant tubes are straight, free-flowing tubes. The fins are given a pattern to break up any smooth laminar air flow (flow staying in layers) that would

keep the air from contacting the surface of the fin. This produces a turbulent air flow that will increase the heat-transfer rate, but it will also add air resistance. Care is taken to design a radiator that will provide maximum cooling with minimum air resistance. With a given frontal area, radiator capacity may be increased by increasing the core thickness, packing more material into the same volume, or both. The radiator capacity may also be increased by placing a shroud around the fan so that more air will be pulled through the radiator. A fan shroud can be seen in Figure 21-7.

Radiator headers and tanks that close off the ends of the core are made of sheet brass 0.020 to 0.050 in. (0.5 to 1.25 mm) thick. These are fitted with tubular brass hose necks. The supporting sides of the core are usually steel. The filler neck and the drain boss are brass. When a transmission oil cooler is used in the radiator, it is placed in the outlet tank, where the coolant has the lowest temperature.

Radiators may be of the down-flow (Figure 21-8) or cross-flow (Figure 21-9) designs. In down-flow designs, hot coolant from the engine is delivered to the top radiator tank and cool coolant is removed from the bottom tank. In cross-flow designs, hot coolant goes to a tank on one side of

Figure 21-7. A typical fan shroud on a V-type engine.

Figure 21-8. A down flow radiator (Courtesy of Dow Chemical Company).

Figure 21-9. A cross flow radiator (Courtesy of Modine Manufacturing Company).

the radiator and flows across the radiator through the coolant tubes to the tank on the other side. In the down-flow designs, the coolant reserve tank is located on the top, or inlet side. In the cross-flow design, the reserve tank is placed on the outlet side. Neither type is more or less efficient than the other. Available space generally dictates which of the two designs is to be used in an automobile.

Pressure Cap. The filler neck is fitted with a pressure cap. The cap has a spring-loaded valve that closes the cooling system vent. This causes the cooling pressure to build up to the pressure setting of the cap. At this point, the valve will release the excess pressure to prevent system damage. Figure 21-10 illustrates pressure cap operation.

Excess pressure usually forces some coolant

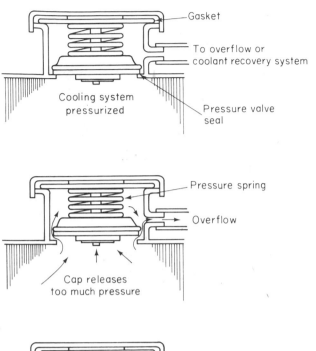

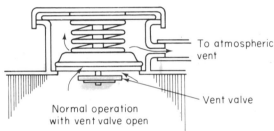

Figure 21-10. A line drawing illustrating the operation of a typical pressure cap.

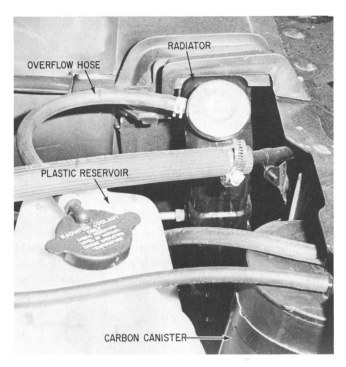

Figure 21-11. Parts of a typical cooling system with a plastic reservoir to hold excess coolant while the system is hot.

from the system through an overflow. The overflow is a tube leading out below the radiator, where coolant is lost. Most modern cooling systems connect the overflow to a plastic reservoir to hold excess coolant while the system is hot. An example of this system is shown in Figure 21-11. When the system cools, the pressure in the cooling system is reduced and a partial vacuum forms. This pulls the coolant from the plastic container back into the cooling system, keeping the system full. Because of this action this system is often called a coolant recovery system. The filler cap used on a coolant system without a coolant saver is fitted with a vacuum valve. This valve allows air to reenter the system as the system cools, so that the radiator parts will not collapse under the partial vacuum. SAE standard latching notches on the cap and neck are sized so that a high-pressure cap will not fit a low-pressure cooling system.

Automobile engine cooling systems are pressurized to raise the boiling temperature of the coolant. The boiling temperature will increase approximately 3 °F (1.6 °C) for each pound increase in pressure. At standard atmospheric pressure, water will boil at 212 °F (100 °C). With a 15 lb/in.² (100 kPa) pressure cap, water will boil at 257 °F (125 °C), which is a maximum operating temperature for an engine. The lubricant does not operate properly above this temperature. The high coolant-system-temperature serves two functions. First, it allows the engine to run at an efficient temperature, close to 200 °F (93 °C), with no danger of boiling the coolant. Second, the higher the coolant temperature, the more heat the cooling system can transfer. The heat transferred by the cooling system is proportional to the temperature difference between the coolant and the outside air. This characteristic has led to the design of small, high-pressure radiators that are capable of handling large quantities of heat. It can be seen that for proper cooling, the system must have the right pressure cap correctly installed.

A problem that sometimes occurs with a high-pressure cooling system involves the coolant pump. To function, the inlet side of the pump must have a lower pressure than its outlet side. If inlet pressure is lowered too much, the coolant at the pump inlet could boil, producing vapor. The pump will then spin the coolant vapors and not pump coolant. This condition is called *pump cavitation.*

Fan. Air is forced across the radiator core by a cooling fan. On most engines, it is attached to a fan hub that is pressed on the coolant pump shaft. An example of this is pictured in Figure 21-12. Most installations with transverse engines drive the fan with an electric motor, as shown in Figure 21-13. The fan is designed to move enough air at the lowest fan speed to cool the engine when it is at its highest coolant temperature. The fan shroud is used to increase the cooling system efficiency. The horsepower required to drive the fan increases much faster than the increase in fan speed. Higher fan speeds also increase the fan noise. Thermoregulating and viscous fan

Figure 21-13. A typical electric cooling fan.

Figure 21-12. A typical water pump shaft and fan hub with the fan removed.

drives (Figure 21-14) have been developed to drive the fan only as fast as required. They also limit the maximum fan speed. This reduces both the power requirements of the fan and the fan noise. Fans with flexible plastic or flexible steel blades have also been developed. These fans have high blade angles that will pull a high air volume when turning at low speeds (Figure 21-15). As the fan speed increases, the fan blade angle flattens (Figure 21-16), reducing the horsepower required to rotate it at high speeds.

Extra Heat Loads. Cooling systems have an added heat load when air conditioning is used. The air conditioning condenser is usually located ahead of the radiator. It operates quite warm and this raises the incoming air temperature 10 to 20°F (6 to 11°C). Because of this, air conditioned cars are usually equipped with a larger-capacity radiator and a higher-capacity fan than cars without air conditioning. High-capacity cooling systems like this are also used on cars equipped for trailer towing.

Figure 21-14. Thermo-regulating and viscous fan drives. A silicone drive coupling is on the left with thermostatic spring valves in the center and the drive couplings on the right.

Figure 21-16. A flexible fan blade has a lower angle at high engine speeds. The photographs in Figure 21-15 and 21-16 were taken with a high speed flash.

Emission controls that retard the spark increase heat rejection to the cooling system. There may be as much as 25% more heat produced at idle as a result of these emission controls. In traffic, the additional engine heat may become critical. To help relieve this critical situation, many engines are equipped with a temperature-sensing vacuum valve similar to the one shown in Figure 21-17. When the coolant reaches a critical temperature, engine vacuum is used to advance ignition timing. This improves combustion but increases exhaust emissions; however, it lowers the amount of heat the engine produces so that the engine does not overheat.

Most of the heat absorbed from the engine by the cooling system is wasted. Some of this heat, however, is recovered by the vehicle heater. Heated

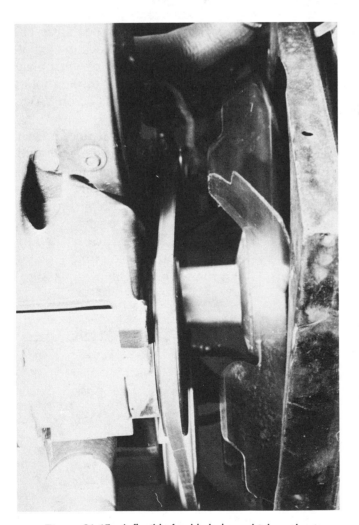

Figure 21-15. A flexible fan blade has a high angle at the engine idle speed.

Figure 21-17. A typical temperature sensing vacuum valve mounted in a cooling system passage of the intake manifold.

coolant is bypassed through tubes in the small core of the heater. Air is passed across the heater fins; then it is sent to the passenger compartment. In some vehicles, the heater and air conditioning work in series to maintain vehicle compartment temperature.

21-4 Cooling System Maintenance

The cooling system is one of the most maintenance-free systems in the engine. Normal maintenance involves an occasional check on the coolant level. It should also include a visual inspection for signs of coolant system leaks and for the condition of the coolant hoses and fan drive belts.

The coolant level should only be checked when the engine is cool. Removing the pressure cap from a hot engine will release the cooling system pressure while the coolant temperature is above its atmospheric boiling temperature. When the cap is removed, the pressure will instantly drop to atmospheric pressure, causing the coolant to boil immediately. Vapors from the boiling liquid will blow coolant from the system. Coolant will be lost and someone may be injured or burned by the high-temperature coolant that is blown out of the filler opening.

The coolant antifreeze mixture is renewed at periodic intervals. When the coolant system is empty during the coolant change is a good time to replace hoses and to check thermoststs. Typical fan drive belt conditions are shown in Figure 21-18.

Coolant system problems are indicated by leaks, excessive engine temperature, and by low engine temperature. The cause of a cooling problem is primarily determined through knowledge of the system operation and by a good visual inspection. This is supplemented by tests to determine the coolant temperature and coolant pressure.

Drain and Refill. Manufacturers recommend that a cooling system be flushed and that the antifreeze be replaced at specified intervals. Some recommendations specify yearly antifreeze replacement. Others recommend replacement every 2 years. Draining coolant when the engine is cool eliminates the danger of being injured by hot coolant. The radiator is drained by opening a petcock in the bottom tank and the block is drained by opening plugs or petcocks located in the lower part of the cooling passage, one on an inline engine and two on V-type engines, one on each side.

While the coolant is out of the engine, preventive maintenance should be done. Flushing the cooling system is a good preventive maintenance practice. Water should be run into the filler opening while the drains remain open. Flushing should be continued until only clear water comes from the system. In some cases, the drains are closed and the radiator is filled with a flushing solution added to the water. After a run to thoroughly warm and cir-

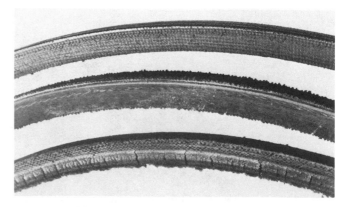

Figure 21-18. Section of fan belts. The top belt is a normal used belt, the middle belt is glazed, and the lower belt is cracked.

culate the solution, it is drained and the system flushed with clear water.

The volume of the cooling system must be determined. It is specified in the owner's handbook and in the engine service manual. The antifreeze quantity needed for the protection desired is shown on a chart that comes with the antifreeze. Protection is required against both freezing and boiling. On automobiles with factory air conditioning, the antifreeze keeps the coolant from freezing in the heater core during maximum cooling. The correct amount of antifreeze is put into the radiator, followed by enough water to completely fill the system. The coolant recovery reservoir should be half filled with the correct antifreeze mixture. In most systems, small air pockets can occur. The engine must be thoroughly warmed to open the thermostat. This allows full coolant flow to remove the air pockets. The heater must also be turned to full heat. In some cases, it is necessary to slightly loosen the upper heater hose to release air trapped in the heater. After the engine has been thoroughly warmed and allowed to cool, the system should be topped off with water to complete the refill job.

Hoses. Coolant system hoses are critical to engine cooling. As the hoses get old, they become either soft or brittle and sometimes they swell larger in diameter. Their condition depends upon the material from which they are made and from the engine service conditions. If a hose breaks while the engine is running, all coolant will be lost. A hose should be replaced anytime it appears to be abnormal. It is good preventive maintenance practice to replace hoses at regular periods, such as every 2 years.

When hoses require replacement, the coolant must be drained. If the same coolant is to be reinstalled in the engine, it will have to be caught in a drain pan. If the coolant is to be replaced, the old coolant should be dumped into a waste drain. It is only necessary to drain coolant from the petcock located at the bottom of the radiator to remove sufficient coolant for hose removal. Hose clamps are loosened and slipped off the portion of the hose that is on the hose neck. Hose clamps found on coolant hoses are a wire spring clamp and a band tightened with a screw. These are shown in Figure 21-19. The wire-spring-clamp type is easily removed

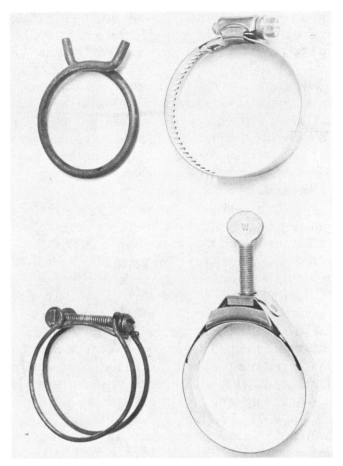

Figure 21-19. Types of hose clamps used on coolant hoses.

and replaced with *special* pliers. The pliers have a slot in the jaw to keep the clamp from slipping. Common slip-jaw pliers can be used, but it is much more difficult to keep them from slipping from the clamp. The band-type clamp is removed and installed using only a screwdriver. With the clamp loose, the hose can then be worked free from the hose neck. Care should be taken to avoid bending the soft metal hose neck on the radiator. The hose neck should be cleaned before a new hose is slipped in place. The clamp is placed on the hose; then the hose is pushed fully over the neck. Hose clamps are secured over the hose at a point from ¼ to ½ in. (6 to 12 mm) from the end of the hose. If the clamp is too close to the hose end, it may slip off. If it is placed too far onto the hose, the edge of the hose may curl and cause a leak. The hose should be cut so the clamp is close to the bead on the neck. This is especially important on aluminum hose necks to avoid corrosion. A typical hose clamp installation is shown in Figure 21-20. When the hoses are in place and the drain petcock closed, the cooling system can be refilled with the correct coolant mixture.

Thermostat. An overheating engine *may* result from a faulty thermostat. An engine that does not get warm enough *always* indicates a faulty thermostat. The thermostat is located in a housing on the engine outlet near the top front of the engine. The thermostat must be removed to check its correct operation.

Coolant will have to be drained from the radiator drain petcock to lower the coolant level below the thermostat. It is not necessary to completely drain the system. The upper hose is removed from the thermostat housing neck; then the housing is removed to expose the thermostat. If the engine operates cold, the thermostat will be stuck open. If the thermostat is closed, a 0.015 in. (0.39 mm) feeler gauge should be forced in the opening so that the thermostat will hang on the feeler gauge. The thermostat is then suspended by the feeler gauge in a bath along with a thermometer, as shown in Figure 21-21. The bath is heated until the thermostat opens enough to release and fall from the feeler gauge. The temperature of the bath *when the*

Figure 21-21. Equipment used to check the opening temperature of a thermostat.

thermostat falls is the opening temperature of the thermostat. If it is within 5°F (4°C) of the temperature stamped on the thermostat, it is satisfactory for use. If the temperature difference is greater, the thermostat should be replaced.

The gasket flanges of the engine and thermostat housing are cleaned. A new gasket is used. One of the flanges may be coated with gasket sealer. The thermostat is placed in the engine with the sensing pellet toward the engine. The thermostat housing is carefully fitted in place with the new gasket. Make sure that the thermostat position is correct. The bolt threads are coated with sealer and tightened in place. Then the upper hose is installed and the system is refilled.

Cleaning. Overheating problems may be caused by deposits that restrict coolant flow. These can often be loosened by back flushing. Back flushing requires the use of a special gun that mixes low-pressure air with water so that it will not damage the cooling system. The upper radiator hose is removed

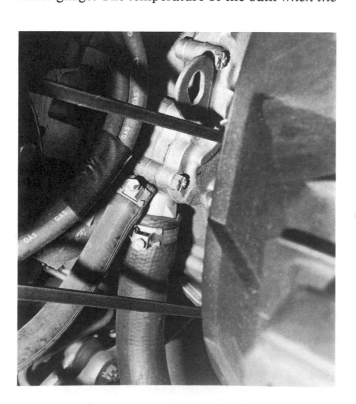

Figure 21-20. A typical hose clamp installation.

from the radiator and the lower hose is removed from the engine. Heater hoses should be removed and the openings plugged. The gun is fitted into the lower hose so that the radiator flushes upward. A long upper hose can be attached to deflect the flushing water from the engine compartment. This setup is illustrated in Figure 21-22. Deposits will come out of the filler opening and out of the hose connected to the upper hose neck. The engine block is back-flushed by fitting the flushing gun into the upper hose, as shown in Figure 21-23. The thermostat is removed and lower hose must be removed

from the radiator. The air and water mixture is forced backward through the engine. The deposits will come out of the lower hose neck. If, after flushing, some deposits still plug the inner portion of the radiator core, the radiator will have to be removed and sent to a radiator repair shop for cleaning. After cleaning the radiator, hoses, drains, plugs, and petcocks are properly assembled. The system is then refilled with coolant.

Overheating can result from exterior radiator plugging as well as internal plugging. External plugging is caused by dirt and insects. This type of plug-

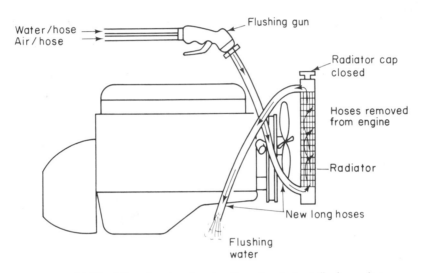

Figure 21-22. A line drawing showing the set up to back flush a radiator.

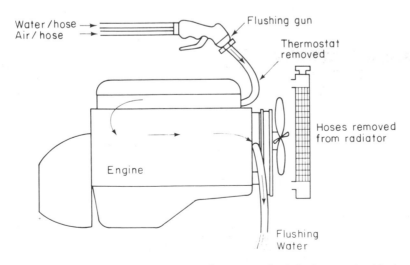

Figure 21-23. A line drawing showing the set up to back flush an engine block.

ging can be seen if you look straight through the radiator while a light is held behind it. The plugged exterior of the radiator core can usually be cleaned with water pressure from a hose. The water is aimed at the *engine side* of the radiator. The water should flow freely through the core at all locations. If it does not, continued use of water pressure and some air pressure will usually loosen the material that is plugging the core. If this does not clean the core, the radiator should be removed for cleaning at a radiator shop. Never push a wire into the core. It may rupture the thin tubes that carry coolant so that the radiator will have to be repaired.

Thermal Fan. The thermal fan is driven by a belt from the crankshaft. It turns faster as the engine turns faster. Generally speaking, the engine is required to produce more power at higher speeds. Therefore, the cooling system will also transfer more heat. Increased fan speed aids in the required cooling. Engine heat also becomes critical at low engine speeds in traffic where the vehicle moves slowly. The fan also turns slowly, so the engine cooling capacity is too small under these conditions. To make conditions worse, the air conditioning places an additional heat load on the system. Thermal fans have been developed to help these high-heat conditions. Drive pulleys turn the thermal fan faster for better radiator cooling during low-engine-speed operations.

The thermal fan is designed so it uses little power at high engine speeds and minimizes noise. The simplest type of thermal fan is one made with flexible blades, as shown in Figures 21-15 and 21-16. A second type of thermal fan has a silicone coupling fan drive mounted between the drive pulley and the fan. A third type of thermal fan has a thermostatic spring added to the silicone coupling fan drive. The thermostatic spring operates a valve that allows the fan to freewheel when the radiator is cold. As the radiator warms to about 150°F (65°C), the air hitting the thermostatic spring will cause the spring to change its shape. The new shape of the spring opens a valve that allows the drive to operate like the silicone coupling drive. When the engine is very cold, the fan may operate at high speeds for a short time until the drive fluid warms slightly. The silicone fluid will flow into a reservoir to let the fan speed drop to idle. A section view of a thermostatic fan drive is shown in Figure 21-24.

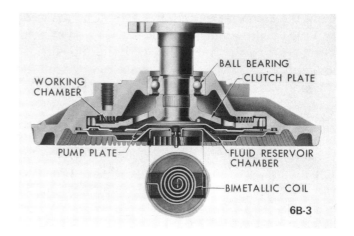

Figure 21-24. A section view of a thermostatic fan drive. (Courtesy of Buick Motor Division, General Motors Corporation).

In some automobiles with a transverse engine, the cooling fan is driven with an electric motor. Temperature switches in the electric circuit of the motor prevent fan operation when the engine is cold. As the engine warms, the fan begins to run at a speed that provides the required cooling. It turns off and on, as needed, to help maintain the required cooling. This also saves energy, to improve fuel economy.

When all other conditions are checked, an overheating problem may result from an improperly operating thermal fan drive. This can be checked by using a stroboscopic light, similar to a timing light. The stroboscopic light will measure fan speeds under different engine operating conditions. Unfortunately, these stroboscopic lights are not usually available in the service shop. When a thermal fan drive malfunction is suspected, the standard procedure in a typical service shop is to replace the drive coupling with a new one and try it out.

Water Pump. The fan is mounted on the water pump shaft on most engines, so the pump speed is governed by the fan pulley size and engine speed. The rate of coolant flow through the system is therefore controlled by the size of the pump impeller. It is sized to match the speed of the water pump to the cooling requirements of the engine. Replacement pumps must have the correct impeller for proper cooling system operation.

Three parts of the water pump may cause trouble and lead to pump failure. Pump seal failure is the most common malfunction. The pump seal is a spring-loaded carbon-face seal that rides against a ceramic seal surface on the impeller. This can be seen in Figure 21-25. A bleed hole to the pump exterior is located between the seal and the shaft bearings to allow leaking coolant to run out of the engine before it can contaminate the pump shaft bearings. This hole can be seen in Figure 18-11.

A second water pump problem is impeller breakage or impeller slippage on the pump shaft. The impeller may be plastic, cast iron, cast aluminum, or stamped steel. A damaged impeller will not pump enough coolant through the system. The only way to be sure that the impeller is satisfactory is to remove the pump and examine it visually. In some engines, a cover will also have to be removed from the pump.

The third problem is the pump bearings. These are sealed for life. If they begin to get noisey, they will have to be replaced with new bearings.

Many service manuals describe water pump overhaul procedures; however, parts departments seldom stock the required parts. The reason for this is that mechanics and technicians are seldom able to make the satisfactory repairs with the tools and equipment normally available in the automotive service shop. The biggest problem in water pump rebuilding is that great care is required to make sure that the pump shaft seal is installed properly. During manufacture, the pumps are assembled in an air-conditioned "clean room" at an exact assembled height to give the seal the correct preload. The sealing surfaces are never touched. In spite of this care, one will occasionally leak. The automotive service technician does not have much of a chance to make a satisfactory water pump repair in a typical service shop. It is a common service practice when a pump is faulty to purchase a new or factory-rebuilt pump rather than run the chance of a "come-back" water pump job.

Radiator Service. Unless they are physically damaged, most radiators are serviced by keeping the radiator exterior and interior clean. Vibration will sometimes loosen some of the soldered joints. These can often be resoldered with the radiator in place, especially open joints that frequently occur around the tanks. Care must be exercised in soldering a radiator because the heat necessary to make a repair may loosen other soldered joints. In most cases, it is advisable to remove the radiator and take it to a specialized radiator repair shop.

The radiator is easily removed. First, the coolant is drained and the hoses are disconnected. If the radiator contains a transmission cooler, these lines will also have to be disconnected from the radiator. The radiator is held in the vehicle with sheet-metal steel side supports or it is clamped in a rubber cushion. These must be unbolted to free the radiator. The radiator can then be lifted straight up from the vehicle. After repairs, the radiator assembly is replaced in the reverse order.

21-5 Cooling System Troubleshooting

The most common cooling system problem is a loss of coolant. Coolant can get out of the system through an overflow pipe. Leaks are usually visible and can be seen after a thoroughly warm engine is turned off. Most cooling systems are equipped with a pressure cap to limit the maximum system pressure. The pressure increases as the coolant temperature increases. While the engine is operating, the coolant temperature is kept below the maximum temperature by coolant circulation and radiator cooling. When the hot engine is turned off, cylinder heat will transfer to the coolant that is not being circulated. This added heat raises the coolant system pressure. Leaks will generally show

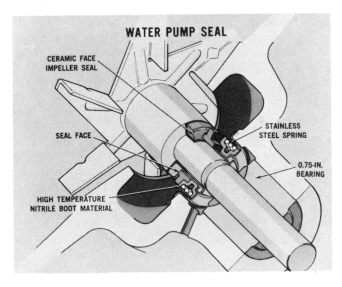

Figure 21-25. A section view of a water pump showing the parts of the seal.

up when this high pressure exists shortly after the engine has been turned off.

Sometimes, coolant leaks can be located by placing a cooling system pressure tester in place of the radiator cap. A typical tester is shown in Figure 21-26. The system is pressurized to the cap pressure setting. If no leakage occurs, the pressure will hold steady. If leakage occurs, the pressure will gradually fall. Leaks can usually be observed visually while the test pressure is on the cooling system.

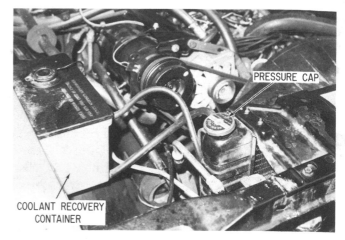

Figure 21-27. A typical coolant recovery system connected to the radiator overflow.

radiator from the recovery container. This prevents coolant loss. Coolant will be lost if the cap is defective on systems that do not have a coolant recovery system. Caps are checked with a cooling system pressure tester that measures the cap release pressure. The cap should hold its design pressure without leakage. Faulty pressure caps should be replaced. Figure 21-28 pictures a typical pressure cap tester in use.

Coolant overflow will also occur if combustion gases leak into the coolant system. This happens when the head gasket does not seal properly or when there is a crack in the head. Sometimes, internal engine leaks allow the coolant to enter the interior of the engine and mix with the motor oil. Internal leakage is usually indicated by an apparent increase in oil quantity on the oil dipstick. The oil

Figure 21-26. A pressure tester being used to pressure test the cooling system for leaks.

Leaks can occur at the thermostat flange, at coolant hose joints, in the radiator, in the heater core, or in engine soft plugs. Any leak should be immediately repaired before loss of coolant causes the engine to overheat and does costly damage.

Coolant loss can occur as the coolant expands while it is being heated. This expanded coolant is forced out through the pressure cap and the overflow pipe. Loss is prevented on most emission-controlled engines by connecting the overflow pipe to a plastic container reservoir similar to the windshield-washer-fluid bottle. Hot coolant overflows into the coolant recovery container. This can be seen in Figure 21-27. When the coolant cools and contracts, the coolant is drawn back into the

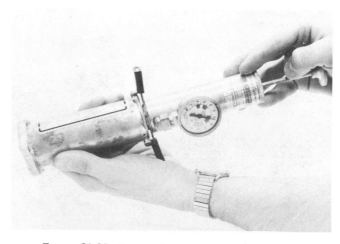

Figure 21-28. A typical pressure tester being used to test the radiator cap pressure.

will have a brown or white foamy appearance. Internal leaks should be corrected as soon as possible to minimize engine damage.

Combustion leakage and air leaking on the suction side of the water pump can often be observed by filling the coolant system clear up into the filler neck. With the cap off and the engine running, leakage is indicated by bubbles forming in the coolant at the filler neck. Another method is to place the radiator pressure tester on the filler neck in place of the filler cap, then start the engine. Combustion or a suction air leak will cause the cooling system pressure to rapidly increase. If a hydrocarbon–carbon monoxide tester is available, combustion leakage will be indicated by hydrocarbons at the filler neck when the tester pickup is held *above* the coolant.

The coolant pump, as well as the fan, is driven by a belt. Belts, like hoses, deteriorate. They stretch, wear, and crack. These conditions can be checked visually as pictured in Figure 21-18. A loose belt will slip and will not drive the pump or fan at the speed required for proper cooling. Belt tension is critical for correct performance. The use of a tension gauge is the best method of measuring belt tension. Figure 21-29 shows a typical belt tension gauge on a belt. In some installations, a square hole is provided in the idler bracket so that a torque wrench can be used to adjust the belt tension. Another is designed with a square boss so that an open-end

Figure 21-29. One type of belt tension gauge being used to measure the tension of a fan belt.

wrench can be used. Other methods used to measure the belt tension are not reliable.

If the belt breaks, the water pump stops turning so that coolant will not flow. This allows the engine to overheat. In many automobiles, the same belt that drives the fan and water pump also drives the alternator. When the belt breaks, the alternator stops charging. This gives the driver an *immediate* visual notice of failure. Either the charge light comes on or the ammeter discharges. It is advisable to examine the belts frequently and to replace them at regular periods, just as hoses are replaced, to provide trouble-free service.

REVIEW QUESTIONS

1. What must the cooling system do? [21-1]
2. What happens when the engine operating temperature is too low? [21-1]
3. What happens when the engine operating temperature is too high? [21-1]
4. How does moisture get into the oil? [21-1]
5. What limits the maximum operating temperature of an engine? [21-1]
6. Why would it be an advantage to use water as a coolant? [21-2]
7. What would be the advantage of using straight antifreeze as a coolant? [21-2]
8. Why is a 50/50 mixture usually recommended as a water–antifreeze mix for use as an engine coolant? [21-2]
9. What is done to keep radiator cost as low as possible? [21-3]
10. Describe radiator construction. [21-3]
11. When is aluminum used for radiators? [21-3]
12. What type of radiator core is generally used in automobiles? [21-3]

13. Where is the main limitation to heat transfer in a cooling system? [21-3]
14. What determines the heat-transfer capacity of a radiator? [21-3]
15. How much heat must the radiator be able to remove from the engine? [21-3]
16. Compare the heat transfer efficiencies of the cross-flow and the down-flow radiator designs. [21-3]
17. Describe the operation of a pressure cap. [21-3]
18. Describe the operation of the coolant recovery system. [21-3]
19. What is the advantage of using a 15 lb/in² pressure cap? [21-3]
20. What is cavitation of the coolant pump? [21-3]
21. What is the advantage of thermo-regulating cooling fans? [21-3]
22. How does a flexible blade fan operate? [21-3]
23. When are high capacity cooling systems used? [21-3]
24. When should the coolant level be checked? [21-4]
25. What causes coolant to come out of a hot engine when the pressure cap is removed? [21-4]
26. How are cooling system problems located? [21-4]
27. How should the cooling system be drained and refilled? [21-4]
28. What preventative maintenance should be done while the coolant is out of the engine? [21-4]
29. Why is the cooling system topped off with water after a refill and engine warm up? [21-4]
30. Describe the procedure that should be followed to replace a cooling system hose. [21-4]
31. What causes an engine to always run cool? [21-4]
32. How is a thermostat checked and replaced? [21-4]
33. How is the radiator back flushed? [21-4]
34. How is the engine back flushed? [21-4]
35. How is the exterior of the radiator cleaned? [21-4]
36. How does the thermal fan operate? [21-4]
37. How is the water pump sized to fit an engine? [21-4]
38. Name three water pump problems. [21-4]
39. What is the cooling system's most common problem? [21-5]
40. When are cooling system leaks most likely to show up? [21-5]
41. How can the technician tell that some coolant has found its way into the lubricating oil? [21-5]
42. How can coolant leakage into the combustion chamber be discovered? [21-5]
43. How are fan belts properly tightened? [21-5]
44. How does the operator know that the fan belt has broken? [21-5]

Ignition System Service

The ignition system forces a spark or an electrical arc across the spark plug electrodes to ignite the air/fuel charge in the combustion chamber. The electrical arc must have enough energy to increase the temperature of the air/fuel charge around the spark plug to the kindling point. At this point, combustion becomes self-sustaining. **The voltage necessary to overcome the resistance of the spark plug gap is called the required voltage.**

It makes no difference what type of ignition system is used. The ignition system may be a conventional breaker-point design, it may use high-energy transistor switching, or it may use a capacitive discharge system. The spark must be delivered to the spark plug with enough energy to ignite the charge.

It is not only important for the charge to ignite, but it must ignite at the correct instant so that the burning charge will produce maximum useful energy as the hot gases build up pressure within the combustion chamber. The spark arc is timed so that the maximum combustion chamber pressure will occur when the crank pin is from 5° to 10° after top center. The ignition firing or timing point needs to be adjusted for changes in the quality of the air/fuel charge. The quality of the charge affects the rate of combustion. Timing is adjusted so that maximum pressure will always occur at the correct crankshaft angle under all operating conditions requiring engine power.

22-1 Ignition System Operation

The ignition system uses the battery as the source of electrical energy while the engine is being started. After the engine is running, the charging system provides the electrical energy. Electrical power is carried through wires, switches, and resistors to an ignition coil. The coil transforms low primary voltage to high secondary voltage. The secondary voltage is delivered through a distributor to the spark plug located in the combustion chamber, where it ignites the combustion charge.

The ignition coil is the source of the energy to produce a spark or an electric arc across the spark plug electrodes. The ignition coil consists of two coil windings, primary and secondary. They are wound around a soft-iron core and placed within a case using connections and insulators. In a breaker-point type of ignition system, the primary coil winding is connected in series with the battery, resistor, ignition switch, and breaker points. In a solid-state ignition system, the primary coil winding is connected in series with the battery, ignition switch, and control transistor. An inductive pickup signals the control transistor when to turn on and off. The secondary coil winding is connected in series with the distributor rotor, distributor cap, and spark plug.

In operation, battery voltage pushes current through the coil primary winding. This current flow builds up a magnetic field around the primary winding and in the soft-iron core, as illustrated in Figure 22-1. When the breaker points open or the control transistor is turned off, the current flow through the primary winding stops, causing the magnetic field to collapse. During collapse, the magnetic lines of force very rapidly cut through the secondary windings as they collapse. This rapid relative motion between the magnetic lines of force and conductor induces a high voltage in the secondary windings of the coil with enough energy to force an arc to flash across the spark plug electrodes. This entire sequence of ignition system events must occur each time a spark plug fires. In an eight-cylinder engine running at 4000 rpm, there are 266 spark plug firings each second. It is impossible to follow this action with a meter because the meter cannot move fast enough, so an engine scope (cathode ray oscilloscope) is used to show this constantly changing voltage.

22-2 Distributor Operation

The ignition distributor is driven from the engine crankshaft at one-half crankshaft speed. The breaker points, inductive pickup, condenser, rotor,

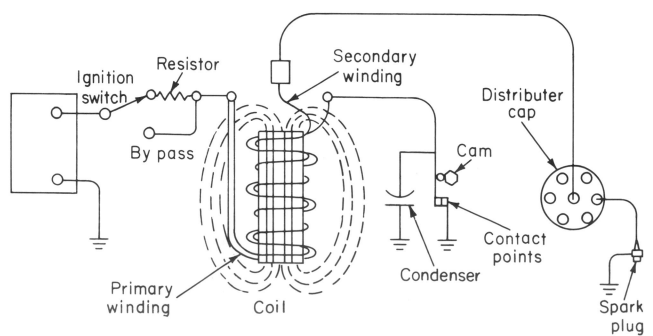

Figure 22-1. The magnetic field around the primary winding of the coil is shown by the dotted lines.

cap, and timing advance mechanisms are located in the distributor. This arrangement puts all the ignition system's moving parts into a single unit. The solid-state module and coil are included with the distributor in HEI ignition systems. This can be seen in Figure 22-2 that shows the major sections of the system.

The distributor gets its name from the part that directs the electrical energy of the secondary coil winding to the spark plugs. It distributes the energy in the correct sequence to match the engine firing order. The electrical energy of the secondary coil is put into the center distributor cap *tower*. It goes through the tower to a *button* inside the cap. A spring clip on the distributor rotor contacts the center button in the cap. The *rotor* has a conductor plate from the spring clip to an extended tip that comes close to the distributor cap *electrodes* as the rotor turns. As the distributor rotates the rotor tip passes near each distributor cap electrode, in sequence, as the breaker points open. This directs the secondary electrical energy through each spark plug cable leading to the spark plug in the correct firing order.

It is convenient to locate the breaker point cam on the same shaft as the rotor because these two parts must always maintain their relative position to each other for rotor and electrode alignment. A breaker-point contact set is attached to the breaker plate within the distributor. There are adjustments to move the points closer to the cam for a larger point gap or away from the cam for a smaller point gap. Figure 22-3 shows breaker points with the rubbing block on the high point of the cam. This holds the points fully open.

There is one cam lobe for each cylinder. Four-cylinder cam lobes are spaced at 90°; even firing six-

Figure 22-3. Breaker point clearance is adjusted when the rubbing block is on the high point of the cam as shown here.

Figure 22-2. The coil and module used in a HEI ignition system can be seen in this illustration.

cylinder cam lobes are spaced at 60°; and eight-cylinder cam lobes are spaced at 45°. Within each of these cam angles, the points must be closed long enough to store electrical energy in the coil as magnetism. The points must open far enough to minimize point arcing. Normally, the points are closed from 65 to 70% of the cam angle. This gives enough time to saturate the coil with magnetism. **The number of degrees the cam rotates while the points are closed is known as the dwell.** Any change in the breaker-point gap will change the dwell approximately 1° for each 0.001 in. (0.025 mm) change in point gap. This is illustrated in Figure 22-4.

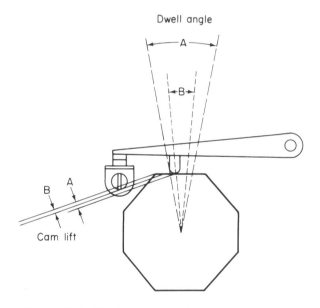

Figure 22-4. This line drawing shows that as the point gap increases the dwell angle also decreases.

The distributor cam opens the breaker points and a breaker-point spring closes the points. A weak breaker-point spring will allow the points to be thrown clear of the cam. When this happens, they will *float* and bounce as they close. Floating points cannot close fast enough. This shortens the dwell period when the points are closed. The available energy stored in the coil will, therefore, be reduced. Excessively high breaker-point spring tension, on the other hand, will cause rapid rubbing-block wear. This, in turn, will cause the point gap to be less. A small point gap will also reduce the amount of electrical energy available at low speeds.

Original-equipment distributor points are designed to maintain constant breaker-point gap

and dwell. The contact surfaces of normal breaker points will gradually burn. This tends to enlarge the gap. Gap enlargement is neutralized by the breaker-point rubbing block wear on the cam. This tends to close the gap. These two wear conditions in service are designed to counteract each other to keep the breaker-point gap nearly constant. When wear causes the breaker-point gap to *decrease,* the engine's basic timing angle will be *retarded.* When wear causes the gap to *increase,* the timing will *advance.*

The breaker-point cam is driven by the distributor shaft through a mechanical advance mechanism. The mechanical advance mechanism consists of centrifugal flyweights that are retained by springs. An exploded view of the mechanical advance parts from a typical distributor is shown in Figure 22-5. As the distributor rotates faster, the flyweights swing outward against spring pressure.

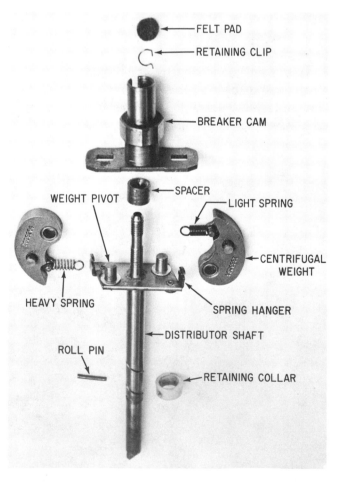

Figure 22-5. An exploded view of the mechanical advance parts of a distributor.

Cam surfaces on the flyweights advance or move the breaker-point cam position forward *in the direction* of cam rotation on the distributor shaft. The pulse inductor of the inductive pickup is attached to the distributor shaft in the same way the breaker point is attached. The mechanical advance moves the pulse inductor in the same way the cam is moved. The pulse inductor can be seen in Figure 22-6. The distributor shaft drive is timed to a specific crankshaft angle with the distributor advance mechanisms in full retard. This is called *basic timing.* The distributor *shaft* always maintains this position in relation to the crankshaft. The distributor advance mechanism moves from the basic timing. The mechanical advance is only sensitive to engine speed. It advances the ignition timing as the engine speed increases. Flyweights and springs control the amount of timing change at any specific engine speed by rotating the cam or pulse inductor in the direction of shaft rotation. When this happens the contact points or the control transistor turns the primary current off sooner. In service, the amount of advance is checked on a distributor machine or by advance controls on a timing light. The actual advance of the distributor is compared against specifications. Corrections in the mechanical advance rate are made on some types by changing the counterweight shaft assembly springs. On others, it can be changed by adjusting the spring hanger position. HEl types require changing the entire shaft and mechanical advance mechanism.

The breaker points and inductive pickups are mounted on a movable plate within the distributor housing. A link from the diaphragm in the vacuum advance unit holds the movable plate position. This is identified in Figure 22-7. The outside part of the vacuum diaphragm is connected by tubing to sense vacuum at a port within the carburetor. Vacuum pulls the movable plate in an advance direction *against the direction* of cam rotation. This opens the contact points or turns the control transistor off sooner. The engine runs with a less dense lean mixture that burns slowly when the ported vacuum is great. This requires a high advance to complete combustion at the correct crankshaft angle after top center. As the throttle is opened, manifold vacuum is reduced. This lowers ported vacuum, which, in turn, allows the vacuum advance mechanism to retard the movable plate. The combustion charge mixture is more dense, and it is usually richer as the manifold vacuum is reduced. This condition requires less advance to complete combustion at the proper crankshaft angle.

The vacuum advance mechanism is sensitive to *ported vacuum*. This, in turn, comes from manifold vacuum. Manifold vacuum results from the engine load and throttle position.

Mechanical and vacuum advance mechanisms work together to provide the engine with the advance required to give the most efficient combustion and lowest practical emission level at each operating condition. Any change in the basic timing

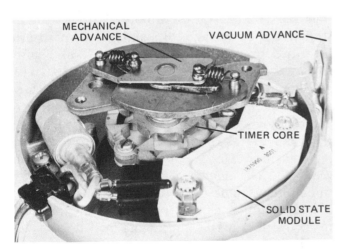

Figure 22-6. The pulse inductor timer core of an HEI ignition system can be seen in this illustration.

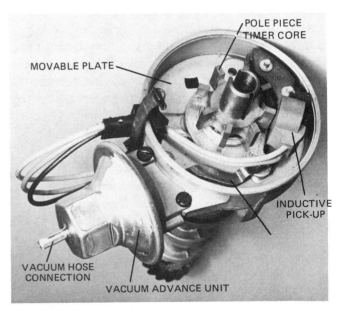

Figure 22-7. The vacuum advance unit and link can be seen in this illustration.

or advance mechanisms will reduce normal performance. If, on the other hand, the engine is modified from the manufacturer's original design, timing and advance curves should also be modified to produce the most efficient performance. The emission decal or references on emission controls should be consulted for the control devices that are used to modify the vacuum advance to minimize exhaust emissions.

22-3 Breaker-Point Replacement

Breaker points are checked for gap, dwell, and point resistance. The breaker cam is shaped to open the points a set amount of gap. As the breaker-point gap is adjusted wider, the rubbing block must start to open the breaker points sooner in order to be wide open at the cam high point. Gap changes will, therefore, affect both dwell and engine timing. If the breaker cam is not worn, point gap will be correct when dwell is correct. Breaker cams seldom wear and when they do, the wear is visible.

As breaker points operate, they develop a frosty appearance on their contact surfaces. This type of surface makes a good electrical contact. The breaker points should be checked for resistance to make sure that the electrical contact is good.

Points and Condenser. Two methods can be used to replace the points and condenser. The amateur mechanic will usually change them while the distributor remains installed in the engine. It is done in this way to keep from changing the basic timing. The professional automotive technician will usually remove the distributor from the engine because the breaker points and condenser can be replaced better and faster. It also provides an opportunity to examine the rest of the distributor for potential problems.

Breaker-point replacement starts by removing the distributor cap. It is fastened with spring clips, screwdriver lock clips, or screws. The cap is loosened and placed at the side of the distributor with all the ignition cables remaining in the cap towers. This is shown in Figure 22-8. If the distributor is to be removed from the engine, the vacuum advance unit and rotor positions should be noted and marked on the distributor housing. This will allow the distributor to be replaced in the same position after the points and condenser are installed. The primary

Figure 22-8. The distributor cap removed from the distributor.

lead is removed from the coil and the vacuum hose is removed from the vacuum advance unit nipple. The distributor hold-down clamp can then be removed so that the distributor can be pulled straight out of its engine opening. These parts are identified on the distributor pictured in Figure 22-9. Heavy internal engine deposits will sometimes make distributor removal difficult. Careless prying can break the distributor housing and bend the shaft. The distributor shaft will turn a small amount as the distributor is removed from most engines. This is caused by the angled-drive gear teeth. The amount of rotation should be noted to ease reinstalling the distributor. The following breaker-point changing procedure will be the same if the distributor is in the engine or if it is on the bench.

The distributor *rotor* is removed. The rotor is merely pulled off from the shaft on distributors

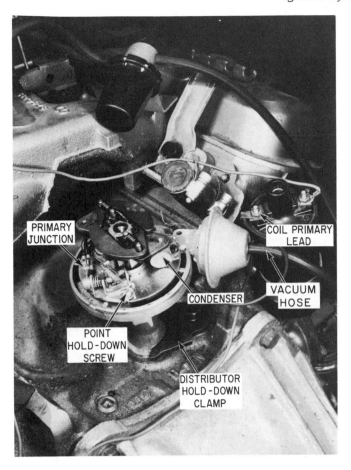

Figure 22-9. Identification of parts of a typical distributor. These parts are used when replacing the breaker points.

Figure 22-10. A condenser combined with the breaker points is used to reduce radio noise.

having the centrifugal mechanical advance *below* the breaker points. Two screws will have to be removed from the rotor to free the rotor from distributors having the mechanical advance mechanism *above* the breaker points. If the rotor tip or distributor cap electrodes show excessive burning, they should be replaced with new parts, especially if the engine has been misfiring prior to the distributor service. Contact breaker points are exposed when the rotor is removed on most distributors. Some distributors require the removal of a metal shield cover to expose the points. The metal shield is used to reduce radio noise on vehicles having the antenna in the windshield glass. Late models combined the points and condenser, as shown in Figure 22-10, for the same reason.

The primary wire and condenser are connected to the breaker points at a single junction. The junc-

tion may have a screw, a bolt and nut, or it may be held by a spring pressure fit. The junction is separated and the screws holding the points to the breaker plate are removed so that the points can be lifted from the distributor. If the condenser is to be replaced, it can also be removed by taking out one hold-down screw. After wiping the cam, a new condenser and breaker point set can be installed following the reverse order used to remove them. If a screw is dropped into the distributor, the screw *must* be retrieved. If it jams within the distributor, it will break the distributor or distributor drive. Cam lubricant should be wiped smoothly over the cam surface and a drop of oil should be put on the felt plug inside the cam (Figure 22-11). Excessive lubrication will get on the points and cause them to burn. The distributor bearings should be oiled if an oiler is used. Most modern distributor bearings have permanent lubrication or they are lubricated with motor oil from within the engine.

When a distributor is in good mechanical condition, the dwell will be correct if the breaker-point gap is correct. A breaker-point gap between 0.016 and 0.018 in. (0.04 and 0.045 mm) will allow the engine to start. Some four-cylinder engines require a breaker-point gap of 0.025 in. (0.055 mm). This point gap will usually place the dwell within the correct range. The point gap is adjusted while the breaker-point rubbing block is on the highest part of one of the breaker cam lobes. An example of this is shown in Figure 22-12. Most technicians adjust the points to the correct dwell. Putting a feeler gauge between the points may accidentally put dirt between the points.

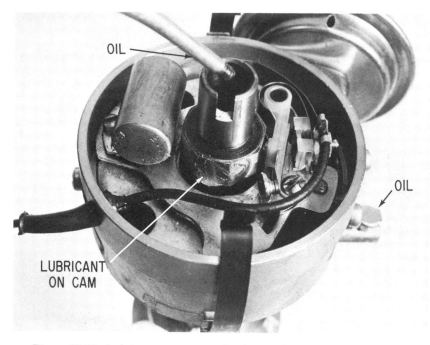

OIL

OIL

LUBRICANT
ON CAM

Figure 22-11. Lubrication points in a distributor when the breaker points are replaced.

The breaker point contact set must be adjusted properly for maximum service. It should be checked for contact point alignment and for spring tension. Contact points can be aligned by carefully bending the stationary contact to align it with the movable contact. Point spring tension of 19 to 23 oz is checked with a spring scale similar to the one shown in Figure 22-13. Tension can be increased by pushing the point spring into the holding screw, and it can be reduced by sliding the spring from the holding screw. The dwell should be rechecked after making any point adjustment. In a professional shop, the centrifugal and vacuum advance mechanisms would be checked at this time.

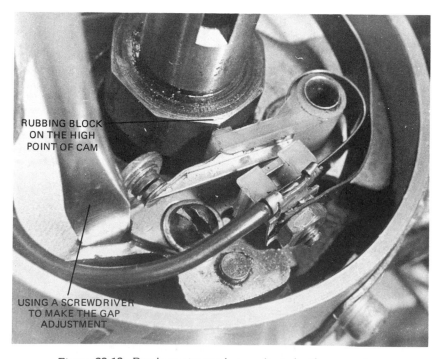

RUBBING BLOCK
ON THE HIGH
POINT OF CAM

USING A SCREWDRIVER
TO MAKE THE GAP
ADJUSTMENT

Figure 22-12. Breaker point gap being adjusted with a screwdriver.

Figure 22-13. Measuring the breaker point spring tension.

The distributor is reinstalled and the basic timing is mechanically set as discussed in Section 12-7. If the ignition cables have been removed from the distributor cap, they can be easily reinstalled in the correct cap tower. Number one cable goes into the tower toward which the rotor points when the timing marks line up as number one cylinder is on its compression stroke. Following around the cap in the direction the rotor turns, the ignition cables are installed according to the engine firing order. The cylinder firing order is usually shown in raised cast numbers on the manifold. If it is not, it will have to be found in a specification book. Moving in the direction of distributor rotation, place the spark plug cable leading to the next cylinder in the firing order into the next distributor cap tower. Follow this by inserting each succeeding spark plug cable into the distributor cap towers in the proper firing order sequence. If the engine is not going to be timed with a timing light, reinstall the vacuum hose. With the parking brake set and the transmission in neutral, start the engine.

Setting Dwell. Dwell is only checked on breaker point ignition systems. If a dwell meter is available, it should be attached to the distributor side of the coil and the dwell checked. Dwell can be easily adjusted on an external-adjustment distributor with the engine idling, as shown in Figure 22-14. If dwell is incorrect on the other types of distributors, the

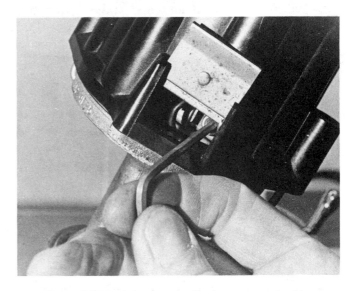

Figure 22-14. Adjusting dwell on an external adjustment distributor.

engine will have to be stopped and the distributor cap removed. The breaker point gap is changed only enough to correct the dwell. Some dwell meters can be used to check dwell while the engine is being cranked. If one of these meters is used, the rotor must be removed. The dwell can then be adjusted as the engine is cranked. Basic ignition timing should *always* be reset *after* adjusting dwell.

Timing Light. Connect the timing-light pickup cable to the number 1 ignition cable and connect the light to its proper power source, either the vehicle battery or a 110-volt electrical outlet. It is critical on most emission-controlled engines to attach this timing light pickup to the number 1 spark plug cable, even if it is not as accessible as one of the other spark plugs. Set the parking brake and place the transmission in neutral; then start the engine. Let the engine run slowly at curb idle speed with the vacuum line to the distributor removed and plugged so air does not go into the carburetor. With the timing light aimed at the timing marks and the distributor hold-down clamp slightly loose, adjust the distributor until the timing mark lines up at the specified degree. Tighten the hold-down clamp and recheck the timing. If the timing changes, readjust the timing as necessary; then reconnect the distributor vacuum line.

Ignition Advance Test. In addition to basic timing, the timing light can be used to determine if the distributor mechanical advance and vacuum advance mechanisms are operating. To do this, remove and plug the distributor vacuum hose so that air does not enter the carburetor port. With the timing light aimed at the timing marks, gradually increase the engine speed. The timing marks should appear to move in the advance direction when the mechanical advance is operating. With the engine running and held at approximately 1200 rpm, observe the timing mark with the timing light as the vacuum line is reconnected. The timing mark will again move in the advance direction if the vacuum advance mechanism operates. This test will indicate that the distributor advance mechanisms are operating. It does *not* indicate that the advance mechanisms are operating correctly.

Some timing lights are made with a built-in advance meter. When doing basic timing, the technician must be sure to have the meter turned *off* so that it will operate as a simple timing light, as pictured in Figure 22-15. After basic timing is set, the engine rpm is adjusted to a specified test speed. At this speed, the advance meter control is turned until the timing mark appears to be in the same position as it had when the basic timing was set. At this point the meter reading indicates the number of degrees the distributor has advanced. This advance check can be done with the vacuum line off to measure mechanical advance. The vacuum line is reconnected to measure the total advance produced by

Figure 22-15. Using a timing light to set the basic ignition timing.

both mechanical and vacuum advance mechanisms. The distributor vacuum advance hose must be connected at the end of this test. If the timing advance is not correct, the distributor will have to be removed and tested on a distributor machine.

22-4 Spark Plugs

The entire ignition system is designed to produce an arc between the spark plug electrodes. If the correct spark plug is not used, ignition will malfunction, resulting in a misfire or in damaged pistons. The spark plug changes the electrical energy in the ignition secondary to thermal energy. The electrical arc across the spark plug electrodes will ignite enough of the intake charge so that the remaining charge will burn in a normal manner.

Spark Plug Construction. The spark plug consists of three major parts, the shell, the insulator, and the electrodes. These can be identified in Figure 22-16. The *shell* supports the insulator and it has threads that screw into the head. The thread portion must be long enough to allow the electrodes of the spark plug to enter the combustion chamber. This length is called the spark plug *reach*. If the reach is too long, it can damage the valves or piston. Threads on the spark plug have metric 14- and 18-mm threads. The shell seals the combustion chamber spark plug hole. Some shells seal with a tapered spark plug seat. Others seal with a metal spark plug gasket (Figure 22-17).

The ceramic spark plug insulator is sealed inside the shell, so that it makes both a pressure and a thermal seal. Much of the spark plug development work has been concentrated on the insulator. It must be able to operate at high temperatures and high pressures as it insulates the high secondary voltage. During manufacturing, the insulator materials are formed to a puttylike consistency. The putty is formed in the insulator shape, then fired in a furnace. The heat causes the insulator to shrink and harden. The finished insulator is close to diamond hardness. It is placed in the shell with sealing material; then the shell is crimped around the insulator to produce a gastight seal.

The center electrode is placed in a hole down through the center of the insulator. Modern electrodes are made from two pieces. The spark plug cable is connected to one. The other extends into the hot combustion chamber. Using two pieces allows

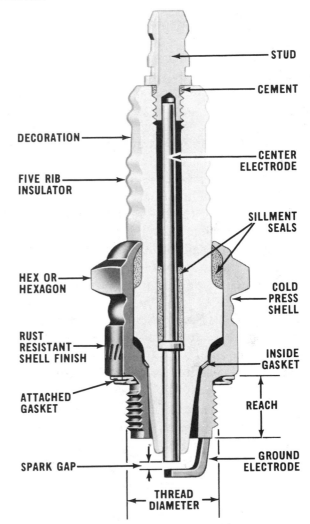

Figure 22-16. Parts of a typical spark plug (Courtesy of Champion Spark Plug Company).

Figure 22-17. Typical automotive spark plugs.

the spark plug manufacturer to select the type of metal that will best meet the operating requirements. The center electrode is sealed with a gastight electrical conducting seal material.

Some spark plug types have a resistor installed between the two sections of the electrode. Its 10,000-ohm resistance changes the oscillating frequency of the ignition secondary voltage at the instant the electrical arc is established across the electrodes. This changes the frequency of the electrical radiation from the ignition system so that it does not interfere with television and radio reception. The resistor also increases the life of the spark plug electrode by cutting down peak current that would flow across the electrode gap in the arc. Figure 22-18 is a section view of a resistor spark plug.

Spark Plug Heat Range. Spark plugs must operate within a specified temperature range. If the spark plug operates too cold, it will *foul* with deposits. These deposits will leak secondary electrical energy to ground, so the spark plug will not fire. If the

spark plug operates too hot, the electrodes will erode rapidly and will cause preignition. Preignition will lead to physical engine damage, usually burned pistons. The minimum spark plug temperature for nonfouling operation is 650°F (340°C). Above 1500°F (815°C), preignition will generally occur. This is shown in Figure 22-19. Spark plugs must operate within these temperatures under all normal operation conditions.

When an engine is running under heavy loads, such as sustained high-speed driving, the combustion chambers become hot. A cold heat range spark plug is required when operating at these high-temperature operating conditions to prevent overheating the spark plugs. Spark plugs that have a hot-heat range are used in engines that have a tendency to run cold. Cold-running spark plugs foul from oil or from light-duty operation. Hot heat range spark plugs keep the temperature of the spark plug nose high enough to eliminate fouling. Drag-racing operation is such a short-time operation on each run that cold heat range spark plugs are seldom necessary.

A cold heat range spark plug transfers heat from the spark plug nose through the shell faster than does a hot heat range spark plug. It can be seen in Figure 22-20 that the heat must travel farther in a hot heat range spark plug. The spark plug heat

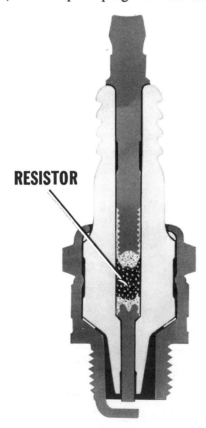

Figure 22-18. A section view of a resistor spark plug (Courtesy of AC Division, General Motors Corporation.)

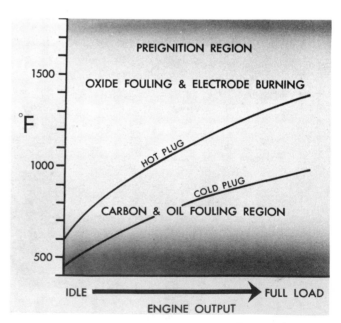

Figure 22-19. A graph showing the operating temperature limits of a spark plug (Courtesy of Champion Spark Plug Company).

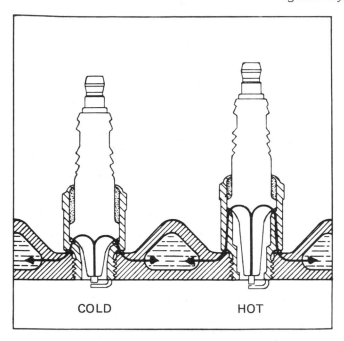

Figure 22-20. Heat transfer through a spark plug (Courtesy of Champion Spark Plug Company).

COLD   HOT

range selected for replacement spark plugs should match the manufacturer's specifications. It should be noted that the spark plug heat range does *not* affect the temperature of the electrical arc. Modified engines may require a different heat range spark plug than those specified for the original engine. Spark plugs for modified engines should be selected by working up from cold spark plugs. If fouling occurs, the next-hotter heat range should be used. The correct heat range is reached when the spark plug does not foul. In this way it is possible to avoid preignition while selecting the correct heat range spark plug.

It is interesting to note that the center electrode of automotive spark plugs has *negative* polarity. Hot bodies are known to have increased electron activity. The center electrode is the hottest part of the spark plug and, therefore, requires the lowest voltage to force free electrons to form an arc. The hot negative voltage center electrode helps to keep required ignition voltage low. If the polarity is reversed, required voltage will increase. Secondary polarity can be changed by reversing the primary coil leads, so care is required to see that coil connections are made correctly.

Spark Plug Service. The spark plug is the end point in the ignition system. It is located in the combus-

tion chamber, where it is subjected to combustion deposits, erosion, and corrosion. Spark plugs require periodic service, regardless of the type of ignition system used.

Spark plugs are removed by first removing the spark plug cable. This is done by carefully twisting and pulling *the boot* that fits over the spark plug. In this way, the cable can be removed without damaging or stretching the spark plug cable. Blow air around the spark plug to remove any loose material before removing the spark plug. Use a good spark plug socket, with an internal cushion, to remove the spark plug without damage. Generally, the spark plugs are removed and laid out for inspection. A better removal practice should be followed when a compression test is to be run. Loosen the spark plugs about one-quarter turn, and reattach the spark plug cables. The engine is then started and sped up two or three times. Loosening the spark plugs will break any carbon at the junction between the combustion chamber surface and the spark plugs. The carbon chips will be blown from the engine when it is run so that they will not get under a valve and cause faulty compression test readings.

When the spark plugs are removed, they should be examined critically. This examination is often called *reading the plugs*. The condition of the electrodes and the type of carbon on the spark plug nose indicate how that cylinder has been operating.

The most obvious spark plug condition is the type of carbon on the spark plug nose. Examples are shown in Figure 22-21. Normal spark plugs will have a light tan to gray deposit, depending on the additives in the gasoline and oil that has been used in the engine. If there are almost no deposits and these are white and the electrodes are badly eroded, the spark plug has been running very hot. Heavy sooty deposits indicate rich air/fuel mixtures, and heavy wet deposits indicate high oil consumption. Dry, fluffy carbon deposits indicate incomplete combustion. Some engines, especially ones with high mileage, tend to develop heavy white deposits that bridge the spark plug gap. This results from oil getting into the combustion chamber to form the deposits. Spark plug manufacturers supply full-color pictures of these conditions so that the technician can compare the spark plug appearance and diagnose the problem.

Figure 22-21. Examples of deposits on the spark plug nose (Courtesy of Prestolite Company).

Spark plugs *can* be cleaned and serviced; however, it is the general practice to install new spark plugs when a tune-up is done on a customer's engine. The customer usually plans to get over 10,000 miles (18,500 km) before the next tune-up and so new spark plugs are expected. New spark plugs at specific service intervals are specified for automobiles equipped with a catalytic converter.

22-5 Ignition System Testing

The ignition system is one of the interrelated engine systems. Input voltage to the coil primary comes from the battery while cranking. It comes from the voltage regulator in the charging system when the engine is running. Ignition system output is the *voltage required* by the spark plug. Any change in combustion chamber conditions, such as compression pressure, temperature, and air/fuel mixture ratios will affect the required voltage. These must be considered when testing ignition systems.

A cathode ray oscilloscope is one of the best instruments to observe the overall characteristics of the complete operating ignition system. Normal or abnormal operating conditions are indicated by the scope trace.

The ignition oscilloscope trace displays secondary voltage on a time base. The scope sweep is usually triggered by the number 1 spark plug cable; however, any secondary cable could be used. As the voltage increases in the number 1 spark plug wire, a spike is produced at the right edge of the pattern. This is the trigger that immediately shifts the sweep to the left edge to begin a new trace. Figure 22-22 shows a trace with several faults.

Anything in the primary that has an effect on available voltage is reflected in the secondary scope pattern. Parts of the ignition that the scope pattern shows to be faulty must be rechecked with a specialized tester to verify the malfunction. Details of the ignition oscilloscope and its use are beyond the subject matter of this book.

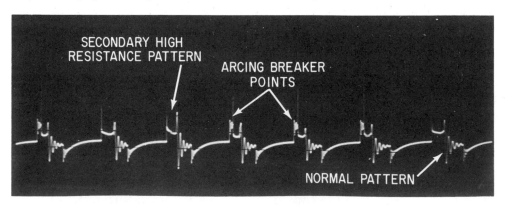

Figure 22-22. An oscilloscope trace showing several ignition faults.

REVIEW QUESTIONS

1. Define "required voltage" in the ignition system. [INTRODUCTION]
2. How is the ignition firing point determined? [INTRODUCTION]
3. What does the coil do? [22-1]
4. What must happen for the secondary voltage to build up enough to make an electrical arc at the spark plug? [22-1]
5. Why does the distributor rotate at one-half of the crankshaft speed? [22-2]
6. How does the ignition secondary energy get through the distributor? [22-2]
7. Why is the breaker-point cam or pulse inductor mounted on the same shaft as the rotor? [22-2]
8. How many cylinders are fired by each breaker cam lobe? [22-2]
9. How is energy stored in the coil? [22-2]
10. What causes the breaker points to bounce? [22-2]
11. What causes excessive rubbing block wear? [22-2]
12. What happens to the basic ignition timing as the point gap decreases from rubbing block wear? [22-2]
13. Which way do the flyweights move the breaker-point cam or pulse inductor? [22-2]
14. What is the starting point for ignition advance? [22-2]
15. What direction, in relation to the distributor rotation, does the mechanical advance move the breaker-point cam? [22-2]
16. What direction, in relation to the distributor rotation, does the vacuum advance move the breaker plate? [22-2]
17. What is indicated by a frosty surface on the contact point face of the breaker points? [22-3]
18. How is the distributor rotor removed? [22-3]
19. How are the breaker points replaced? [22-3]
20. How should the point gap be set correctly? [22-3]
21. How should the point spring tension be set? [22-3]
22. Describe the procedure to reinstall the distributor and set the basic ignition timing. [22-4]
23. How should the ignition cables be installed in a distributor cap? [22-3]
24. What should be done after the dwell is adjusted? [22-3]
25. How is timing set with a timing light? [22-3]
26. How can a timing light be used to see if the mechanical and vacuum advance units are operating? [22-3]
27. How can the amount of distributor mechanical and vacuum advance be checked on a running engine? [22-3]
28. Name two types of spark plug seals. [22-4]
29. What is the purpose of a spark plug resistor? [22-4]
30. What happens when the heat range of the spark plug is incorrect? [22-4]
31. How is the correct heat range spark plug selected for an engine? [22-4]
32. How can secondary voltage be reversed? [22-4]
33. How should the spark plugs be removed? [22-4]
34. What is "reading the plugs"? [22-4]
35. What conditions affect required voltage? [22-5]

CHAPTER
23

Electrical System Service

The electrical system is required to supply electricity to the ignition system, starter, some emission control devices, and instruments. The battery supplies the electricity when the engine is not running. When the engine is running, the charging system supplies enough electricity to operate all electrical systems and to recharge the battery.

Fortunately, the modern engine electrical system needs little service. A majority of modern batteries require no maintenance. Starters and generators (alternators) are designed to last the life of the engine in normal use. Both of them are mechanical devices, so the bearings and brushes will wear. Because of this they will have to be replaced in time. Switches will wear out, too. The electrical wiring can break (open), two wires can touch together causing a copper-to-copper contact (short), or the wire insulation can fail. When this happens, the wire can contact engine or chassis metal. This makes a copper-to-iron contact (ground). Shorts and grounds cause fuses to blow.

The most common electrical problem is loose or corroded wire terminal connectors. Terminal connectors that are clean and tight are not likely to cause trouble. In many cases, a special silicone insulating grease is used on the terminal junction to keep moisture out. This nearly eliminates corrosion problems at the terminals.

23-1 Battery Care

In order for an engine to start, the battery must be in good condition to provide enough electrical power for cranking the engine and for ignition. To ensure dependable service, the battery must be serviced during each tune-up, and sometimes more frequently.

During a tune-up, the battery is checked to make sure it is in good physical condition and it is installed correctly. The battery should be clean and it should be securely mounted in the carrier with proper hold-down clamps. Two typical hold down methods are pictured in Figure 23-1. The cables and terminals must be clean and tight. Water should be added to the battery fluid, called *electrolyte,* to fill it to the indicator level, as shown in Figure 23-2. It is a

Figure 23-2. Electrolyte level as seen through the cell cover.

good practice to check the battery electrolyte level each time the engine oil level is checked. Only distilled water is recommended for addition to the electrolyte. Drinking water is generally satisfactory for use as battery water, but it is not recommended by battery manufacturers.

If the exterior of the battery is moist and dirty, electricity will leak across the battery case between the battery terminals. This will gradually discharge the battery. Leakage can easily be determined by placing voltmeter lead terminals in the moisture on the top of the battery. If leakage is occurring, a voltage reading will be observed on the voltmeter. An example of this can be seen in Figure 23-3.

Hydrogen and oxygen gases are formed inside the battery as it is discharged and recharged. This gas combination is *very* explosive. Care should always be taken to avoid any flame or electrical spark near the battery. All electrical switches should be turned off before disconnecting the battery terminals. The negative battery cable should be removed first because there will be no spark if the wrench removing the negative cable accidentally touches the vehicle metal. After the negative cable is removed, the battery is electrically insulated so that there will be no sparks as the positive cable is removed.

23-2 Battery Testing

Testing the battery will help prevent vehicle problems that result from battery failure. The battery is tested to determine its state-of-charge and to determine how well it either produces or accepts electrical current. The battery is faulty if it cannot be charged, if the voltage of a fully charged battery is low while discharging, or if the voltage is either too high or too low while charging.

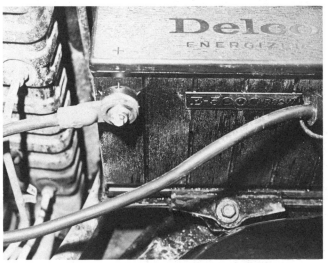

Figure 23-1. Typical battery hold downs.

Figure 23-3. Electricity leaking across the top of the battery as shown on the voltmeter. The voltmeter leads are placed in the moisture on the top of the battery case.

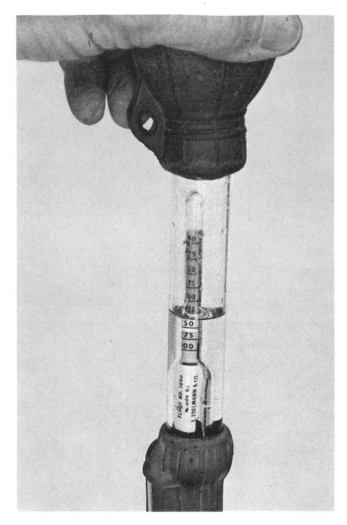

Figure 23-4. A hydrometer being used to measure the state-of-charge of the battery cell.

State-of-Charge. As electricity is used from the battery, the chemical reaction reduces the acidity of the electrolyte. This causes the electrolyte to thin or reduce its specific gravity. **Specific gravity is measured with a hydrometer.** This is done by withdrawing a sample of the electrolyte from the cell with the hydrometer. If the specific gravity is low, the electrolyte is thin and the hydrometer float low. When the cell is charged, the specific gravity is high and the hydrometer floats high. The acid concentration in the electrolyte, as indicated by the hydrometer float level, is an indication of the battery state-of-charge (Figure 23-4).

Batteries with sealed covers often have an indicator in a small window. The indicator will show green (ball is up) when the battery is charged. It will show black (ball is down) when it is discharged. The electrolyte is below the indicator if the indicator is either white or yellow. This type of hydrometer is illustrated in Figure 23-5.

A fully charged battery will have a hydrometer reading of 1270 (indicating a specific gravity of 1.27 times that of water). A completely discharged bat-

Figure 23-5. A hydrometer that is built into one battery cell.

tery will have a hydrometer reading of 1070 (indicating a specific gravity 1.07 times that of water). The use of a hydrometer to measure the specific gravity of the electrolyte is the *best* means of checking a battery state-of-charge. Two precautions should be considered when using a hydrometer. If water has just been added to the cell electrolyte, it will remain on top of the more dense electrolyte. Checking this cell with a hydrometer will produce a reading that is *lower* than the actual battery state-of-charge. The second precaution has to do with battery temperature. If the battery is hot, the electrolyte will be thin and the hydrometer will give a false *low* reading. If the battery is cold, the electrolyte will be thick and give the hydrometer a false *high* reading. Four points are added to the hydrometer reading for each 10 °F (5.5 °C) that the electrolyte temperature is above 80 °F (27 °C). Four points are subtracted for each 10 °F (5.5 °C), the electrolyte temperature is below 80 °F (27 °C). Good-quality battery hydrometers have thermometers built into them with the correction factors indicated on the thermometer scale. The scale of one of these hydrometers is pictured in Figure 23-6. For normal service a battery should be replaced when the hydrometer reading of the cells are more than 50 points between the highest and lowest.

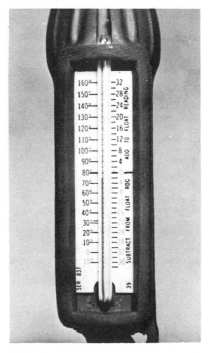

Figure 23-6. A temperature correction scale on a good quality hydrometer.

Capacity. The capacity test or load test of a battery measures the ability of the battery to rapidly convert chemical energy to electrical energy. This is done by drawing a heavy current from the battery while observing the voltage at the battery terminals. When current is drawn from the battery faster than the chemical action can occur within the battery, the battery terminal voltage is lowered. In the battery capacity test, an open variable carbon pile is connected in series with the battery and a high-reading ammeter. A voltmeter is also connected across the battery terminals. The ammeter, voltmeter, and carbon pile are usually built into a battery-starter test unit. This test circuit in a battery-starter tester is illustrated in Figure 23-7.

The battery capacity test is only used to test batteries with a state-of-charge over 1225 or on batteries having a green charge indicator. Voltage is noted 15 seconds after the cabon pile is adjusted to produce a current that is three times the ampere-hour capacity of the battery. **Ampere-hour capacity is a rating used to indicate the maximum potential electrical power or size of the battery.** It is marked either on the battery or given in specification books. If the ampere-hour capacity cannot be determined, use a test amperage that is half the cold cranking rating of the battery. For a healthy, fully-charged battery at 80 °F (27 °C) or above, the battery terminal voltage should not drop below 9.5 V at the end of 15 seconds while the test current is flowing. When the battery is at 30 °F (−1 °C), the minimum battery voltage needed for this test is only 9.0 V or above.

If this test were to be performed on a battery with a hydrometer reading below 1225 or the charge indicator is black, the terminal voltage at the end of the capacity test would be very low. This voltage reading serves no useful purpose in determining the battery condition. The capacity test is only valid when performed on a battery that has a state-of-charge above 1225. A battery with a low state-of-charge should be recharged before a capacity test is performed.

A battery with a low state-of-charge is not able to supply the amount of electrical current required for cranking the engine and still have enough voltage to produce a good ignition spark. The battery should be checked to be sure that it is at least

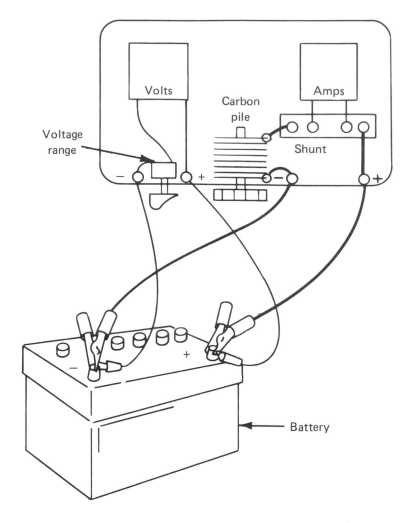

Figure 23-7. A tester used to check the capacity of a battery.

three-fourths charged (1225) and that it has satisfactory capacity before any other electrical system problem is diagnosed. If the battery does not meet the standards, it should either be charged or temporarily replaced with a good fully charged battery. The rest of the electrical system can then be tested.

23-3 Battery Charging

A battery will charge when a voltage is applied to the battery that is higher than the terminal voltage. The higher voltage forces electricity backward through the battery to produce a reverse chemical action within the battery. The higher the charging voltage is, the faster the battery internal chemical action will occur. Excess charging voltage placed across the terminals, will force chemical action to occur so fast that the battery will gas excessively. It will also cause the plates to buckle, the active

material will break loose from the plates, and the battery will be destroyed. To minimize damage, battery temperature during charging should never be allowed to exceed 110°F (49°C). At higher temperatures, the battery quickly loses its resistance to the charging voltage, allowing the charging rate to go very high at a normal charging voltage. This will rapidly heat the battery and it will be ruined.

A battery charger should be *turned off* when it is connected to or disconnected from a battery. If the charger is on, a spark will occur when the charging cable clips are being attached to or removed from the battery terminals. This spark can ignite the gases in the battery cell and cause an explosion. The explosion will break the battery and throw acid out, usually on the person who is carelessly attaching or removing the cable clips.

Before charging the battery, check the electrolyte to make sure that it is at the proper level.

Connect the red charger cable clips to the positive battery post. The positive terminal is always the larger of the two battery posts and it is usually marked with a (+) sign or with red paint. The black charger cable is connected to the negative terminal. The charger is always turned to the lowest charging rate when it is first turned on. The charging rate is then increased to the desired rate. This prevents an excess voltage surge that could damage the battery.

A normal battery can be fully charged by using a charger that will force from 3 to 5 A through the battery. Chargers are equipped with ammeters that show the charging rate. A battery will be fully charged when three successive hydrometer readings taken at 1-hour intervals show no increase in readings.

A battery that has a hydrometer reading below 1225 will accept a higher charging rate than a battery that is near full charge. A fast charger can be used for as long as 30 minutes at rates as high as 50 A on a partially charged 12-V battery. A fast charger should only be used long enough to get the battery charged for engine starting. A normally operating vehicle charging system will finish charging the battery after the engine is running. If the vehicle charging system does not recharge the battery, the charging system will have to be repaired to provide satisfactory electrical operation.

It is helpful in understanding battery operation if you connect a voltmeter across the battery to watch terminal voltage during both charging and discharging. The battery terminal voltage will be near 14.0 V during slow-charging and it will be over 14.5 V during fast-charging. After the charger is turned off, the terminal voltage will gradually drop to the normal battery voltage, as the chemical action works into the plates. During discharge, the voltage drops below normal open-circuit terminal voltage. The higher the discharge rate, the lower the terminal voltage will be. After a heavy discharge, the terminal voltage will remain temporarily low. When the battery is allowed to stand, the terminal voltage will gradually increase because the chemical action within the plates works to the surface to produce normal terminal voltage.

23-4 Starter

The battery provides electrical energy to crank the engine and to fire the spark plugs. The positive ter-minal of the battery is connected to the starter through heavy insulated cables and switches. The negative terminal is connected to the engine block with another heavy cable or flexible strap. A heavy-duty starter switch is operated from the driver's control. This control is usually built into the ignition switch.

The starter switch is used to complete the battery circuit by sending current from the battery terminals through the starter, engine block, and cables. When the starter switch is open, no current will flow and the starter will not operate.

The cranking speed of the starter results from the starter design, the voltage supplied by the battery through cables, and the engine cranking load. The technician can do nothing about the starter design or the engine load but can make sure the starter gets the maximum possible voltage from the battery. This is done by connecting a fully charged battery to the starter with proper-size cables using clean, tight junctions.

Cranking Voltage. The starting system can be given an overall check by measuring the cranking voltage. To keep the engine from starting during this test, the coil secondary cable can be removed from the distributor cap center tower and placed against the engine block. An alternative method is to connect a jumper wire from the *distributor* side of the coil primary to ground. The wire connected to the BAT terminal is removed from the HEI distributor to keep the engine from starting. The negative voltmeter lead (black) is connected to the engine metal for ground. The positive voltmeter lead (red) is connected to the battery cable terminal of the starter. The cranking voltage should be above 9.5 V when the starter is cranking at normal speeds. Cranking should be limited to the shortest cranking period necessary to get a steady voltage reading. An alternative location for voltmeter connections that is more accessible is to connect the positive voltmeter lead to the switch side of the coil. This would be the wire that was removed from the BAT terminal on the HEI distributor. The minimum cranking voltage at the coil should be 9.0 V. In some cases, cranking voltage is measured across the battery terminals while the engine is cranking. With this connection, cranking voltage should be above 9.6 V. Manufacturers' specifications should always be followed. In a normal warmed-up engine in good operating condition, the cranking voltage will usually be 10.0 V or slightly more. An automotive service technician should make a habit of placing the transmission in neutral and setting the parking brake when cranking the engine, especially when a remote-control starter switch is being used.

Low cranking voltage indicates problems with battery cables and their connections or with the starter itself. Details of the test procedures used to pinpoint these problems are beyond the scope of this book.

23-5 Charging System

With the engine running, the charging system is designed to supply all of the electrical current required by the electrical units that are continually being used in normal vehicle operation and to charge the battery. The battery supplies the occasional extra electrical demand that exceeds the capacity of the charging system. This may occur at idle speed when a large number of accessories and lights are turned on. It also supplies the high electrical demand of power windows, horns, power antennas, and so on.

The charging system consists of a belt-driven generator, a regulator to limit maximum voltage, and electrical wiring to connect it into the automobile electrical system.

In 1960, *diode rectified* generators, usually called *alternators,* were first installed as standard equipment on some domestic automobiles. By the mid-1960s, all domestic automobiles were using them. The use of the alternator did not change the rest of the electrical system. It produced the same type of pulsing direct current that the older commutator rectified generator had produced. An alternator dces, however, produce current at lower engine speeds, it has lighter construction, operates safely at higher speeds, and is less expensive than the old-style generator. All of these were good reasons to use an alternator.

23-6 Testing Charging Circuits

Many automotive service technicians do not understand charging systems. When there is a charging system problem, many of these technicians change parts until the system operates. This method

is time-consuming and the parts are expensive for the customer. Technicians correcting problems in this way have helped to give the automotive service trade a poor name. Modern test equipment can be used to test the charging system quickly and simply. The use of test equipment together with a thorough understanding of operating principles of the charging system will lead to quick, accurate diagnosis. Only the malfunctioning part will need to be repaired, thus saving time, parts cost, and "comeback" that results from improper repairs.

As in any electrical test work, the first thing to check is the battery. It should be tested to see that it is in good condition and charged. If the battery is faulty, it should be temporarily replaced with a good battery before further tests are run. The alternator belt tension should be checked and adjusted as necessary. Figure 23-8 shows a belt tension gauge.

Figure 23-8. A typical belt tension gauge used to check the generator drive belt.

Alternator Tests. The battery test is followed by an alternator *output test.* In this test, a test ammeter is connected in the charging circuit, either at the alternator or at the battery. The best test ammeters use an inductive pickup that clamps around the wire coming from the alternator BAT terminal. This type of pickup is pictured in Figure 23-9. The alternator field wire is removed and a jumper wire is connected between the field terminal and battery ter-

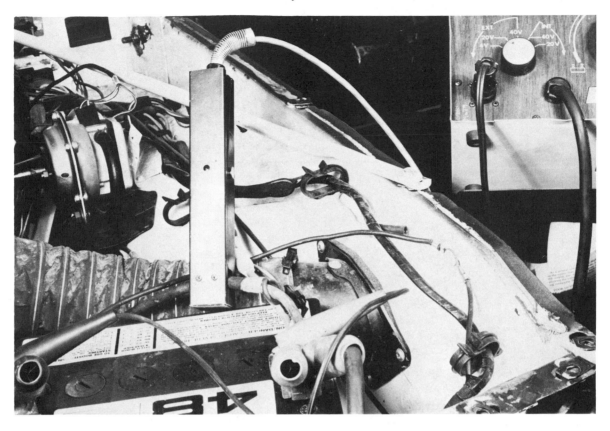

Figure 23-9. An inductive pick-up used on one type ammeter.

minal on mechanically regulated internally grounded types (Figure 23-10). This type of connection is often called a B circuit. On externally grounded types, the jumper wire is connected between the field terminal and ground. This type of circuit is often called an A circuit. A heavy carbon-pile tester is placed across the battery. It is adjusted as necessary to keep the charging system voltage at the maximum voltage given in the applicable specification, approximately 14 to 15 V. This test connection is illustrated in Figure 23-11. The engine is started, and adjusted to the required speed, about 1500 rpm. The carbon-pile rheostat is tightened to control the maximum charging voltage to the specified voltage. Alternator output is the read on the ammeter while the engine speed and voltage are operating at their specified settings.

Alternators having the regulator built inside the case have their fields connected like an externally grounded charging system. The service technician can bypass the regulator by inserting a screwdriver through a hole in the alternator case. The screwdriver blade should touch a tab on the grounding brush at the same time that it touches the edge of

Figure 23-10. The jumper connection used to measure the alternator output of an internally grounded type.

the case hole, as shown in Figure 23-12. This completes the field circuit to ground, to produce maximum field current.

Some service manuals do not recommend the use of a test ammeter placed in the charging circuit

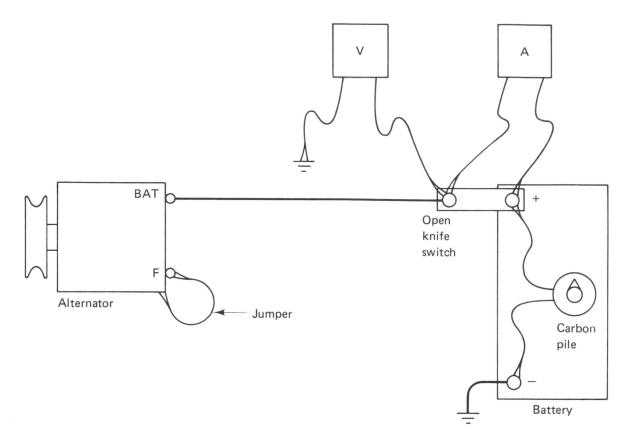

Figure 23-11. The test connections to measure the alternator output of an externally grounded type.

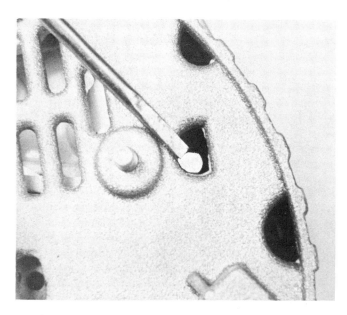

Figure 23-12. A screw driver blade is used to ground the tab in this type of DR alternator to produce maximum field current.

to check alternator output. The charging system is checked while it is fully connected in its normal manner. If the test ammeter connection should accidentally separate during an output test, a high-

voltage surge would be sent through the electrical system. This voltage could be high enough to damage solid-state control units used in other vehicle systems. When an ammeter is not used, output of the generator is checked by measuring the electrical system voltage. The voltmeter is connected to the battery terminals. Battery voltage is noted before starting the engine. After starting, the engine speed is increased and held at 1500 rpm. When all electrical systems except ignition are turned off, the voltage should gradually increase and hold at 14.0 to 14.5 V. With the engine speed held at 1500 rpm, all of the electrical systems are turned on. This should cause the voltage to drop to about 13.0 to 13.5 V. If these voltages are not reached or if they are exceeded, the charging system requires a thorough diagnosis to determine the problem. Details of these diagnostic tests of the charging system test procedures are beyond the scope of this book.

If the alternator output is low, further alternator tests are required. One of the best test instruments for alternator problem analysis is the oscilloscope. Many engine scopes have connections and circuits that can be used for alternator testing.

The primary test leads will usually work for this. The pattern displayed on the scope, similar to Figure 23-13, will indicate a normal alternator. If the pattern is not normal, the alternator will have to be removed for repair or replacement. Serious alternator malfunctions will also show up on test meters.

Voltage drop Tests. All conductors must be free of abnormal resistances if the charging system is to function correctly. Voltage drops throughout the charging circuit are checked while the alternator is producing 20 A in the circuit. This is done using the same test connections as used to check alternator output. The insulated charging circuit should have a voltage drop of less than 0.7 V on systems with an ammeter, or 0.3 V on systems with a charge indicator lamp. The grounded side of the circuit

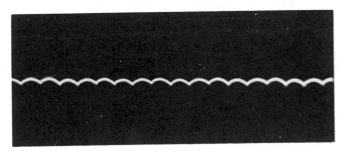

Figure 23-13. An oscilloscope pattern of normal alternator output.

should have a voltage drop of less than 0.1 V. Test connections are shown in Figure 23-14.

When the alternator operates properly and no excess resistance is found in the charging circuit, the jumper wire or grounding screwdriver can be removed and the regulator connected so that the regulated voltage can be checked.

Checking the Regulator Voltage Setting. The voltage regulator limits the maximum voltage in the charging system. If the regulator setting is too low, the battery will not fully charge. A setting too high will shorten the service life of the battery, ignition points, electric motors, radio, lights, and so on, by forcing excessive current through them. High-voltage regulator settings may be first recognized as an unusual amount of headlight flare when the engine accelerates from idle. The battery will also use excessive water as a result of excessive gassing.

Charging system voltage increases when there is no place for the electrical current to flow. The voltage decreases when there is a demand for electrical current. There must be enough resistance in the system to produce a high system voltage when a voltage regulator is being tested. This voltage will be in the regulator controlling range. The system voltage can usually be increased by increasing the engine speed to 1500 rpm with all accessories turned off. A ¼-ohm series resistor can also be inserted in

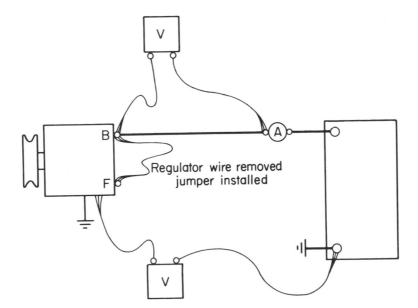

Figure 23-14. Typical test connections for measuring charging circuit voltage drop.

the charging circuit to add enough circuit resistance to force system voltage to increase to the regulating range. Most volt–ampere testers have a built-in ¼-ohm fixed resistance, as illustrated in Figure 23-15. Specific test settings and specification procedures are provided by each vehicle manufacturer. These should be followed when checking the specified voltage regulator settings. Usually, the specified range is between 13.8 and 14.5 V. Test

procedures vary from manufacturer to manufacturer, from model to model, and from year to year. Generally speaking, alternators using mechanical regulators have internal field grounds (B circuits), while those using solid-state regulators have external field grounds (A circuits). It is important to use the correct procedures and test specifications for the vehicle and for the test equipment being used to determine the regulator setting.

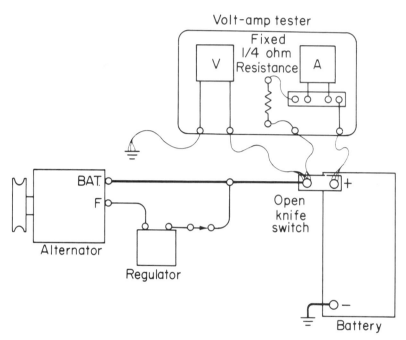

Figure 23.15. Electrical connections used to check the voltage regulator setting.

REVIEW QUESTIONS

1. How do electrical opens, shorts, and grounds differ? [INTRODUCTION]
2. What is the most common electrical problem? [INTRODUCTION]
3. What is added when the battery electrolyte level is low? [23-1]
4. When should the battery electrolyte level be checked? [23-1]
5. Why is it important to keep open flames and electrical sparks away from a battery? [23-1]
6. What procedure should be used to remove the cables from a battery? [23-1]
7. When is a battery faulty? [23-2]
8. What indicates a battery's state of charge? [23-2]
9. List two precautions that should be considered when using a hydrometer. [23-2]
10. How is the hydrometer reading corrected for temperatures of the electrolyte? [23-2]

11. What is measured during a capacity test of a battery? [23-2]
12. When is the battery capacity test used? [23-2]
13. How much current is used to make the battery capacity test? [23-2]
14. What is the minimum terminal voltage when running the battery capacity test? [23-2]
15. What should be done before electrical tests are run after the battery does not pass the test standards? [23-2]
16. When does a battery charge? [23-3]
17. What happens when excessive charging voltage is placed across the battery terminals? [23-3]
18. What precautions should be observed when the battery is turned off and on? [23-3]
19. How should the battery be charged? [23-3]
20. How should a fast charger be used? [23-3]
21. How can the technician make sure the starter gets the maximum possible battery voltage? [23-4]
22. How is the engine kept from starting while the cranking voltage of the starter is being checked? [23-4]
23. What do high and low cranking voltages indicate about the starter and the cranking system? [23-4]
24. What is the first thing to check when there is an electrical problem? [23-5]
25. What is done to an alternator to make an output test? [23-5]
26. How can satisfactory alternator output be checked with a voltmeter? [23-5]
27. How is a voltage drop test done on the charging system? [23-5]
28. What is the purpose of a voltage regulator? [23-5]
29. How is a high voltage regulator setting recognized? [23-5]
30. How is a low voltage regulator setting recognized? [23-5]

CHAPTER
24

Fuel System Service

The automotive fuel system includes the gasoline tank, fuel pump, fuel filter, connecting lines, and carburetor. Gasoline is stored in the tank under the rear floor pan in front-engine automobiles. Rear-engine automobiles usually mount the fuel tank in front of the passenger compartment. The tank is vented to allow vapors and air to move in and out. This is necessary to keep pressures nearly equal as the vehicle goes up and down hills, as atmospheric pressures change, and as fuel is drawn from the tank. On emission-controlled automobiles, the tank is vented through a vapor separator and a carbon canister to keep the gasoline vapors from reaching the atmosphere. The filler neck of emission-controlled automobiles is fitted with a sealed cap. The cap has a pressure valve and a vacuum valve that allow *excessive* pressures to equalize. This is necessary to prevent damage to the tank if the normal evaporative emission-control vent fails to operate properly. The filler neck has a smaller opening on automobiles that are designed to use no-lead gasoline.

A diaphragm-type fuel pump, usually mounted on the engine, moves the gasoline from the tank to the engine. Gasoline is pumped through a filter to the carburetor or fuel injection system. Sometimes, an electrically driven turbine-type pump is used in the fuel line or in the gasoline tank. To function correctly, the fuel system must deliver liquid gasoline having enough volume and pressure to keep the carburetor bowl full of clean gasoline, regardless of vehicle speed or maneuvering.

When the fuel system malfunctions, it is up to the technician to locate and correct the cause of the malfunction. The only decision that the automobile operator can make is to have the proper grade of gasoline put into the tank.

24-1 Fuel System Operation

The fuel pump moves the gasoline in the fuel system. It transfers gasoline from the tank to the carburetor or fuel injection system. Engine-mounted diaphragm-type fuel pumps are operated by an eccentric lobe on the camshaft or distributor shaft. A spring-loaded lever arm of the fuel pump is held against the eccentric lobe, contacting it at all times. In some systems, a short pushrod is fitted between the eccentric lobe and the spring-loaded arm. A lobe on the camshaft moves the arm. This action pulls the diaphragm *away from* the fuel chamber side to increase its volume and this draws fuel from the gasoline tank. A spring on the lever side of the diaphragm pushes against the diaphragm as the eccentric lobe movement relaxes its pull on the arm lever. The force of this spring on the diaphragm makes the fuel pressure. The movement of the diaphragm moves the fuel toward the carburetor or fuel injector. The fuel pump is fitted with two check valves. The one on the tank side of the pump will only allow gasoline to go into the pump (inlet check valve). The check valve on the engine side of the pump will only allow gasoline to leave the pump (outlet check valve). As the pump fuel chamber is made to increase in volume *by the linkage,* gasoline is drawn into the pump from the tank through the inlet check. As the pump fuel chamber is made to decrease in volume *by the spring,* gasoline is pushed toward the engine through the outlet check. The fuel pump operation is illustrated in Figure 24-1.

Connecting fuel lines are made of steel or synthetic rubber hose. Steel tubing uses standard flare-type fittings to connect between the parts of system and beads to connect to hoses. Synthetic rubber fuel line hoses are fitted snugly over the beads on the ends of the tubing and nipple at the system components. They are usually secured to the tubing and nipple with a hose clamp. In some cases, the hose is fitted with a flare-type metal fitting end. The tubing or hose must be in good condition so that the gasoline does not leak out. They must also keep air

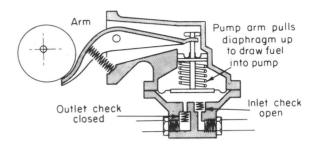

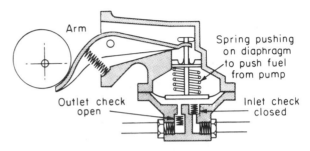

Figure 24-1. The operations of a typical engine driven fuel pump.

from leaking into the system on the tank side of the pump during its intake stroke.

The gasoline tank has a vapor separator with several lines that is designed to vent the tank while the automobile is standing in any position. It is located within the fuel tank or within a space in the automobile body, where it is difficult or impossible to see without extensive work. It rarely causes a problem so it is not considered to be a part of normal vehicle service.

24-2 Fuel System Testing

The most important characteristic of gasoline is that it must be able to start the engine. This sounds extremely fundamental, but its importance can best be demonstrated if one tries to start a cold spark-ignited automobile engine on diesel fuel. Fuel must be able to *vaporize.* **The ability to vaporize is called volatility.** A basic fact to be remembered is that the fuel must be in the vapor form to burn. It must also be mixed with air in the correct proportions, which is the function of the carburetor or fuel injector. Finally, the fuel vapors must be thoroughly mixed with the air. This is accomplished in the intake manifold, head port, and combustion chamber.

389

All of the gasoline should be burned in the combustion chamber. If it is not, some of the partially burned fuel can get past the piston rings and into the engine oil. The remaining partly burned fuel will go out with the exhaust gases to form unburned hydrocarbon exhaust emissions.

Fuel Filter. Correct carburetor or injection system operation requires clean gasoline. A fuel filter is placed in the gasoline supply line to trap any particle that may have been accidentally put in the gasoline tank at the service station or any particle that develops within the fuel system of the vehicle.

Periodic fuel filter replacement is recommended by all automobile manufacturers. If the fuel filter replacement recommendation is followed, the operator is unlikely to have a problem from a plugged fuel filter. When the filter becomes plugged, the engine will not produce its usual power or speed. The fuel filter can be checked by running a fuel pump capacity test on the carburetor side of the fuel filter. This test can only be done when the fuel filter is located outside the carburetor.

Fuel Pump. The fuel pump capacity test is one of the easiest tests to perform, even when a pressure-volume fuel pump tester is not available. The fuel line is disconnected from the carburetor and directed into a measured container. If a shop towel is held around the fitting as it is loosened from the carburetor, the towel will absorb any fuel leakage. The gasoline saturated towel must be removed from the engine compartment before starting the engine. This will reduce the fire hazard that is always present around gasoline. An extension hose is slipped over the fuel line to make it easier to direct the flow of fuel. The engine is started and allowed to idle for 30 seconds as the fuel is being collected in a measured container. This procedure is pictured in Figure 24-2. The engine is then turned off. There is usually enough gasoline in the carburetor to operate the engine at idle for 30 seconds. The fuel pump specifications give the amount of gasoline that the fuel pump should deliver in 1 minute. The amount collected in the 30-second test must be doubled to determine the pump capacity per minute. For example, if one pint of gasoline is delivered in 30 seconds,

Figure 24-2. Running a fuel volume test. A pressure gauge is also included with this specific piece of test equipment.

the fuel pump has the capacity to deliver 2 pints in 1 minute. Fuel pumps will have a minimum delivery of approximately 1 pint per minute on small-displacement engines and approximately 2 pints per minute on large-displacement engines. The appropriate specifications should be checked for the engine being tested.

If the fuel system capacity is normal when tested on the carburetor side of the fuel filter, no further test is required on the fuel filter. If the capacity is low, the test should be repeated on the fuel pump side of the filter. The filter is plugged if the test on the fuel pump side of the fuel filter is normal while the test on the carburetor side of the filter is low. If both tests are low, the problem is with the fuel pump operation or with the fuel lines.

In addition to providing fuel volume, the fuel pump is required to produce enough pressure to open the carburetor float valve and keep the carburetor bowl full of fuel. Fuel pump pressure is checked by placing a pressure gauge on the outlet

side of the fuel pump, usually at the carburetor end of the fuel line, and measuring the fuel pressure as the engine idles. Figure 24-3 pictures one type of fuel pressure gauge being used. Fuel pressures will range from 5 to 10 lb/in.^{2} (35 to 70 kPa), depending upon the engine. Fuel pressure specifications that apply to the engine being tested should be consulted. A fuel pump may pass the pressure test, but fail the pump capacity test. If the fuel pump eccentric is normal, low fuel pressure indicates that a new pump is needed.

When the gasoline volume produced by the capacity test is too low, a fuel pump vacuum test should be run. To run this test, a vacuum gauge is fastened to the tank side of the fuel pump. The fuel lines coming from the tank will have to be plugged during the test to keep gasoline from leaking out of the line. A normal fuel pump will pull about 10 in. (25 mm) of vacuum at idle. If the vacuum is less, the pump or pump cam is faulty. A vacuum check at the tank end of the fuel line will indicate air leaks in the

Figure 24-3. Fuel pressure being measured while the engine idles.

line that could cause low pump capacity while still producing the required pressure and vacuum. A broken rocker arm return spring on the fuel pump linkage produces a tapping sound, but will not normally affect pump operation.

Air Filter. The air coming into the engine must be clean for the engine to continue to function properly for long periods of time. Engines have air filters that trap abrasives, dirt, and other contaminants. As the particles of these materials are trapped, they plug the small openings in the filter. Excessive plugging restricts air flow into the engine and this upsets calibration of the carburetor. Air filter plugging makes a rich fuel mixture, which, in turn, reduces engine power, increases fuel consumption, and increases both carbon monoxide and unburned hydrocarbon exhaust emissions. Badly plugged air filters will limit engine power and speed.

Some manufacturers recommend cleaning the air filter at specific mileage intervals by blowing air *from the inside* toward the outside. Others recommend replacing the air filter at set mileages. These schedules should be followed for maximum engine service. Air filter changes are required more frequently when unusually dirty operating conditions are encountered. Some equipment companies have developed test methods to indicate air filter plugging. These pieces of test equipment are not commonly used by the automotive service trade.

One of the more accurate air filter checks involves the use of the hydrocarbon–carbon monoxide emission tester. If the engine produces more unburned hydrocarbons and carbon monoxide with the air filter installed than it does with the filter removed, the filter is partly plugged and should be cleaned or preferably replaced. Air filter removal is pictured in Figure 24-4.

24-3 Carburetors

A carburetor mixes the correct amount of fuel into the incoming air to give the engine a combustible intake charge. The charge has more fuel in proportion to the air (rich mixture) at idle speeds, during acceleration, and at full throttle than it does when operating at cruising speeds. This change in the mix-

Figure 24-4. Air filter removal.

ture ratio is necessary to provide *drivability,* a term used to describe acceptable automobile operation.

As the engine idles, the air flow through the carburetor and manifold is slow, so some of the fuel collects on the manifold walls. A rich idle mixture is needed to make sure that the air/fuel charge mixture reaching the leanest cylinder has a combustible mixture. Engine modifications for emission control have improved manifold and carburetor designs. Engines with these modifications will operate satisfactorily and provide drivability with much leaner mixtures than those used on pre-emission-controlled engines.

Engine power results from the pressure in the combustion chamber as the charge expands during combustion. A rich mixture carries more fuel in proportion to the incoming air, so it will release more energy to produce more power. This is required during acceleration and full throttle. The charge has more than enough fuel to use all of the available oxygen contained in the charge air. If this same rich mixture were used for cruising speeds, the engine would give very poor gasoline mileage.

At cruising speeds, fuel economy is desirable. The carburetor delivers a lean mixture containing more than enough oxygen in the charge air to burn all the fuel in the mixture. This uses all the available fuel energy to give economical operation. The graph in Figure 24-5 shows how the air/fuel ratio changes at different vehicle speeds.

Excessively rich mixtures waste fuel and produce large amounts of exhaust emissions. Excessively lean mixtures may not ignite or they may be only partly burned in the combustion chamber. These unburned or partly burned combustion products go out of the engine as unburned hydrocarbon emissions. Emission control carburetors will make the correct air/fuel ratios for all engine operating conditions. It is up to the automotive service technician to correctly maintain and adjust them for continued proper operation. The adjustments are done when the engine is given a tune-up.

In most spark-ignited engines, the air and fuel are mixed and delivered to the combustion chamber as a combustible intake charge mixture. Combustible air/fuel mixtures range from an 8:1 to a 20:1 air/gasoline ratio *by weight.* The ideal mixture has the correct amount of air that is required to burn the entire quantity of fuel. This is called a *stoichiometric* mixture. **The mixture is stoichiometric when the substances have the exact chemical proportions that are required to complete the reaction.** Internal combustion engines operate best on air/fuel ratios between 11.5:1 (rich) and 15:1 (lean), except at idle, where engine mixtures may go as rich as 9.5:1 for cars built before 1968. For emission-controlled engines, idle mixtures are approximately 14:1 to 16:1, with cruising mixtures as lean as 18:1 when operated near sea level.

It makes no difference if an engine is carbureted, throttle-body-injected, or sequenced fuel-injected—the engine still needs the same fuel mixture ratios. Fuel-injected engines that discharge their fuel close to the intake valve deliver mixture ratios that are more equal between cylinders than carbureted or throttle-body-injected engines. This occurs because fuel injection does not depend upon the air velocity and the mixing of the air and fuel in the manifold to deliver the fuel to the cylinder.

24-4 Carburetor External Adjustments

The most common carburetor adjustment is engine idle. It is done by setting the *idle mixture screw* to make a smooth idle and by setting the *idle speed screw* to give the curb idle speed on a warm engine. Adjustable emission-control carburetors have limiters that prevent excessively rich idle adjustments and, therefore, it may be difficult to adjust them rich enough for a smooth idle. Before an idle adjustment is made, all engine systems should be functioning correctly and the engine should be warm enough so that the choke will fully open. This will release the fast-idle cam so that the engine can operate at curb idle.

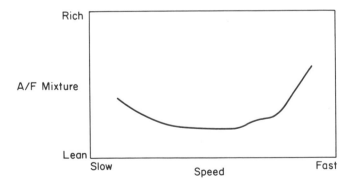

Figure 24-5. A graph showing the air/fuel ratio change at different vehicle speeds.

Positive Crankcase Ventilation. Before idle adjustment is attempted, the positive crankcase ventilation (PCV) system must function correctly. The PCV valve can be checked by slipping it from the engine. Before the engine is started shake the valve. If the valve rattles it is moving freely in its case. The PCV hoses should be in good conditon and should be attached correctly. A normally operating PCV system allows some crankcase vapors with air to enter the intake manifold. With the engine running the PCV hose is pinched closed to stop vapors and air from flowing through the PCV system into the manifold. This will reduce idle speed 50 rpm or more. If the engine speed does not drop, the PCV system is plugged and should be serviced. The PCV problem will be either a stuck valve or plugged hoses.

Curb Idle Adjustment. Other conditions must be met before curb idle can be correctly adjusted. Depending on the engine make and model, these conditions include things like having correct ignition timing; placing the transmission in neutral with parking brake set; turning the headlights on; turning the air conditioning on; removing the vapor vent line; and removing the distributor vacuum line and then plugging it while the idle adjustment is being made. The conditions to be checked are usually indicated on an emission decal located on an inside fender pan, on a radiator support, or on the inside of the hood within the engine compartment. Idle is adjusted with the air cleaner installed and other vacuum hoses connected in their normal position. When the conditions that apply are met, the engine curb idle can be correctly set.

Idle settings involve two adjustments an *air adjustment* that primarily controls idle speed and a *mixture adjustment* that controls idle smoothness. Figure 24-6 shows typical idle adjustments. The speed is adjusted by limiting the amount of throttle plate closing with a throttle stop screw located on the carburetor linkage. Air going around the slightly open throttle plate supplies the air required for engine idle. Increasing air flow into the engine will increase engine speed. Idle mixture is controlled by idle needle screws, usually located near the base of the carburetor, as shown in Figure 24-7. In some carburetor models, this screw is located in the idle air-bleed passage. The screws control the amount of

Figure 24-6. Typical idle adjustments on a carburetor.

fuel delivered through the carburetor idle system to the incoming air. Emission-control carburetors have limiters sealed in the idle circuit passages that limit rich idle mixtures. The adjustable mixture screws can lean the mixture from this preset rich mixture limit point. In most carburetor models each carburetor primary barrel has an idle mixture screw. Some carburetors, however, are designed to use a single mixture screw to control the mixture on two primary barrels. Starting in 1980, some carburetors have their idle mixture screws capped to prevent tampering or improper adjustment.

Curb idle is set by *first* adjusting the idle mixture screws, within the range of the limiter stops, to make the engine run smoothly at the highest rpm. High vacuum gauge and tachometer readings help determine this point. Some manufacturers also recommend the use of a hydrocarbon–carbon

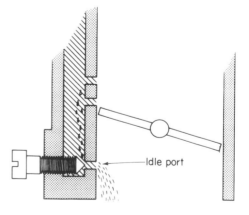

Figure 24-7. A line drawing of a typical idle mixture needle screw located near the base of the carburetor.

monoxide tester when setting idle. Idle mixture is set to the lowest hydrocarbon readings. *Each* mixture screw should be rotated $\frac{1}{16}$ of a turn at a time, alternating between screws. About 10 seconds should be allowed for the engine speed to stabilize before making the next adjustment. Idle mixture is adjusted by rotating the idle mixture screws *clockwise to lean* the mixture until the engine begins to slow down, then rotating them *counterclockwise to enrich the mixture.* This should cause the engine to begin to speed up again. Continued rotating counterclockwise past the point of highest speed will again slow the engine speed as the screws are rotated too far. Some idle screws have an internal limit-stop. In these carburetors, the screw will break if the service technician tries to force it to rotate past the stop. The screws are readjusted to the point where the engine will idle smoothly at the highest speed. Idle-mixture adjustment is done on each mixture screw, rotating each the same degree. If rotating the mixture screws has no effect on engine idle, the idle system is malfunctioning. To correct this condition, the carburetor will have to be opened and the idle system cleaned to give proper operation. The external limiter stops can be removed if the engine does not run smoothly when it is adjusted within the range of the idle mixture limiters. This will allow the idle mixtures screws to be adjusted

further to give the smoothest and fastest possible idle. Some states have laws that allow only service technicians with licenses for emission-control service to remove the limiter caps and make the adjustments. They will use a hydrocarbon-carbon monoxide tester to make the adjustments.

With the engine running smoothly, idle speed is set by turning the throttle stop screw or air screw to provide the required engine curb idle speed. In some model emission-control carburetors, the idle speed is set by making an adjustment on an idle-stop solenoid (Figure 24-8). The mixture screw setting should be checked *after* the idle speed has been set to make sure it is still at the best mixture setting.

A simple idle adjustment procedure is given for emission-controlled engines by some manufacturers. This method follows the normal idle adjustment procedure just previously described. However, the idle speed is set at about 75 rpm above specified idle speed. After this adjustment is made, the correct curb idle speed and and emission mixture are obtained by further leaning the mixture screws equally $\frac{1}{16}$ turn at a time, *without touching* the speed screws.

When more than one mixture screw is used, alternate adjustments between mixture screws until the idle speed is correct. This leans the idle mixture, slowing the engine. Using this adjustment pro-

Figure 24-8. A typical idle stop solenoid on the throttle linkage.

cedure will lean the mixture to meet the emission specification. Specific idle adjustment procedures are shown on the emission decal placed in the engine compartment of emission-controlled engines.

In 1975, Ford introduced an idle mixture adjustment procedure using *propane enrichment*. By 1978, all domestic automobile manufacturers recommended using some form of this procedure. A controlled amount of propane is allowed to flow into the incoming air. The propane enrichens the intake mixture and this increases the engine speed a specified number of rpms when the idle is correctly adjusted. This occurs because the idle mixture on emission-controlled engines is lean to control carbon monoxide. If the rpm's increase too much with propane, the idle mixture is set too lean. If they increase too little, the idle mixture is set too rich. A 1-lb propane tank with a hose and two valves, as shown in Figure 24-9, is used for propane enrichment.

Propane is fed through a tube into the air horn, air filter, or a carburetor port, as specified on the emission decal. The propane is adjusted to produce maximum rpm's on a warmed-up engine running at

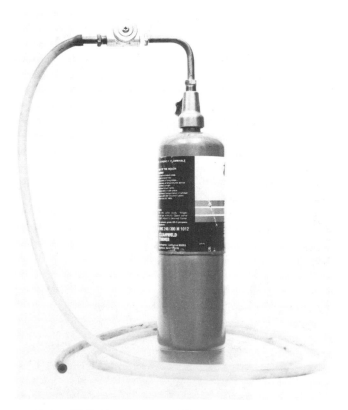

Figure 24-9. Equipment used for propane enrichment idle mixture adjustment.

curb idle. With propane flowing, adjust the idle-speed screw to give the specified propane-enriched idle speed indicated on the emission decal. Turn the propane off. Without changing the idle-speed screw, set the carburetor curb idle speed by adjusting only the idle mixture. The mixture setting should be rechecked using the propane enrichment procedure. The engine speed should increase the correct rpm.

Emission-control carburetors are adjusted lean to help keep exhaust emissions low. It is best to use the propane enrichment method or a hydrocarbon-carbon monoxide (HC-CO) tester when adjusting curb idle. Using the tester, the idle mixture screw is adjusted to give as smooth an idle as possible while keeping the meter reading below the maximum rich-mixture specification. This is less than 1% CO on emission-control carburetors. If the idle mixture limiter caps were removed during idle adjustment, they should be replaced with colored replacement-type caps.

It might be noted here that during calibration of school instruments for the Plymouth Trouble Shooting Contest, the instruments used by schools were frequntly found to be incorrect. Tachometers were as much as 200 rpm incorrect and dwell meters 5° incorrect. Instruments this far off from calibration are worse than useless for adjusting modern emission-controlled engines. It is a good practice to have instruments calibrated at regular intervals.

Linkage Adjustment. Other external carburetor controls are positioned by linkages. Linkage adjustments can be made on the carburetor without disassembly when it is off the engine. Many of the adjustments can be made while the carburetor remains installed on the engine. The linkages seldom have to be adjusted. They will have to be adjusted if someone has tried to correct the carburetor operation by bending the linkages or if the carburetor has been in an accident. All carburetor linkages must move freely with no binding. They operate without lubrication. Lubrication would collect dirt, which will generally cause the linkages to stick and wear.

Each carburetor make, model, and application has its own specific linkage adjustment. The name of the *carburetor make* is cast on its exterior surface. For each carburetor make, there are a number of carburetor models. Sometimes, the *model iden-*

tification is cast on the exterior. Most models are quite different, so it is easy to identify carburetor models with illustrations of carburetors in reference books. A single model of a carburetor may be used on a large number of different engine types requiring different metering and different operating linkages. Numbers and letters stamped on the carburetor or on an attached tag give the carburetor *setting number*. The setting number indicates which specific carburetor–engine combination the carburetor is set for. Details of linkage adjustments for all carburetor makes, models, and settings are beyond the scope of this book. Service manuals and instruction sheets accompanying repair parts kits that apply to the specific engine and carburetor will have to be consulted for accurate carburetor adjustment. General linkage adjustment procedures will be discussed.

A common linkage adjustment method is to move one end of the link to an extreme position, then measure the position or the amount of travel at the other end of the link. Measurement may be made with a scale, a numbered drill shank, or a special gauge (Figure 24-10). A second method of linkage adjustment is to move the operating members on each end of linkage to one of their extreme positions. The linkage should then drop into place or match a notch or an indicating mark. Still another method is to measure the length of the linkage directly (Figure 24-11). If the linkage is incorrect on carburetors built before 1980 it can be

bent at points provided in the link to give the correct linkage adjustment (Figure 24-12). Most carburetors built after 1980 have hardened linkages that will break if any attempt is made to bend them. The latest designs use a protractor with a level that is attached to a lever or plate with a magnet. The protractor will indicate the number of degrees of

Figure 24-11. A typical method of measuring the linkage position.

Figure 24-10. Choke opening being measured with the shank of a numbered drill.

Figure 24-12. Bending a linkage to correct the linkage adjustment.

angular movement (Figure 24-13). After adjustment, the linkage should be rechecked to see if it
operates freely.

Linkages are held on the carburetor by the way
they are bent, by an upset tang, with cotter pins,
with an E ring, with special spring-type carburetor
clips, or by screws or nuts. During carburetor service, the special clips often fly off and are lost unless
the technician takes *special care* to hold them as
they are removed. It is usually difficult to find the
correct replacement for lost clips.

Figure 24-14. One type of bowl vent that directs the
gasoline vapors to the canister that traps the vapors.

Figure 24-13. A protractor being used to indicate the
number of degrees of angular movement.

Bowl Vent. Carburetor bowls have special vents
that open to allow pressure from gasoline vapors to
escape from the carburetor when the engine is stopped. One type is pictured in Figure 24-14. If the
vapors were trapped in the float bowl, they would
force liquid fuel from the discharge nozzle into the
manifold after the engine is turned off. This type of
action is called *percolation*. The bowl vent allows
the vapors to escape, leaving liquid fuel in the carburetor bowl. In emission-controlled cars, the carburetor bowl vapors are directed to a *canister* that
traps vapors; in non-emission-controlled cars,
vapors were vented directly to the atmosphere.

The bowl vent adjustment is checked with the
throttle closed and the engine off. If the vent opening is incorrect, its operating linkage can be adjusted to provide the proper amount of vent opening.

Fast Idle. Fast idle settings can be made while the
carburetor is off the engine as well as when it is on
the engine. When checking fast idle with the carburetor off the engine, the usual procedure is to
place a specific-size drill bit shank or gauging tool
betwen the throttle plate and carburetor bore. The
throttle plate is closed against the gauge and the
choke fully closed. The idle-speed screw is then adjusted to touch the choke-controlled fast idle cam at
a specified point. Figure 24-15 shows this point on
one carburetor. To find this point on some carburetors, it may be necessary to place a drill bit
shank or a gauging tool between the choke plate and
carburetor bore as shown in Figure 24-10.

Fast idle can also be checked and adjusted on a
running engine. The adjustment is made on a warm
engine. The fast-idle cam is positioned so that the
fast-idle stop screw is resting on a specified step of
the fast-idle cam. This position may require gauging
the amount of choke plate opening. If the low cam
stop is specified, the speed will be over 2000 rpm.

On some emission-control carburetors, it may
be necessary to disconnect some of the emission
conrols during fast-idle adjustment. In general, the
emission controls must be made to operate in the
same way they would operate when the engine is

Figure 24-15. One type of fast idle cam positioned for fast idle adjustment with the carburetor off the engine.

Figure 24-16. One example of a choke unloader is shown by the arrow.

cold. It is necessary to follow specific carburetor–engine instructions because there are so many different combinations used in modern emission-control systems.

Unloader. The choke is closed while the engine is cranking. If the engine does not start immediately, because of weak ignition, for example, the engine will pull too much fuel into the manifold, causing an overrich or *flooded* condition. Carburetors have an unloader to overcome this condition. The unloader is a tab or tang on the throttle control lever that contacts the fast-idle cam when the throttle is opened fully. One example is shown in Figure 24-16. The unloader moves the cam, which, in turn, opens the choke plate through the fast-idle linkage. The amount of choke plate opening is measured with a drill bit shank or gauge between the choke plate and carburetor bore. If the opening is incorrect, the tab or tang on the throttle control lever can be bent to open the choke the correct amount.

Acceleration Pump. A linkage from the throttle control lever is used to mechanically operate the acceleration pump. When the throttle is closed, the acceleration pump fills. Opening the throttle pushes fuel from the acceleration pump discharge nozzle. Two points are critical on the acceleration pump linkage adjustment. The pump must be in the correct starting position to make sure that it will be able to deliver the full quantity of fuel. This is usually checked by measuring the distance between the pump end of the linkage and a fixed point on the

carburetor as shown in Figure 24-17. The distance of the pump starting point is correctly positioned by bending the link or by adjusting a screw between the throttle control lever and the pump. A second adjustment that needs to be checked is the acceleration pump stroke. Pump stroke is the measured distance the pump travels from closed throttle to fully opened throttle as shown in Figure 24-18. If it is incorrect, most acceleration pump linkages are provided with several different linkage position holes for longer or shorter pump strokes.

Figure 24-17. An example of the location used to check the acceleration pump linkage starting position and travel of the stroke on a carburetor.

Figure 24-18. Measuring the travel of the acceleration pump stroke on the carburetor.

Choke. The automatic choke uses a heat-sensitive thermostatic spring. It may be built integral with the carburetor or it may be fastened on the manifold using a crossover link. The link is connected to the carburetor choke lever. When the thermostatic spring is integral, it has an insulated heat tube running be-tween a manifold heat stove and the choke spring housing. Figure 24-19 shows one of these. Engine vacuum, pulling on an internal vacuum choke piston or through a special passage in the choke spring housing, draws fresh air through the choke stove, where it is warmed by exhaust heat. The warmed air flows through the insulated tube to the choke spring housing so that the spring can sense engine temperature. Some integral choke springs are warmed with an engine coolant passage or an electric heating coil that turns on with the ignition. An example is pictured in Figure 24-20. As the engine coolant or electric heater becomes warm, it heats the choke spring to allow choke opening.

When the heat sensing coil spring is not directly on the choke shaft, it is connected to the choke shaft by a linkage. An example is shown in Figure 24-21. For adjusting, the choke spring is positioned or is held against its closed extreme limit and the length of the link is adjusted to give proper choke plate closing tension. Service manuals should be consulted for specific adjustments when the choke does not function correctly.

Vacuum applied to the choke piston or to a diaphragm opens the choke. When a piston is used, it is usually located in the choke thermostatic spring

MANIFOLD HEAT CHOKE STOVE

CONNECTION FOR ELECTRIC CHOKE HEAT

Figure 24-19. A typical integral choke and insulating heat tube.

Figure 24-20. An example of a choke with an electric heating coil.

Figure 24-21. A choke heat sensing coil in the manifold with a linkage connected to choke shaft.

housing. It is attached to the choke shaft with a link (Figure 24-22). The piston is seldom adjustable and it must move freely. The vacuum diaphragm, sometimes called a choke breaker or vacuum kick, is mounted independently from the thermostatic spring, and is connected to the choke operating arm by a link. Some carburetors use two choke breakers. The breaker is adjusted by retracting the vacuum diaphragm fully with an external vacuum source. The link is then adjusted to give the correct amount of choke opening. The amount of choke opening is measured with a drill bit shank or a gauge.

An improperly operating choke breaker or vacuum kick will cause engine performance problems during warm-up. The mixture will be either too rich or too lean, depending on the specific fault. This may be the result of bent linkages, improperly set choke opening, leaking diaphragm, or leaking vacuum hoses.

The thermostatic choke springs, which are remotely mounted in a manifold well, are set for the specific carburetor model used. They are not or-dinarily designed for adjustment. Integral thermostatic coil choke springs are adjusted by turning the spring housing cover in the direction that will close the choke plate. In Figure 24-23 the alignment

Figure 24-23. The choke spring housing cover is turned to align the notches on the choke housing and cover.

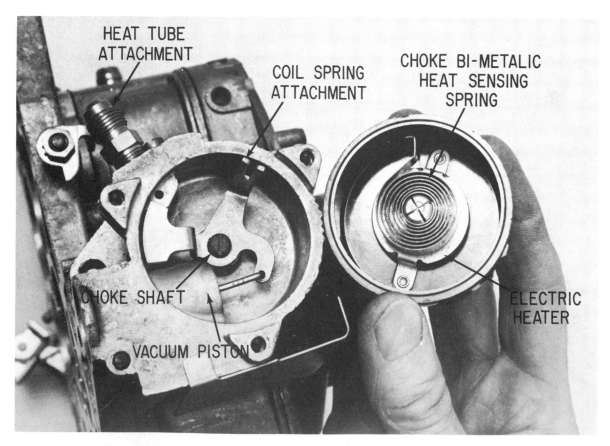

Figure 24-22. An integral choke with a choke opening piston attached to the choke shaft with a link.

notches on the choke housing and cover can be seen. The choke is set to the specified number of rich or lean notches.

Secondary Throttle. Secondary throttles should start to open after the primary throttles are about two-thirds open. This can be seen in Figure 24-24. Mechanically operated secondary throttle linkages

are checked by opening the primary throttle plates a measured amount. At this point, the secondary throttle plates should start to open when the choke is open. Early model carburetors with vacuum-operated secondary throttles have an adjustable stop screw that requires positioning. Late-model carburetors do not require adjustment on vacuum-operated secondary throttles.

Some carburetors with mechanically operated secondary throttles use air valves above the discharge cluster. The air valve may be held closed with a balance weight or with spring tension. The balance weight closing requires no adjustment. The spring closure type must be adjusted. To do this, the spring tension is released until the air valve is open. The spring tension then is gradually tightened until the air valve just closes. The spring is tightened still further, approximately two turns, to give the air valve the correct closing tension for normal operation.

The preceding adjustments are the ones most commonly found on modern carburetors. Some additional specific adjustments will be found on carburetors. These are described in service manuals and on instruction sheets that are included with carburetor repair kits. Typical carburetor kit instructions and gauges are shown in Figure 24-25.

Figure 24-24. The primary throttle plate is opened just to the point when the secondary throttle plate will start to open.

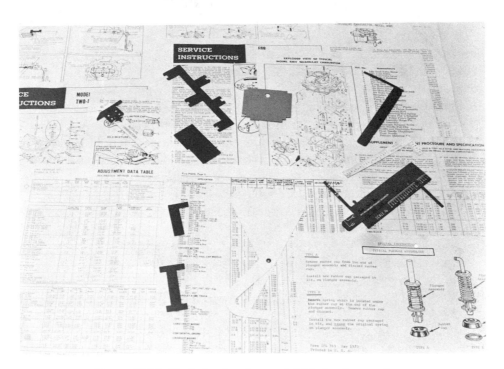

Figure 24-25. Typical carburetor repair kit instructions and gauges.

REVIEW QUESTIONS

1. Through what parts does the fuel flow from the tank to the carburetor? [INTRODUCTION]
2. What must a fuel system do to function correctly? [INTRODUCTION]
3. What part of an engine operates the engine-driven fuel pump? [24-1]
4. What part of the fuel pump provides the force to make the fuel pressure? [24-1]
5. Why does the fuel pump need two check valves? [24-1]
6. What are the most important characteristics of gasoline? [24-2]
7. In what form must gasoline be in to burn? [24-2]
8. How should the operator know that the fuel filter was plugged? [24-2]
9. How is the capacity of the fuel system checked? [24-2]
10. How can the fuel system capacity test indicate a plugged fuel filter? [24-2]
11. What is indicated by low fuel pump pressure? [24-2]
12. When should a fuel pump vacuum test be run? [24-2]
13. How would the operator know that the air filter was plugged? [24-2]
14. How can a service technician check an air filter for plugging? [24-2]
15. During what operating conditions is the intake charge rich? [24-3]
16. What does drivability mean? [24-3]
17. What does the carburetor do to give the maximum fuel economy at cruise? [24-3]
18. What is the range of combustible air/fuel mixtures? [24-3]
19. What is a stoichiometric mixture? [24-3]
20. How does the air/fuel ratio delivered to the cylinders differ between carbureted and fuel-injected engines? [24-3]
21. What adjustments are made to set idle speed and mixture? [24-4]
22. How is the operation of a PCV system checked? [24-4]
23. What conditions must be met before the engine idle can be adjusted? [24-4]
24. How is idle speed controlled? [24-4]
25. How is idle mixture controlled? [24-4]
26. How is idle speed adjusted? [24-4]
27. How is idle mixture adjusted? [24-4]
28. Describe the propane enrichment idle adjustment procedure. [24-4]
29. What causes the carburetor linkages to require adjustment? [24-4]
30. Why is the carburetor setting number important? [24-4]
31. List the ways the position of linkages are measured. [24-4]
32. List the ways that linkages are adjusted. [24-4]
33. What causes percolation in a carburetor? [24-4]
34. How is fast idle set? [24-4]
35. Why do carburetors have an unloader? [24-4]
36. How is the acceleration pump linkage checked? [24-4]
37. Name two types of chokes. [24-4]
38. What problems will be caused by an improperly operating choke breaker? [24-4]
39. What adjustment is made on a choke? [24-4]
40. How is the secondary air valve adjusted on a carburetor? [24-4]

CHAPTER 25

Emission System Service

The best means of maintaining low emission levels is by giving the engine a good periodic tune-up. On emission-controlled engines the tune-up should include a check of emission-control devices. When these devices are not functioning properly, the vehicle will produce excessive vehicle emissions. In addition, faulty emission control devices can affect vehicle drivability and increase fuel consumption. The positive crankcase ventilation system affects engine idle and crankcase contamination. Ignition timing affects all operations. The air preheat helps drivability during warm-up. It can also cause abnormal combustion after warm-up if it continues to heat the air. Backfiring may result from a faulty air injection system. Rough unstable idle and backfiring can be caused by improper operation of the exhaust gas recirculation system. Evaporative control system problems can affect engine idle and fuel flow from the tank to the fuel pump.

25-1 Positive Crankcase Ventilation

The valve and hoses of the PCV system (positive crankcase ventilation) are simple and easy to service, but they are often neglected, just as the exhaust heat riser is neglected. The PCV system allows filtered air to enter the crankcase, usually through the rocker cover. If the PCV inlet filter, located in the air cleaner, is dirty, it should be replaced. If a visual inspection of the hoses, either inside or outside, shows deterioration, they should be replaced.

The PCV valve is located between the carburetor base and the rocker cover or intake
manifold. The parts of the system are intercon-
nected with hoses, as illustrated in Figure 25-1. The
PCV supplies the engine with some air at idle. Pinch-
ing the hose until it is closed stops this air flow so
that the engine idle speed should drop about 50
rpm. If you have a hydrocarbon tester, its pickup
can be placed at the oil fill cap. It should indicate
hydrocarbons when the engine is not running.
When the engine is started, the hydrocarbon level
should immediately drop. The hydrocarbon tester
can also be used in its normal tailpipe connection. If
the PCV valve is disconnected, the hydrocarbon
content of the exhaust should decrease.

When the PCV valve does not operate satisfactorily according to one of the above tests, it should
be serviced. Take-apart PCV valves can be
disassembled and cleaned in a bath of carburetor
cleaner. This takes time and labor time and it is expensive to the customer. New PCV valves are low-
cost items, so inoperative PCV valves are usually
replaced with new valves. The proper valve should
be installed because different engines are designed
to operate with different PCV air flows. The valves
are manufactured with different plungers and springs
even when they may look alike. The interior of the
PCV hose should also be checked. This hose frequently collapses inside to block the PCV system.
After a new PCV valve is installed, the carburetor
idle should be readjusted as described in Section
24-4.

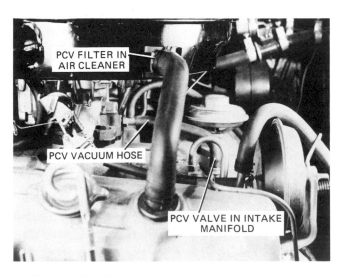

Figure 25-1. Parts of a typical crankcase ventilation
system on a V-type engine.

25-2 Ignition Timing Control

Basic timing on emission-controlled vehicles is
retarded from the best timing so it is necessary to
have the throttle open further to provide the same
engine idle speed. This added air allows the engine
to run leaner and, together with the higher
temperatures that result from retarded timing, helps
reduce HC and CO emissions. The amount of exhaust emissions produced by a normally operating
engine results from the effects of carburetion and
the ignition timing on combustion. Basic engine
timing must be correct. It is checked with a timing
light while the engine runs at curb idle with the distributor vacuum hose removed and temporarily
plugged so that no air flows into the carburetor.
These procedures are discussed in Section 22-4. Ignition mechanical advance is based on engine
speeds. In engines without emission controls
operated at full throttle, there is *no* manifold
vacuum. The mechanical advance is the only advance operating when there is no manifold vacuum.
This is also true on emission-controlled engines.
Because of this, emission controls that modify the
distributor vacuum advance signal have no effect
on the full throttle power produced by the engine.
Distributor vacuum advance modified by emission
controls will only affect part throttle and idle operation. This, in turn, affects fuel economy. The ignition timing advance is set for low exhaust emission.
This is done by modifying the vacuum signal applied to the distributor vacuum advance unit. The
distributor vacuum unit in most vehicles is connected to the carburetor sensing port through a
temperature-operated bypass valve. This type of
valve is called by a number of names: thermostatic
vacuum switch (TVS), ported vacuum switch
(PVS), orifice spark advance control (OSAC), or
distributor vacuum control valve. The distributor
vacuum unit receives its signal from a carburetor
vacuum port located just above the closed throttle
plate. The hose connecting the port to the
distributor vacuum system will usually include an
advance delay, and this may be temperature-
compensated. At engine idle, the throttle is closed,
and there is no vacuum advance. When the engine
overheats during prolonged idle, the temperature
operated bypass valve closes the carburetor vacuum
port. At the same time, it applies full manifold
vacuum to the distributor vacuum-control unit.

This fully advances the distributor vacuum unit, which improves the combustion efficiency. The efficient combustion increases engine speed, thereby lowering the engine temperature. When the engine cools to normal engine temperature, the valve returns to its normal position. This again connects ported vacuum to the distributor vacuum advance mechanism. A typical temperature-operated bypass is pictured in Figure 25-2.

Figure 25-3. A typical vacuum solenoid valve in the transmission controlled spark (TCS) system.

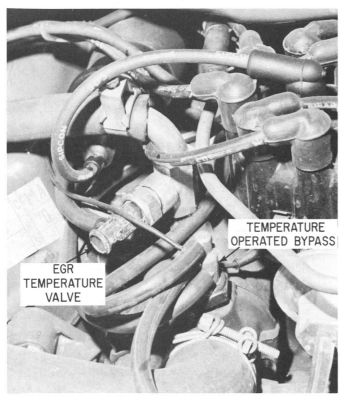

Figure 25-2. A typical temperature operated by-pass valve.

The temperature-operated bypass valve can be checked by applying a test vacuum source to the manifold vacuum port of the valve while heating it with a heat gun, such as a hair dryer. A thermometer held next to the sensor will indicate the operating temperature when the port opens to release vacuum.

Another approach to the control of emissions is to allow vacuum advance in direct drive only. This is done by putting a switch in the transmission. The switch controls a vacuum solenoid valve in the distributor vacuum advance hose. This valve can be identified in Figure 25-3. The solenoid valve allows ported vacuum to reach the distributor vacuum unit

only when the transmission is in direct drive. In a manual transmission, the switch is mechanically operated. In an automatic transmission, it is operated by direct-drive transmission fluid pressure.

When the ignition switch is turned on, the solenoid closes the vacuum hose that connects the distributor vacuum advance unit to the carburetor vacuum port. At the same time it opens the hose to the atmosphere to prevent any vacuum advance. When the vehicle shifts into direct drive, the transmission switch will open the electrical circuit of the solenoid. The solenoid valve will then close the atmospheric vent. At the same time, it connects the distributor vacuum advance hose to the carburetor vacuum port hose in its normal manner. This system is called a Transmission Controlled Spark (TCS) system by one manufacturer. It is called a Transmission Regulated Spark (TRS) control system by another manufacturer.

The TCS system can be checked by jacking the drive wheels from the ground and connecting a vacuum gauge to a tee placed in the distributor hose. The vehicle is started and accelerated through the gears. Vacuum should be observed on the vacuum gauge when the transmission shifts into direct drive with engine speed at approximately 2000 rpm. If no vacuum is observed, the vacuum hose should be visually checked and the electrical components checked for opens, shorts, or grounds.

In some vehicle applications, the system may also include a temperature switch to advance the ignition when the engine becomes too warm, regardless of which drive gear is being used. To do this, both the TCS and TRS systems use the Thermostatic Vacuum Switch (TVS) or Ported Vacuum Switch (PVS) previously described. The TRS system also uses an electrically operated thermal switch that is mounted on the door pillar to sense outside air temperature. This switch is connected parallel with the transmission switch so that it can bypass the normal operation of the transmission switch. The thermal switch remains closed below approximately 55 °F (13 °C) to hold the electrically operated vacuum solenoid valve closed. This keeps the ignition retarded on the cold engine even in direct drive so that the engine will warm up rapidly. Above this temperature, the thermal switch opens to allow the transmission switch to control the vacuum solenoid valve. A thermometer and voltmeter are required to check the proper operation of the temperature switch. Cooling will close the switch and warming will open it. It can be cooled by holding ice wrapped in plastic against the switch. It can be warmed with a hair dryer or a light bulb held close to the switch. The voltmeter is used to indicate switch operation and the thermometer held next to the switch will indicate the operating temperature.

In another application, Ford used a speed sensor in place of the transmission switch. It is called an Electronic Spark Control (ECS) system. This is not to be confused with the Chrysler computer-controlled spark advance having the same name. The speed sensor, consisting either of a rotating magnet and a stationary field winding in a small case, or of flyweights that operate a switch. This unit is connected in series in the speedometer drive cable, as shown in Figure 25-4. As vehicle speed increases, the voltage frequency of the speed sensor increases or the switch turns on. The voltage frequency signal may be fed into a small solid-state amplifier in certain applications where the electrical output of the speed sensor is small. The speed sensor electrical output operates a vacuum advance solenoid valve. The valve connects the distributor vacuum advance hose to the carburetor vacuum port above 23 to 35 mph (36 to 56 km/h), depending on the specific vehicle application. Below this speed, the distributor vacuum advance hose is vented to atmosphere to prevent ignition vacuum advance. The electronic spark control (ESC) system

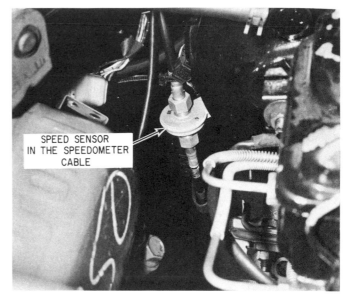

Figure 25-4. A speed sensor located in the speedometer cable.

has approximately the same effect as the transmission-regulated spark (TRS) system, but it will advance the ignition at the control speed setting even when the transmission is in a gear below direct drive.

The component units of the ESC system are not repairable. If they do not function properly, they must be replaced. They can be checked by placing a vacuum gauge in the distributor hose and operating the vehicle with the drive wheels raised. If vacuum is not observed at the proper time, the system can be checked using a jumper wire placed around the thermal switch. A voltmeter can be used to check the speed control and amplifier output to determine the speed at which they function. Disconnecting the solenoid wire will break the electrical circuit. The open circuit causes the solenoid to move to a position that will allow ported vacuum from the carburetor to reach the distributor. This will be accompanied by a slight increase in engine speed.

The typical mechanical and vacuum advance units in the distributor are slow to respond to changes in engine speed and load. Because of this, they could not change the ignition timing fast enough to meet the emission and economy standards set by Congress. It was necessary to develop computerized controls to provide the precise timing required. In 1976, Chrysler's first version of a computerized ignition timing control was called the Lean Burn system. Figure 25-5 shows a picture of

Figure 25-5. The first version of Chrysler's computerized ignition advance.

this system. The 1977 Oldsmobile Toronado was equipped with the first GM computer-controlled ignition advance system. It was called the MISAR system. The 1978 Lincoln Versailles introduced a computer to control both the ignition advance and the EGR valve. This system was called the EEC-I. Many improvements have been made since the introduction of these systems. The names of the systems have also been changed. Now microcomputers control most of the varible functions of the modern automotive engine.

Sensors on the engine feed information into the microcomputer. They sense the crankshaft position, crankshaft rpm, throttle position, throttle rate of movement, coolant temperature, inlet temperature, manifold pressure, and barometric pressure. The pulse inductor in the distributor also sends a signal to the computer. The computer compares all of these data to the standard data stored in

the computer. It then signals the ignition module to fire the coil at the precise instant ignition is needed. Electrically, the computer is located between the pulse inductor of the distributor and the ignition control module of a standard solid-state ignition sytem. If the computer-controlled ignition system fails, all of the standard items should be checked *before* condemning the computer. These items include the spark plugs, ignition cables, ignition coil, and timer core coil.

25-3 Carburetor Controls

Most emission-controlled automobiles preheat the inlet air using a thermostatic air cleaner to shorten the warm-up time. Air is drawn from the space between a shield and the exhaust manifold. The warm air is carried through a tube to the air control damper on the air cleaner assembly. This can be seen in Figure 25-6. Once the engine is warm, the

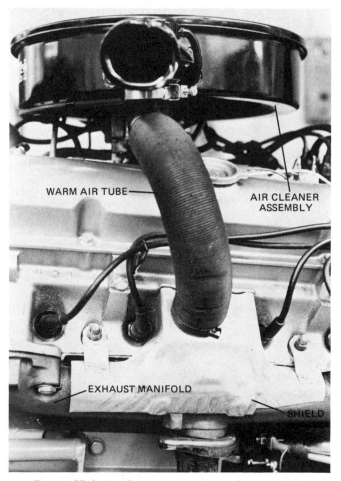

WARM AIR TUBE

AIR CLEANER ASSEMBLY

EXHAUST MANIFOLD

SHIELD

Figure 25-6. A tube carries warm air from a shield around the exhaust manifold to the air cleaner.

preheat is turned off so that it has no effect on engine operation or performance. The operation of the air control damper is illustrated in Figure 25-7. Correct operation can be checked visually. When the engine is cold, before starting, the control damper should be open to deliver cold air. Immediately after the engine starts, the damper should close to deliver warm air. As the engine warms, the damper will gradually open to mix cold and warm air. It is fully open in the cold air position when the engine is warm. The operating temperature can be checked by fastening a thermometer in the air cleaner inlet. The damper should open when the temperature is slightly above 100°F (38°C). If the control damper does not open, the temperature sensor or the vacuum-diaphragm actuator motor is not operating correctly. The first thing to check is to make sure that the vacuum hoses are in good condition. They must be installed securely on the proper hose nipples. If they are correctly installed and the damper still does not operate correctly, remove the air cleaner assembly and allow it to cool. Apply a 20 in. (50 cm) Hg vacuum source to the vacuum hose nipple of the heat sensor. This should cause the

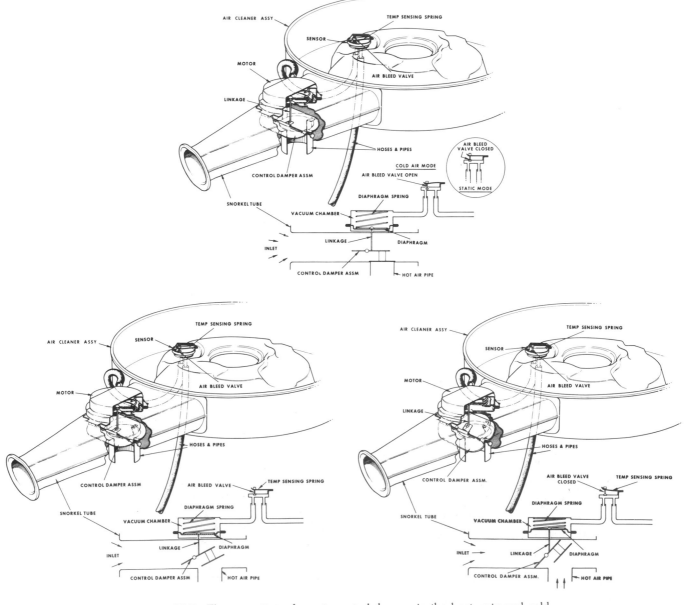

Figure 25-7. The operation of an air control damper in the heat, mix and cold positions (Courtesy of AC Division, General Motors Corporation).

damper to move to the closed position. If it does not, move the vacuum source directly to the vacuum–diaphragm motor. If this causes the damper to operate, the sensor is not opening at a cool temperature. If the damper still does not operate, the vacuum–diaphragm motor will have to be replaced. Sensor operation can be checked by connecting the vacuum source to the engine hose and a vacuum gauge to the diaphragm hose. Heating should cause the sensor vacuum to close the diaphragm below 100 °F (38 °C) and to open it at 135 °F (57 °C).

The carburetor may also be equipped with an antidieseling solenoid or an idle-stop solenoid. One example is pictured in Figure 25-8. When the ignition is turned on, the solenoid plunger is out, contacting the throttle idle stop. It must be in this position when idle speed is adjusted. When the engine is turned off, the solenoid plunger moves back. This lets the throttle plates close completely so that the engine cannot draw intake air.

The solenoid can be checked by placing a jumper wire to connect battery voltage to the solenoid lead. This should cause the plunger to move out. If the plunger does not move out when the ignition is on but moves out when a jumper is used to power it

directly from the battery, there is an open in the electrical feed wire from the ignition switch. This open will have to be located and repaired.

As emission standards became more severe, it was impossible for automobiles to meet the standards at both sea level and in the mountains. Altitude carburetors were developed for use at high elevations. These carburetors were equipped with altitude-compensation devices. The altitude-sensing device in the carburetor consists of an accordion-shaped container which is sealed so no air can get in or out. This sealed container is called an *aneroid*. This unit is compressed by atmospheric pressure at sea level. It expands in lower atmospheric pressures at higher elevation. A valve is moved by one end of the atmospheric-pressure-sensing unit. In one type, the valve restricts the amount of air bleeding into the main well at sea level to enrich the mixture. This type is pictured in Figure 25-9. It allows increasing amounts of air to bleed into the main well as the atmospheric pressure drops when the automobile operates at higher elevations. This increase in bleed air leans the mixture. In another type of altitude compensation, the atmospheric-pressure-sensing unit moves a valve in a part-throttle jet located in the bottom of the float

Figure 25-8. A typical idle-stop solenoid.

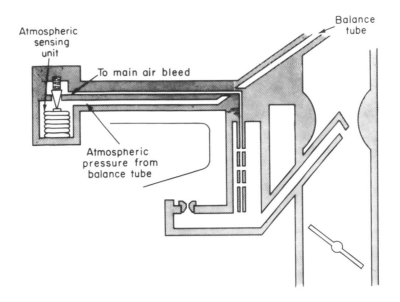

Figure 25-9. A line drawing of an air bleed-type carburetor altitude control.

bowl. Figure 25-10 pictures this type of control. The sensing unit expands at high elevations, pushing the valve into the jet to reduce the effective size of the jet opening. This reduces fuel flow into the main well to lean the mixture. Altitude compensation provides the automobile with low emissions and good drivability at all elevations.

A catalytic converter has been used since 1975 to help vehicles meet the HC and CO emission standards. A typical installation is shown in Figure 25-11. In 1981, HC, CO, and NO_x Federal standards, became so tight that an additional catalyst was required to reduce NO_x. This three-way converter is efficient only when the air/fuel ratio is maintained very close to the chemically balanced stoichiometric ratio. The control was accomplished by developing a computer to control the air/fuel ratio delivered by the carburetor or fuel metering system. A sensor measures the oxygen in the exhaust. This information is passed along to a computer. The computer adjusted the carburetor, as necessary, to maintain the correct air/fuel ratio. In a sense, the system automatically *keeps* the engine in tune.

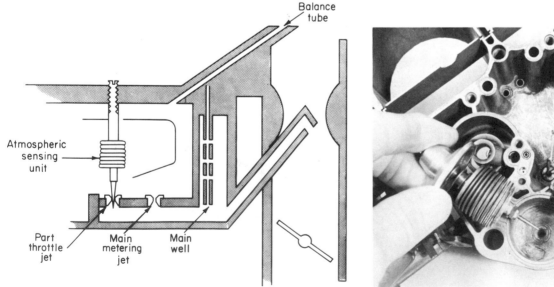

Figure 25-10. A line drawing of a fuel valve-type carburetor altitude control (a) and a picture of the control unit (b).

Figure 25-11. A typical installed catalytic converter.

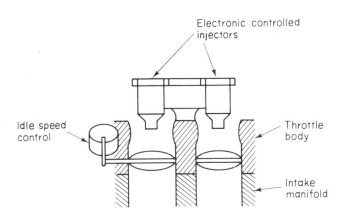

Figure 25-13. A line drawing illustrating throttle body fuel injection.

General Motors introduced the C-4 (Computer Controlled Catalytic Converter) in some early 1980 model automobiles in California in April 1979. This system is illustrated in Figure 25-12. This system sensed the oxygen remaining in the exhaust. Too much oxygen indicated an excessively lean mixture. Too little oxygen indicated too rich a mixture. The computer adjusted the carburetor to correct the air/fuel ratio. Computer throttle body injection was introduced on some other 1980 GM automobiles, as illustrated in Figure 25-13. These systems were also controlled by the oxygen sensor to maintain the correct air/fuel ratio.

25-4 Air Injection System

The air injection system called Thermactor or Air Injection Reactor (AIR) supplies air under pressure to each cylinder exhaust port to help burn combustible products remaining in the exhaust gases. The system used on a V-type engine is illustrated in Figure 25-14. A belt-driven vane-type pump supplies the air. Some of the first pumps were repairable, but the modern air-injection pump is permanently lubricated and requires no periodic maintenance. It is not to be lubricated in any way, even if it squeaks when turned by hand.

A plastic centrifugal inlet filter-fan is located behind the pump drive pulley. If the engine is to be washed down or steam-cleaned, the filter should be

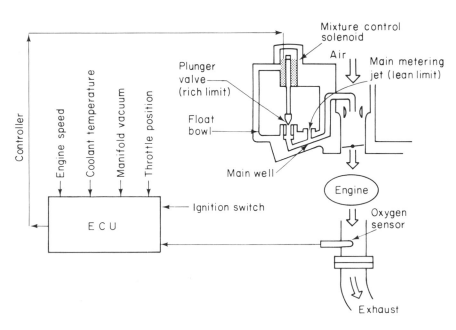

Figure 25-12. A line drawing illustrating a C-4 computer controlled carburetor mixture.

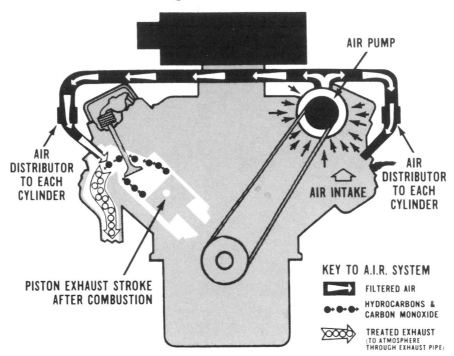

AIR PUMP

AIR DISTRIBUTOR TO EACH CYLINDER

AIR DISTRIBUTOR TO EACH CYLINDER

AIR INTAKE

PISTON EXHAUST STROKE AFTER COMBUSTION

KEY TO A.I.R. SYSTEM

FILTERED AIR

HYDROCARBONS & CARBON MONOXIDE

TREATED EXHAUST (TO ATMOSPHERE THROUGH EXHAUST PIPE)

Figure 25-14. A line drawing illustrating an air injection reaction system (Courtesy of General Motors Research).

masked off to keep all cleaning agents out of the pump. The centrifugal filter-fan is the only part of the pump to be serviced. The pulley is removed, then the plastic filter-fan is taken off with a pliers, as pictured in Figure 25-15. The filter-fan will usually break up as it is being removed, so care must be taken to keep any plastic chips from entering the pump inlet. It is a recommended precaution to wear safety glasses to prevent particles from getting into the eyes as the plastic breaks. A new filter-fan is installed by drawing it into position with the pulley assembly bolts.

The air pump must have its belt adjusted correctly to operate properly. Air pump operation can be checked by removing the hose feeding air to the exhaust and by feeling the pump outlet air flow. It is normal for the pump to make some noise that becomes louder as engine speed increases. Air leaks in the system can often be felt by hand. They can be checked more accurately by brushing soapy water on the hoses, tubing, and connections. The soap will bubble around any leak.

The check valves can be inspected by removing the air hoses. The operation of the valves can be observed by depressing the check with a probe. It should return freely to its original position.

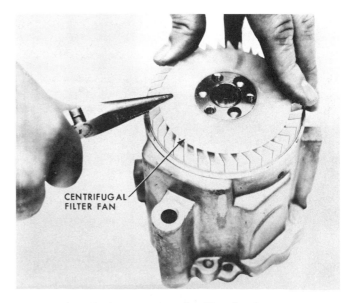

CENTRIFUGAL FILTER FAN

Figure 25-15. Removing the filter fan from an air pump (Courtesy of AC Division, General Motors Corporation).

The operation of the bypass or diverter valve, as shown in the section drawing in Figure 25-16, can be checked by listening at its silencer. One end of a piece of hose held close to the silencer and the other held close to your ear will help you to listen to its operation. Engine speed is brought up to 2000 rpm

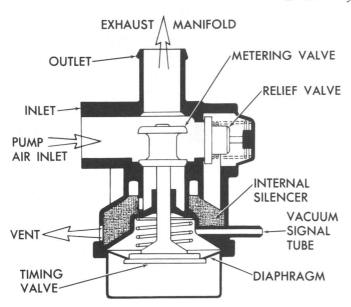

Figure 25-16. A section drawing of an air injection system diverter valve (Courtesy of AC Division of General Motors Corporation).

and held until the speed stabilizes. The throttle is allowed to close suddenly. When it closes you should hear a rush of air coming out of the silencer when the valve is operating correctly. If no rush of air is heard or if an exhaust backfire occurs, the bypass or diverter valve is faulty and must be replaced. The valve is also faulty if air escapes from the silencer at any other time.

After the air injection reactor system is checked, the hoses and tubing should be reconnected and left in good condition. All fittings and connections should be tight and the drive belt tension should be adjusted correctly.

In some cases, an engine puts out too much HC and CO without an air pump. Adding an air pump would supply too much air. These engines are equipped with a *pulsair* or *aspirator system*. This system uses the exhaust pressure pulses to draw some air into the exhaust system. Clean air is taken from the air cleaner to the one-way, spring-loaded diaphragm valve. The diaphragm opens to allow fresh air to mix with the exhaust gases during negative pressure pulses that occur in the exhaust. Two different systems are pictured in Figure 25-17.

The valve is not repairable. When it fails, it must be replaced. The first sign of valve failure is an increase in the underhood noise. Exhaust gases leaking through the valve will cause the connecting hoses to harden. If valve failure is suspected, remove the hose connecting the valve to the air cleaner. Exhaust pulses can be felt through a failed valve inlet fitting while the engine is idling and the noise increases.

25-5 Exhaust Gas Recirculation

Oxides of nitrogen (NO_x) are controlled with a special camshaft that has a long valve overlap. NO_x is also controlled with a system to return some of the exhaust gas back to the combustion chamber. This system is called the Exhaust Gas Recirculation (EGR) system. Flow of exhaust gas in this system is

Figure 25-17. Two typical aspirator systems.

proportional to the difference between intake manifold vacuum and exhaust back pressure. Maintenance of the EGR system recommends the removal of the carburetor and an inspection of the orifices each 12,000 miles (19,300 km).

The most common EGR design uses a vacuum-controlled valve to allow a small flow of exhaust gases to enter the intake. A section drawing of a typical EGR valve is shown in Figure 25-18. In some late models, a computer controls the opening of the EGR valve. A spring-loaded diaphragm holds the valve closed. The vacuum chamber located above the diaphragm is connected to a special signal port in the carburetor. The port connects to a hole or a vertical slot in the carburetor bore or to a port that senses venturi vacuum. A slot port can be seen in Figure 25-19. As the carburetor throttle plate is opened, the port vacuum increases. In some applications, the effect of the vacuum is modified by exhaust pressure. The EGR valve stays closed at idle. It gradually opens as engine speeds are increased. At full throttle, it again closes. The actual flow of exhaust gases into the intake depends upon the EGR valve opening position and upon the pressure difference between the gases in the exhaust manifold and the vacuum in the intake manifold. A vacuum amplifier, as pictured in Figure 25-20, is required to increase the vacuum port signal when venturi vacuum is used to control the EGR valve.

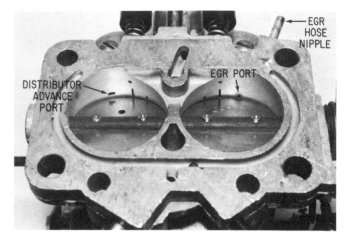

Figure 25-19. The EGR port can be seen by looking into the bottom of the bores of a carburetor.

The operation of this system can be checked by installing a vacuum gauge in the control vacuum hose going to the EGR valve. The start of EGR is delayed with a timer on some engines. Increasing engine speed should increase EGR port vacuum. If it does not increase, the carburetor passage will have to be examined. The operation of the valve can be checked by applying vacuum to the vacuum connection. The operation is pictured in Figure 25-21. The idle speed of the engine will slow as the valve opens. If the valve does not work correctly, it will have to be cleaned or replaced.

The exhaust gas recirculating controls require

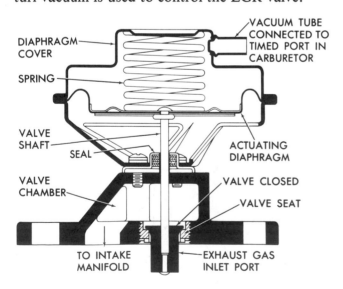

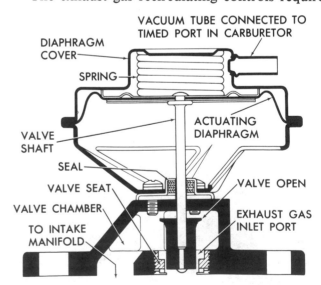

Figure 25-18. Section view showing the operation of a typical EGR valve. (a) Valve is closed. (b) Valve is open (Courtesy of Cadillac Motor Car Division, General Motors Corporation).

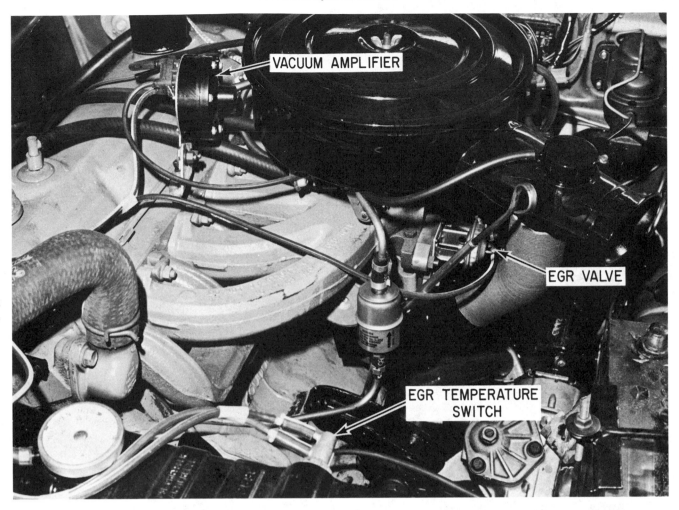

Figure 25-20. A vacuum amplifier used to control the EGR valve.

the use of a temperature switch in the coolant system. The EGR vacuum is blocked off at low coolant temperatures. As the coolant warms, the EGR system vacuum is allowed to go to the EGR valve. The coolant temperature switch is checked in the same manner as the thermostatic vacuum switch.

25-6 Catalytic Converter

Severe emission standards cannot be met entirely by engine modifications, so devices have been developed to finish cleaning the exhaust after it leaves the engine.

Catalytic converters are located in the exhaust pipe between the exhaust manifold and the muffler. They consist of a chamber with a catalyst through which the exhaust gases pass. The catalyst increases the rate of oxidation of the HC and CO in the exhaust. In three-way converters, it also increases the rate of reduction of NO_x at lower temperatures.

Two things can happen to cause premature failure of the catalytic converter. Its case or housing can burn through, allowing the exhaust gas to escape, or the catalyst can become ineffective. This will happen when a cylinder misfires for a period of time. Ineffective catalyst can be determined with a HC–CO tester. In either case, the catalyst or converter will have to be replaced.

Automobiles equipped with catalytic converters *must* use only lead-free gasoline. Lead from the gasoline will contaminate the catalyst and make it ineffective. If this happens, the catalytic converter will need to be replaced to reduce emissions to the required emission standards. Fuel-tank fill necks are designed so that only lead-free gasoline can be used to fill them with the gasoline pump nozzle.

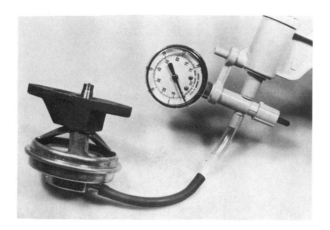

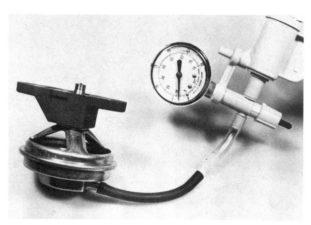

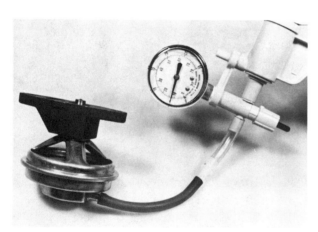

25-7 Evaporative Control System

The evaporative control system requires very little service. Its hoses and lines should be secure and in good condition. If they are replaced, they must be replaced with special fuel-resistant materials marked EVAP or FUEL. Vapor leaks can best be checked with a hydrocarbon test instrument. The tester pickup should be moved along below the lines. Hydrocarbon vapors tend to sink in air. An increase in hydrocarbon readings indicates vapor leaks. The source of any leak must be found and repaired.

The gasoline tank filler cap used on emission-controlled vehicles is a pressure-vacuum cap. If it is replaced, it must be replaced with the correct type of cap to prevent excess pressure or vacuum that could damage the fuel system parts.

The carbon canister is fitted with a filter, as shown in Figure 25-22. Under normal driving conditions, the canister filter should be replaced once each year or each 12,000 miles (19,300 km), whichever comes first. When the vehicle is operated under dirty conditions, the canister filter should be replaced more often, just as is required of the engine filters. If the filter is changed at proper intervals, the activated carbon does not require replacement.

Other parts of the evaporative fuel system are not repairable, but must be replaced if they are damaged, worn, or aged to the point that they are no longer effective.

Figure 25-21. EGR valve movement as vacuum increases from illustration a to d.

Figure 25-22. Replacing the filter on the bottom of a carbon canister.

REVIEW QUESTIONS

1. What emission control systems should be checked during a periodic tune-up? [INTRODUCTION]
2. How is the PCV system tested? [25-1]
3. When should the PCV valve be replaced? [25-1]
4. Why don't the ignition timing controls for emission have an effect on full throttle power produced by the engine? [25-2]
5. What does a temperature-operated bypass valve do? [25-2]
6. How is the temperature-operated bypass valve checked? [25-2]
7. How does the vacuum solenoid valve operate? [25-2]
8. How is the operation of the vacuum solenoid valve checked? [25-2]
9. How does the ESC system differ from the TRS system? [25-2]
10. What brought about the microcomputer controls for engine functions? [25-2]
11. Where does the computer fit into the ignition system? [25-2]
12. How is the thermostatic air cleaner checked for operation? [25-3]
13. What is the purpose of the idle stop solenoid? [25-3]
14. How does the altitude compensation device change the air/fuel mixture? [25-3]
15. Why is it necessary to have a computer-controlled carburetor when a three-way catalytic converter is used? [25-3]
16. What part of the air pump can be serviced? [25-4]
17. How is the operation of the air pump checked? [25-4]
18. How are the air pump check valves inspected? [25-4]
19. How is the air pump diverter valve checked? [25-4]
20. How is the aspirator system checked? [25-4]
21. What does an EGR valve do? [25-5]
22. When is the EGR valve open? [25-5]
23. How is the operation of the EGR valve checked? [25-5]
24. What can cause the catalytic converter to fail? [25-6]
25. What service should be done on the evaporative control system? [25-7]

GLOSSARY

A

Abrasive: A very hard material that will cut or scratch a surface.

Accumulates: Gradually builds up or collects.

Additive: A product added to a material that will change the characteristics of the material.

Adhesion: The characteristic that causes one material to stick to or cling to another material.

Adjacent: Next to or beside.

Adverse: Against or opposed to an action. A reaction.

Aftermarket: The sales market designed for the consumer after the product is purchased from a dealer.

Air/fuel Ratio: A mixture of air and fuel by weight. Pounds of air per pound of fuel.

Agitation: Mixing or stirring.

Alternate: Going back and forth between two things or places.

Amplitude: The extent or size of vibrating motion.

Application: Something that fits on, or is applied to, another thing.

Assembly: A group of parts that are connected together. The process of putting parts together.

Assumes: Take for granted.

Assure: To make certain.

Authorize: The power to give permission.

B

Back Flush: A procedure used to clean a system by forcing flow in the reverse direction from an external source.

Baffle: A sheetmetal plate used for deflecting a fluid or gas.

Balance: Having equal weight on each side of a supporting point. No change in the position of the center of gravity when a part is in motion.

Base Circle: The part of the cam that has the smallest diameter from the center of the camshaft.

Binders: Materials that hold other materials together.

Bonding: Permanent joining together.

Boss: A heavy cast section that is used for support, such as a heavy section around a bearing.

Bypass: Go around some restriction, such as a valve, switch, or circuit.

Burr: Roughness on an edge.

420

C

Calculation: The use of mathematics to solve a problem.

Calibrate: To check against a standard measurement.

Canted: Angled away from another object or from vertical.

Capacity: The volume of a container. The volume that can be pumped in a given period of time.

Case: An enclosure.

Cast: A part made by pouring molten metal into a mold where it cools in the shape of the mold.

Catalytic Converter: A device in the exhaust system used to reduce polluting emissions.

Coefficient of Friction: An expression that relates the force to move an object to the weight or load produced by the object. Force/load.

Characteristic: Distinctive trait, property, or quality that separates.

Check: An examination procedure used to determine the condition of an item or assembly.

Cock: To tip when the part is designed to be straight.

Combustion: Burning fuel to produce heat.

Commercial Standards: The quality of repairs made for the general motoring public. A satisfactory repair made at the lowest cost.

Compatibility: Working together in harmony. Helping each other.

Compensate: To adjust to correct for changing conditions.

Compression Ignition: Diesel engine operation. The heat in compressed air causes the fuel to ignite.

Compromise: Giving up a little of one thing to gain a part of another thing.

Conceal: Cover up or hide.

Concentric: Circles located evenly around a common center point.

Conform: To change the shape of a piece to match the shape of a part with which the piece is in contact.

Connecting Rod Journal: The surface of the crank pin.

Consumption: Using up materials.

Contour: The shape of the surface.

Converted: Changed from one thing to another.

Corrosion: Pits in the surface of metal caused by oxygen, moisture, or acid.

Corrosive: A material that causes parts to corrode.

Crank Pin: The metal pin between the crankshaft cheeks to which the connecting rod is attached.

Crankshaft Cheek: The flange of the crankshaft between the crank pin and the main bearing pin.

Crank Throw: The offset between the crank pin and the main bearing pin.

D

Damage: Injury or harm that reduces the usefulness.

Dampen: Reducing vibration to an acceptable level.

Deck: The flat upper surface of the engine block where the head mounts.

Deflection: Slight bending under a force or load.

Deform: To change the shape slightly.

Degrade: To reduce the useful properties.

Deposit Ignition: Ignition of the intake charge by hot deposits in the combustion chamber.

Deterioration: A gradual decline in the quality.

Detonation: A knock produced by the sudden combustion of the intake charge ahead of the flame front in a combustion chamber.

Diaphragm: A flexible surface on one side of an enclosed chamber.

Disperse: To distribute uniformly throughout a solid, liquid, or gas.

Displace: Move to another place.

Distress: Damage to a part caused by use.

E

Eccentric: Two or more circles, one around the other, and each having different centers.

Economical: The lowest cost for the same result.

Effective: Something that does what it is designed to do.

Efficient: Having the highest output for a given input.

Enclose: To put a cover around a part.

Energy: The capacity to do work.

Equivalence Ratio: The actual air/fuel ratio divided by the stoichiometric air/fuel ratio.

Erode: To wear away with high velocity abrasive material.

Excessive: More than the amount needed or wanted.

Extent: The amount or limit.

F

Fatigue: The breakdown of material through a great number of loading and unloading cycles. Cracks are the first sign of fatigue.

Feedback: Data about the output of a device continuously sensed by a controller.

Fillet: A rounded joint between two surfaces.

Flange: A small surface at an angle to the major surface used to position, reinforce, or fasten the major surface in place.

Flex: To bend.

Flow: To move along in a stream.

Formulation: A mixture of specific materials mixed acording to a formula.

Frequency: The number of occurrences in a second. Called hertz in SI metric units.

Function: To perform a specific action or activity.

G

Gallery: A large passage in the engine block that forms a reservoir for engine oil under pressure.

Glaze: A polished, work hardened surface.

H

Hertz: Cycles per second in SI metric units.

Hot Spot: A spot in the combustion chamber with enough heat to start combustion of the intake charge.

Housing: A supporting enclosure around moving parts.

I

Identical: Exactly the same as another.

Idle: Running freely with no power or load being transferred.

Impinge: To press against the surface.

Indicate: To show the condition or amount.

Inertia: The force that causes a part that is stopped to stay stopped or one in motion to stay in motion.

Inspect: To look over carefully. To compare against standards or specifications.

Instant: A very short amount of time. A specific moment.

Integral: Part of the same piece.

Interchange: To transfer or take the place of another part.

Interference Fit: The hole is slightly smaller than the object forced into the hole.

Interrelated: Working together.

J

Journal: The surface on which a bearing operates.

Junction: A joint or connection.

K

Keyed: Held from rotating on a shaft with a small metal part called a key.

Knock: The sound produced in a combustion chamber as a result of abnormal combustion, usually detonation.

Knurl: A rough surface caused by a tool that displaces metal outward as it pushes into the surface.

L

Lash: Slack in the valve train that must be taken up before movement begins.

Load: The work being done by the engine.

Lugging: Operating the engine at full throttle with a load that will keep the engine rpm low.

M

Maintain: To keep in condition for normal operation.

Malfunction: A failure to operate normally.

Mass: The amount of material in an object or volume of fluid. Mass is measured in pounds or kilograms in this book.

Mature Failure: A part that wears out, as expected, at high mileage.

Mean Effective Pressure: The calculated average pressure in the combustion chamber during the power stroke.

Micropeen: To work harden the surface by hitting it with small hard particles moving at high speed.

Minimize: To reduce as low as possible. To prevent whenever possible.

Modify: To change the form, quality, or function.

N

Nick: A notch, groove, or chip in the surface of a part.

Node: Point of minimum vibration.

Nominal: Normal position, size, or operating condition.

O

Obvious: Easily seen, recognized, or understood.

Occur: To happen or take place.

Octane Number: A rating number that indicates the anti-knock property of gasoline.

Opportunity: A favorable time or occasion.

Option: The opportunity to choose among several things or ways of doing things.

Optional: Leaving something to one's choice.

Oscillation: A vibration back and forth or up and down.

Oxidizing: Combination of a material with oxygen. Heat is usually released in the process.

P

Pad: Flat surfaces on castings upon which parts are mounted.

Parallel: Straight lines or surfaces that are the same distance apart from end to end.

Partial: Only a part of something. Not complete.

Performance: Effectively operating at the maximum designed operating conditions.

Piezoelectric Crystal: A material that produces a very small electrical charge when pressure is put on it.

Ping: A sharp sound like a stone hitting a metal object.

Pit: A small depressed scar in the surface of metal.

Plenum: The space in the manifold below the carburetor from which the intake runners lead.

Polymerized: The combination of molecules to form larger, complex molecules.

Port: An opening through which liquids or gases flow.

Potential: The capability of becoming or doing.

Porting and Relieving: Using a hand grinder to smooth, enlarge, and match the openings of manifolds and ports in the heads.

Position: To put in the proper place in a specific way.

Power: The amount of work done in a given time.

Precisely: Being exactly as it should be.

Precision Standards: Methods of repair that make the finished part as close to perfect size and finish as possible.

Preflame Reactions: Chemical reactions that occur in the end gases ahead of the flame front during combustion.

Preignition: Ignition of the intake charge before the spark forms across the spark plug electrodes.

Preliminary: Steps taken before a major activity.

Premature Failure: Unexpected or early failure of a part before the end of its normal service life.

Pressure: Force on the surface of a unit area. Pressure is measured in pounds per square inch or kilograms per square meter, called newton-pascal.

Preventative Maintenance: Service procedures done before a malfunction occurs to prevent failure.

Primary: Carburetor barrels in use during all engine operation. Heavy coil windings in the ignition coil that are in use during magnetic buildup.

Produce: The ability to cause, create, or rearrange.

Propellant: The high velocity air that carries cleaning beads.

Propelled: Being forced to move.

Properly: Done following established standards.

Proportional: Having the same or constant ratio.

Pulverize: Break up into very fine particles.

Q

Quench Area: A surface in the combustion chamber designed to cool the temperature of the gases to help control the combustion process. Also called the squish area.

R

Radial: In a straight line out from the center.

Rate: The amount of change in a period of time or distance.

Reciprocate: Move back and forth.

Reduce: Make smaller. Remove oxygen from a material.

Related: Associated with or connected to.

Relative Motion: Motion between two parts, even if they are both moving when compared to a third object.

Relatively: Something having a property compared to another, such as relatively high or relatively hot.

Repair: To restore to good operating condition.

Resonate: Vibrate together at a common frequency.

Restrict: Confine or keep within limits.

S

Score: Grooves in a surface resulting from metal rubbing against metal.

Secondary: The barrels of a carburetor that are in use near full throttle. The small ignition coil windings that are only in use as the spark plug fires.

Secured: Fastened so there is no likelihood of failure.

Seize: Metal-to-metal contact that prevents movement.

Sequence: The ordering of one thing after another in a specified succession.

Service: Make adjustments, repairs, or replace parts.

Sensor: A device that converts a temperature or force into a signal to be used for control.

Service Operations: Adjustments, repairs, or replacement of parts done for preventative maintenance, to make the system operate after failure. To make modifications.

Specific: Having a special application or precise qualities.

Specified: The precise values that are given in references.

Squish Area: (*See* Quench area.)

Stoichiometric: The chemically balanced air/fuel ratio. The correct amount of oxygen in the air to combine with the fuel in the mixture.

Surge: Increasing and decreasing engine speed while the throttle is held in one position.

Surrounded: Enclosed on all sides.

Swirling: Moving around and mixing a fluid or a gas.

T

Tamper: To purposely alter or damage a part so that it can be adjusted outside of legal specifications.

Theory: A particular conception or understanding of the way something works.

Thermal Shock: A sudden change in temperature that affects both expansions and contractions, often causing something to crack.

Thrust Surface: A surface upon which a pushing load is applied.

Tolerance: The difference between the specified maximum and minimum levels.

Torque: The twisting force on a shaft.

Trace: An extremely small amount. One that can barely be identified.

Trade-Off: Give up one thing to gain another. (*See* compromise.)

Transverse: In a crosswise direction. Side to side.

Turbulance: Mixing of gasses and/or liquids.

Typical: Something that is representative of other similar designs or methods.

U

Undercut: Machining below the normal surface.

Upset: Bend over.

V

Vapor Lock: Gasoline evaporating in the fuel line on the fuel tank side of the fuel pump.

Variation: Slight differences between similar parts.

Velocity: The relationship between the distance something travels and the amount of time it takes to travel the distance.

Verify: Check to prove that the results are correct.

Vibration: Periodic back and forth motion.

Viscosity: The thickness or fluid body of a liquid.

Voids: Open spaces between material.

Volatility: The tendency to evaporate.

Volume: The amount of space measured in cubic units.

W

Work: Energy used to move an object.

Index